Chemical Mechanisms in Bioenergetics

D. RAO SANADI, Editor
Boston Biomedical Research Institute
Boston, Mass.

ACS Monograph **172**

AMERICAN CHEMICAL SOCIETY

WASHINGTON, D. C.   1976

QH
510
.C47

Library of Congress CIP Data

Chemical mechanisms in bioenergetics.
  (ACS monograph; 172)

  Includes bibliographical references and index.

  1. Bioenergetics. 2. Membranes (Biology). 3. Photosynthesis. 4. Muscle contraction.
  I. Sanadi, D. R. II. Series: American Chemical Society. ACS monograph; no. 172.
QH510.C47   574.1'9121   76-26707
ISBN 0-8412-0274-5   ACMOAG 172 1-272

Copyright © 1976

American Chemical Society

All Rights Reserved. No part of this book may be reproduced or transmitted in any form or by any means—graphic, electronic, including photocopying, recording, taping, or information storage and retrieval systems—without written permission from the American Chemical Society.

PRINTED IN THE UNITED STATES OF AMERICA

GENERAL INTRODUCTION

American Chemical Society's Series of Chemical Monographs

By arrangement with the interallied Conference of Pure and Applied Chemistry, which met in London and Brussels in July 1919, the American Chemical Society undertook the production and publication of Scientific and Technologic Monographs on chemical subjects. At the same time it was agreed that the National Research Council, in cooperation with the American Chemical Society and the American Physical Society, should undertake the production and publication of Critical Tables of Chemical and Physical Constants. The American Chemical Society and the National Research Council mutually agreed to care for these two fields of chemical progress.

The Council of the American Chemical Society, acting through its Committee on National Policy, appointed editors and associates (the present list of whom appears at the close of this sketch) to select authors of competent authority in their respective fields and to consider critically the manuscripts submitted. Since 1944 the Scientific and Technologic Monographs have been combined in the Series. The first Monograph appeared in 1921, and up to 1972, 168 treatises have enriched the Series.

These Monographs are intended to serve two principal purposes: first to make available to chemists a thorough treatment of a selected area in form usable by persons working in more or less unrelated fields to the end that they may correlate their own work with a larger area of physical science; secondly, to stimulate further research in the specific field treated. To implement this purpose the authors of Monographs give extended references to the literature.

American Chemical Society
F. Marshall Beringer
Editor of Monographs

EDITORIAL BOARD

C. T. BISHOP	ERWIN CHARGAFF	JAMES P. LODGE
MICHEL BOUDART	GERHARD L. CLOSS	JAMES J. MORGAN
FRANK A. BOVEY	ROBERT A. DUCE	ALEX NICKON
DARYLE H. BUSCH	EDMOND S. FISCHER	MINORU TSUTSUI
MELVIN CALVIN	THEODORE A. GEISSMAN	VICTOR E. VIOLA
JAMES CARPENTER	MURRAY GOODMAN	JOHN T. YOKE
	HENRY A. LARDY	

Contents

Preface

1. Biothermodynamics — 1
   S. Roy Caplan

2. Biomembranes—Structure and Function — 38
   Donald F. H. Wallach

3. Membrane Transport — 78
   Joseph D. Robinson

4. Chemical Reactions in Oxidative Phosphorylation — 123
   D. Rao Sanadi and Hartmut Wohlrab

5. Mechanisms in Photosynthesis — 172
   Bessel Kok and Richard Radmer

6. The Mechanism of Muscle Contraction — 221
   John Gergely

Index — 263

Preface

BIOENERGETICS DEALS WITH the phenomenon of transformation of one form of energy into another in biological systems. Examples are the conversion of electromagnetic to chemical bond energy in photosynthesis, phosphate bond energy of ATP to chemical gradients across membranes, and ATP energy to mechanical work in muscle contraction. Conversion of the free energy of oxidation in mitochondria to ATP bond energy would be another example, particularly if the synthesis is driven obligatorily by a proton gradient or conformational change.

Most of these energy transduction processes have several common denominators, one being the key role of membranes. Recent advances in membrane biochemistry have increased the scope of the technology in studying the molecular basis of bioenergetics, and the volume of literature is growing profusely. A multidisciplinary effort involving workers engaged in research on the different types of energy transduction phenomena could lead to extensive cross-fertilization and hopefully hasten the solution. With this aim, we have undertaken to produce a volume that could form the basis for an advanced graduate course in bioenergetics and also serve as a combined selected reference source of introductory material to beginning researchers in bioenergetics.

Boston Biomedical Research Institute
Boston, Mass.
November 1975

D. RAO SANADI

Chapter

1

Biothermodynamics

S. Roy Caplan, Laboratory of Membranes and Bioregulation, Weizmann Institute of Science, Rehovot, Israel

THIS CHAPTER PRESENTS in relatively simple and compact terms the basic concepts of nonequilibrium thermodynamics (NET), shows how these concepts can be used for the macroscopic description and evaluation of bioenergetics, and collects together some examples of the application of NET to the experimental study of biological systems. The chemical and physical processes with which we are concerned in biology are irreversible, and generally take place far from equilibrium. Classical thermodynamics is much too limited in scope for situations of this kind, as was made clear by Prigogine in a very illuminating monograph (*1*). The classical thermodynamic approach is based on the consideration of equilibrium states and its methodology depends on the notion of reversible processes, in other words on hypothetical ideal processes that can occur without disturbing equilibrium. It is a far cry from such idealized systems to the living cell—hence, the need for biothermodynamics. This need is exactly what NET, even in its simplest linear form, can often fulfill.

Thermodynamics, in its broadest sense, establishes a logical framework within which observed behavior pertaining to the energy transactions of a system may be organized and tested for self-consistency. Thus correlations are readily predicted among observations, even in the entire absence of any knowledge of the structure of the system. This remains true when we deal with NET, which extends the classical method by the explicit introduction of time, and is therefore able to provide correlations among the all-important kinetic parameters. The essence of the approach is to study, if possible, the processes occurring when the system is in, or very close to, a steady state. By analyzing the effects of the system on its surroundings under these conditions, it becomes possible to determine the number and nature of the degrees of freedom involved. This in turn indicates the choice of constraints which must be applied to the system if it is to reach a well-defined stationary state and the relationships between the constraints. The analysis may also be invaluable in modelling structure in relation to function since it reveals thermodynamic restrictions to which such models must necessarily conform.

For a more detailed treatment of the matters considered in this chapter, the reader is referred to the textbook of Katchalsky and Curran (2) and to the reviews by Caplan (3) and Caplan and Mikulecky (4).

Coupled Phenomena

Dissipation by Irreversible Processes. Figure 1 depicts two typical physical arrangements in which an irreversible process produces a measurable effect on its surroundings. Considering Figure 1a first, we see a vessel labelled

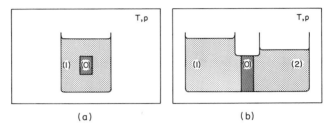

Figure 1. The system: irreversible processes are confined to region (O), which is assumed to be of small capacity compared with the surrounding regions (1) or (1) and (2). In both arrangements (a) and (b), the system is enclosed by rigid adiabatic walls. T = *temperature;* p = *pressure.*

(1) standing in an enclosed space, containing a liquid within which is suspended a "black box" or inner region labeled (0). In the inner region (0), an irreversible process occurs, the process itself being confined to this region although its effects are felt in the surroundings. For example, region (0) might be a bacterium, a mitochondrion, or a fragment of tissue or synthetic membrane with enzymatic activity. In all these cases, substrates present in (1) enter (0), undergo reaction, and are returned to (1) as products. The region (1) may be thought of as a portion of the local environment of (0) large enough to contain completely all possible effects of the process in (0). It is convenient to consider (1) enclosed by rigid "adiabatic walls." In other words, (1) is isolated or sealed off entirely from the world beyond, a situation that is not uncommon experimentally. For simplicity we may take both temperature (T) and pressure (p) to be constant everywhere in (1), which is supposed to be well mixed, as again would often be the case experimentally. Two possibilities arise, (a) and (b).

(a) The region (0) is a closed system—*i.e.,* it can exchange only heat with (1). This possibility excludes the examples quoted above but is nevertheless conceptually useful. Writing the first law for region (0), we have

$$dU = dQ - dW \qquad (1)$$

where, as usual, dU is the internal energy change of (0) as a consequence of its gaining a quantity of heat dQ from (1), performing a quantity of work

dW on (1), and in the process undergoing a change of state. According to the second law, region (0) obeys the inequality

$$dS \equiv \frac{dQ_{\text{rev}}}{T} > \frac{dQ_{\text{irrev}}}{T} \qquad (2)$$

where dS, the entropy gain, is determined by the heat that would be absorbed in a reversible change, dQ_{rev}. This is always greater than the actual heat absorbed in any real change, dQ_{irrev}. Instead of the inequality, we may write for real processes

$$dS = \frac{dQ_{\text{irrev}}}{T} + \frac{dQ'}{T} \qquad (3)$$

where dQ_{irrev} is the actual uptake of heat as before and dQ' is a positive quantity, termed by Clausius the "uncompensated heat." In the notation of NET, which concerns itself explicitly with the uncompensated heat, it is customary to write Equation 3 in the form

$$dS = d_eS + d_iS \qquad (4)$$

The terms on the right of Equation 4 correspond, respectively, to those on the right of Equation 3. Thus d_eS is the exchange contribution to the entropy change in (0) and d_iS is an internal contribution, created by the occurrence of irreversible processes.

(b) The region (0) is an open system, *i.e.*, it can exchange both matter and heat with (1). This is the more interesting of the two possibilities. However, Equations 1 and 2 are now ambiguous since matter carries with it an associated energy and the heat transferred is not well defined. It is necessary to rewrite Equation 1 in a form that takes explicit account of changes in composition; this is the well-known Gibbs equation which we shall use later, obtained by introducing the chemical potentials of the component species. Equation 4 can still be written for region (0), but the terms on the right-hand side are no longer identifiable with those of Equation 3. To see more clearly what the precise meaning of d_iS is, it is worth considering the entropy changes in both regions of the system, bearing in mind that the irreversible process relevant to these considerations occurs only in (0). The total entropy change of the system is evidently given by

$$dS^{\text{total}} = dS^{(0)} + dS^{(1)} \qquad (5)$$

where $dS^{(1)}$ is taken to include entropy changes in the outer container. However, from Equation 4,

$$dS^{(0)} = d_eS^{(0)} + d_iS^{(0)} \qquad (6)$$

$$dS^{(1)} = d_eS^{(1)} = -d_eS^{(1)} \qquad (7)$$

since the entropy change in (1) is solely the result of exchange with (0).

From Equations 5, 6, and 7 it is clear that

$$dS^{\text{total}} = d_i S^{(0)} \tag{8}$$

Thus $d_i S^{(0)}$ represents the total increase of entropy both locally and in the surrounding world from the process taking place in (0).

Intuitively we feel more comfortable with the concept of work than with the concept of entropy. It is readily shown that the irreversible production of entropy is associated with a loss of free energy, *i.e.*, capability for work (*3*). At constant temperature and pressure, the maximum work capability is measured by the Gibbs free energy G. Under these conditions the changes in G in each region are governed by

$$dG^{\text{total}} = dG^{(0)} + dG^{(1)} = -TdS^{\text{total}} \tag{9}$$

Since for any real process dS^{total} is necessarily positive, the free energy of the entire system undergoes a decrease. We shall be concerned with the rate of this loss or dissipation of free energy, which is given by the dissipation function Φ. (The conventional notation $\dot{\chi}$ will be used for the rate of change or time derivative of any quantity χ.)

$$\Phi \equiv T\dot{S}^{\text{total}} = -\dot{G}^{\text{total}} \geq 0 \tag{10}$$

The equality to zero refers of course only to equilibria or hypothetical reversible processes. The dissipation function has the dimensions of power and may be measured in watts if convenient.

For a system in a stationary state, the evaluation of the dissipation function can be carried out readily by using the Gibbs equation to express the internal energy change of the various regions as discussed above. For example, suppose that the process taking place in region (0) is the enzymatic cleavage of a substrate A to give a product B:

$$A \rightleftharpoons 2B \tag{11}$$

Now, when stationarity is reached the properties of (0) do not change with time. The rate at which A is taken up balances the rate at which it is converted to B, and the latter rate just makes up for the loss of B to the surroundings. Hence the internal energy, entropy, volume, and concentrations associated with (0) are all constant. (Stationary states will be reached by region (0) only if it possesses a very small capacity in comparison with the surrounding region (1). The intensive parameters of region (1) should not be appreciably altered either during the approach to stationarity or while the stationary state persists although some parameters must necessarily be altered to a small extent. Such changes are exactly what one measures in characterizing the system.) The Gibbs equation need therefore only be written for (1):

$$dU^{(1)} = TdS^{(1)} - pdV^{(1)} + \sum_i \mu_i^{(1)} dn_i^{(1)} \qquad (12)$$

The summation on the right side accounts for changes in composition in (1) resulting from the reaction's taking place in (0). The quantities $\mu_i^{(1)}$ and $n_i^{(1)}$ are, respectively, the chemical potential and the number of moles of the ith species in (1). Since $dV^{(1)} = -dV^{(0)} = 0$ and $dU^{(1)} = -dU^{(0)} = 0$, we conclude immediately that

$$TdS^{(1)} = -\sum_i \mu_i^{(1)} dn_i^{(1)} \qquad (13)$$

Furthermore, since $dS^{(0)} = 0$, we see from Equation 5 that

$$TdS^{\text{total}} = -\sum_i \mu_i^{(1)} dn_i^{(1)} \qquad (14)$$

The dissipation function is now obtained by taking the derivative of Equation 14 with respect to time:

$$\Phi = T\dot{S}^{\text{total}} = \sum_i \mu_i^{(1)} \dot{n}_i^{(1)} \qquad (15)$$

For a particular system which undergoes the reaction specified in Equation 11, we would have

$$\Phi = -\mu_A^{(1)} \dot{n}_A^{(1)} - \mu_B^{(1)} \dot{n}_B^{(1)} \qquad (16)$$

The velocity of the reaction, v, may be taken to be the rate of disappearance of A in moles/second. If so,

$$\Phi = v(\mu_A^{(1)} - 2\mu_B^{(1)}) \qquad (17)$$

The quantity $\mu_A - 2\mu_B$ (in this example) constitutes the driving force for chemical reaction and was termed the affinity of the reaction by De Donder, who first introduced it as a function of state. More generally the affinity is given by

$$A = -\sum_i \nu_i \mu_i \qquad (18)$$

where ν_i represents the stoichiometric coefficient of the ith species in the reaction, but taken to have a positive sign if the ith species is formed as a product and a negative sign if it is consumed as a reactant. Equation 17 can therefore be conveniently rewritten as

$$\Phi = vA \qquad (19)$$

where A represents the affinity of reaction 11 in region (1). It should be clear from Equation 15 that if r different independent reactions occurred simultaneously we would have

$$\Phi = v_1 A_1 + v_2 A_2 + v_3 A_3 + \ldots + v_r A_r \tag{20}$$

This is the characteristic form of the dissipation function when reactions only are involved.

In the above treatment we have taken into account only the velocities and affinities of chemical reactions, which are scalar processes and hence not associated with any directionality in space. If we consider Figure 1(b), however, we see that the inner region (0) may divide the surrounding vessel into two separate outer regions or compartments, (1) and (2). Thus vectorial flows may occur through (0), which in this case will usually be a membrane separating regions of high and low chemical potential of a given species. However, whatever the combination of processes taking place in a given system, a simple extension of the previous derivation shows that the dissipation function always has the form of a sum of products of conjugate flows and forces (these terms being used in a generalized sense):

$$\Phi = J_1 X_1 + J_2 X_2 + J_3 X_3 + \ldots \tag{21}$$

Here the J's are thermodynamic flows which may include reaction velocities, while the X's are thermodynamic forces, such as reaction affinities and chemical potential differences (both measured in kcal/mole), as well as differences in electrical potential and hydrostatic and osmotic pressure. The number of individual products in the dissipation function corresponds to the number of degrees of freedom of the system.

Relationships between Flows and Forces: The Phenomenological Equations. Several relationships between corresponding flows and forces are well known: for example, Ohm's law for flow of electric current, Fick's law for diffusion, Fourier's law for heat flow, and Poiseuille's and Darcy's laws for fluid flow. In each case the flow is given by the product of a conductance coefficient and the force. In NET, where we usually deal with a set of simultaneous flows, the simple linear relationship is extended to cover all possible cross-effects or interferences. To illustrate this, suppose we have a system that gives rise to a three-term dissipation function, indicating the presence of three independent processes. The phenomenological equations relating the forces and flows would be written

$$\begin{aligned} J_1 &= L_{11} X_1 + L_{12} X_2 + L_{13} X_3 \\ J_2 &= L_{21} X_1 + L_{22} X_2 + L_{23} X_3 \\ J_3 &= L_{31} X_1 + L_{32} X_2 + L_{33} X_3 \end{aligned} \tag{22}$$

This set of linear relations holds good only for relatively slow processes—*i.e.*, for processes sufficiently close to equilibrium. It might be thought that the range of linearity would be too small for a simple linear formulation to be of practical utility. Generally, however, the criterion for closeness to equilibrium is empirical. For vectorial processes, linearity is often observed over a surprisingly wide range of magnitudes of the forces. For chemical reactions the criterion is usually much more restrictive, a problem that will be discussed in the next section. It is seen that each flow J_i is related to its conjugate force X_i through a straight coefficient L_{ii}. It may also be affected by any other force X_j, the coupling or cross coefficient being L_{ij}.

A striking advantage of the formulation now presents itself. Rather than having entirely arbitrary values, the phenomenological coefficients are subject to important thermodynamic restrictions. First, according to Onsager's law, the matrix of coefficients is symmetrical so that

$$L_{ij} = L_{ji} \tag{23}$$

This affords a considerable reduction in the number of coefficients to be measured. Second, since the dissipation function can never be negative (*see* Equation 10), the straight coefficients can likewise never be negative, and the cross coefficients must satisfy the condition

$$L^2_{ij} \leqslant L_{ii}L_{jj} \tag{24}$$

There is an alternative way of formulating the phenomenological equations which makes use of resistance rather than conductance coefficients. As a consequence of the linearity of Equation 22, we could equally well have written

$$\begin{aligned} X_1 &= R_{11}J_1 + R_{12}J_2 + R_{13}J_3 \\ X_2 &= R_{21}J_1 + R_{22}J_2 + R_{23}J_3 \\ X_3 &= R_{31}J_1 + R_{32}J_2 + R_{33}J_3 \end{aligned} \tag{25}$$

The thermodynamic restrictions that apply to this set are completely analogous to those that applied to the previous set.

The phenomenological coefficients cannot be considered to be perfectly constant. They are functions of the parameters of state and will be more or less sensitive to variations in the state of the system. However, they are not functions of the forces or the flows, except indirectly insofar as variations in these may bring about changes in state. It often proves possible, therefore, to carry out experiments over a range of conditions such that the coefficients do in fact remain sensibly constant.

Chemical Reactions

Linearity. For chemical reactions, it will be remembered, the driving force is the affinity given by Equation 18, and the corresponding flow or reaction velocity (in moles or equivalents per second) is

$$v = \dot{n}_i/\nu_i \tag{26}$$

Clearly it is immaterial which of the reactants or products is selected for the estimation of v. For most purposes, in systems of biological interest, the affinity can be identified with the negative of the Gibbs free energy change resulting from the reaction. The Gibbs free energy of the system is given by

$$G = \sum_i n_i \mu_i \tag{27}$$

If temperature and pressure are constant and reaction proceeds to an extent small enough for the chemical potentials of all species to remain essentially constant, then the change in Gibbs free energy (per equivalent of reaction) is evidently, from Equations 18 and 27,

$$-\Delta G = A \tag{28}$$

Under what circumstances is it permissible to consider v linearly proportional to A? Since we are dealing with steady-state enzymatic reactions, it will be advantageous to assume that a reaction scheme of the Michaelis–Menten type applies. Whether or not this is the case, it can be shown (2, 3) that a general condition for linearity is

$$A \ll RT \tag{29}$$

in which case

$$v = L_r A \tag{30}$$

Equation 29 is an extremely restrictive condition, limiting the range of linearity to a region very close to equilibrium. However, Rottenberg (5) has recently shown that in the special case of Michaelis–Menten kinetics the reaction rate is a proportional function of the affinity over a range of more than 2 kcal/mole under appropriate conditions. To this consideration must be added the suggestion by Prigogine (1) that when the affinity of a given chemical reaction is large, the reaction may often be split into a certain number of elementary reactions, each having a sufficiently small affinity to justify the application of linear phenomenological laws. In the stationary state the velocities of all the elementary reactions become identical, and although the total affinity (the sum of the individual affinities) is much greater than RT, the system will remain linear since each individual affinity is much less than RT. Kinetic schemes for enzymatic reactions invariably consist of a set of consecutive steps, and the reaction sequences characteristic of biological systems frequently involve many enzymatic reactions that do not depart far from equilibrium (6). Thus the elemental steps may well be very close to equilibrium. If the reactions are under equilibrium control (7), L_r appears to be insensitive to changes in the concentration of substrates. The above arguments suggest that it is not unreasonable to treat many of the chemical reaction systems of biology as linear, at least as a first approximation.

Coupling and Stoichiometry. In applying NET to an experimental problem, one set of variables often proves to be more convenient than another. In this section a general principle is illustrated—namely, that the set of conjugate forces and flows that appear in a given dissipation function may be replaced by alternative sets, if this is done in such a way as to leave the dissipation function invariant. Mathematically this statement is expressed by writing

$$\Phi = J_1 X_1 + J_2 X_2 + J_3 X_3 + \ldots = J_1' X_1' + J_2' X_2' + J_3' X_3' + \ldots \tag{31}$$

where the unprimed and primed quantities represent equivalent alternative sets. The number of terms in each set need not necessarily be the same, but there is an irreducible minimum corresponding to the number of independent processes taking place, *i.e.*, to the number of degrees of freedom of the system. If we are satisfied that the unprimed forces and flows form a legitimate set (derived by the method described earlier) and that the system behaves linearly, then the alternative primed set gives rise to phenomenological coefficients that still obey Onsager symmetry—providing it is obtained in the following way. The new flows must be linear combinations of the old flows, and the new forces must be the appropriate conjugate linear combinations of the old forces (*2, 3*).

Let us consider a well-known class of biochemical reactions, the synthetic reactions mediated by ATP. Many of these apparently conform to a general stoichiometric equation, which may be written (*8*)

$$XOH + YH + ATP = XY + ADP + HP_i \tag{32}$$

where HP_i is used to represent H_3PO_4. The species XY is synthesized at the expense of ATP. An example of this is the synthesis of sucrose in bacteria from glucose (XOH) and fructose (YH). We may speculate as to whether Reaction 32 could conceivably proceed by the way of the following two reactions:

$$XOH + YH = XY + H_2O \tag{33}$$

$$ATP + H_2O = ADP + HP_i \tag{34}$$

As before, these will occur in a system such as that illustrated in Figure 1a. The appropriate enzymes are bound or otherwise confined to region (0), while region (1) is the pool of reactants and products characterized by the two affinities. We denote the affinity and velocity of Condensation Reaction 33 by A_c and v_c, and the affinity and velocity of the Hydrolytic or Driving Reaction 34 by A_h and v_h. The overall Synthetic Reaction 32 is characterized by A_s and v_s. Suppose that region (0) has an extremely low capacity for water, to which it is essentially impermeable. This might be the case if it were lipid in character. For example, a fragment of cell membrane or endoplasmic reticulum might behave in this way. In such a medium, small changes in water concentration are accompanied by large changes in the chemical potential of water. Suppose further that an ATPase that is functional in a nonaqueous environment and suitable facilitation of sugar transport are present. Then in the steady state, the occurrence of ATP hydrolysis may lower the chemical potential of water within region (0) to such an extent that the local affinity of the condensation reaction becomes appreciable, and it may proceed at a respectable rate. Under these circumstances water plays the role of a common intermediate between the reactions. While this mechanism may be fanciful, it does exemplify an important principle. If region (0) is truly impermeable to water, then the rates of the two reactions will be identical in any steady state, *i.e.*, they will be completely coupled. The stoichiometric ratio (XY produced)/(ATP consumed) will always be unity. However, if region (0) is not truly impermeable to water, it will enter from region (1), and the two reactions will then be incompletely coupled. The degree of coupling obviously depends on the extent to which water leaks in. If we write phenomenological equations for the steady state of this system, *i.e.*,

$$v_c = L_c A_c + L_{ch} A_h \tag{35}$$

$$v_h = L_{hc} A_c + L_h A_h \tag{36}$$

the coefficients characterize the transport properties of region (0) as well as the enzyme kinetics. Indeed the coefficients can be analyzed in terms of these properties (*9*) providing sufficient information is available concerning the nature of region (0). In the case of incomplete coupling, the reaction rates are unequal and the "stoichiometric ratio" will vary with the affinities. (A quantitative measure of degree of coupling will be given later.) On the other hand, if the system is indeed completely coupled, the velocities of the component reactions must be identical and

$$\Phi = v_c A_c + v_h A_h = v_s A_s \tag{37}$$

The foregoing suggests that an integral and constant stoichiometry between component reactions need not be obligatory in biochemical systems. To illustrate this point further, consider oxidative phosphorylation. The sources of uncoupling described above are vectorial, in the sense that they involve the actual leakage or flow of an intermediate from the reaction zone into the surroundings. Scalar sources of uncoupling, involving chemical leaks, are just as readily envisaged. It is commonly held that in oxidative phosphorylation there is a stoichiometric relation between the oxidation and phosphorylation reactions. However, this need not necessarily be so (*10*). For example, the oxidation of acetaldehyde by NAD^+ in microorganisms (*11*) is a well-known case of substrate level phosphorylation. Here, lack of stoichiometry in the oxidation and phosphorylation reactions considered separately results from trivial side reactions for which easy corrections can be made. On the other hand, if hydrolysis of the intermediate linking the two reactions (acetyl co-enzyme A) occurs, then the reactions will be incompletely coupled.

Although the stoichiometry of the principal reactions remains undisturbed, there is a deviation from the "theoretical ratio" relating them to one another.

Transport Phenomena and the Influence of Chemical Reactions

Membrane Characteristics. Membrane processes proved to be a particularly fertile field for the application of NET, but only the barest outline will be presented here, as several extensive treatments exist (2, 4, 12, 13, 14). These processes provide particularly elegant and instructive examples of the value of linear transformation of the dissipation function and the importance of the Onsager relations. The dissipation function is derived by considering the situation represented diagrammatically in Figure 1b. For solutions of electrolytes with passage of electric current, reversible electrodes (rather than salt bridges) must be introduced into each of the compartments. Considering aqueous solutions of a single permeable salt, and denoting the cation and anion by subscripts 1 and 2, respectively, and water by w, the dissipation function takes the form

$$\Phi = J_w \Delta \mu_w + J_1 \Delta \tilde{\mu}_1 + J_2 \Delta \tilde{\mu}_2 \qquad (38)$$

where all fluxes are considered positive in the direction (1) → (2), the electrochemical potential difference $\Delta \tilde{\mu}_i = \tilde{\mu}_i^{(1)} - \tilde{\mu}_i^{(2)}$, and $\tilde{\mu}_i$ is defined by

$$\tilde{\mu}_i = \mu_i + z_i F \psi \qquad (39)$$

z_i being the charge associated with an ion of species i, F the faraday constant, and ψ the electrical potential. This representation of Φ is not, however, convenient. The transformations which lead to operational forms of the dissipation function for membrane processes were studied in a classic series of papers by Kedem and Katchalsky (15, 16, 17, 18, 19). A much more useful form for dilute solutions is the following:

$$\Phi = J_v (\Delta p - \Delta \pi) + J_s (\Delta \pi_s / \bar{c}_s) + IE \qquad (40)$$

Here J_v, J_s, and I are respectively the volume flow, solute (neutral salt) flow, and electric current density, while Δp, $\Delta \pi_s$, and E represent the hydrostatic pressure difference, the osmotic pressure difference attributed to the permeable solute, and the potential difference between the electrodes. The total osmotic pressure difference $\Delta \pi$ includes a contribution from impermeable solutes, $\Delta \pi_i$. The quantity $\bar{c}_s$ is defined by $\bar{c}_s = \Delta \pi_s / \Delta \mu^c_s$, where $\Delta \mu^c_s$ is the concentration-dependent part of $\Delta \mu_s$; in practice $\bar{c}_s$ is essentially the logarithmic mean solute concentration between the two compartments, i.e., $\Delta c_s / \Delta \ln c_s$.

Equation 40 shows that the system as given has three degrees of freedom. If a neutral solute is used, or no electrodes are inserted, the third term will always be zero and the number of degrees of freedom will be reduced to two. Another way of reducing the number of degrees of freedom is to maintain identical solutions on either side of the membrane ($\Delta\pi = \Delta\pi_s = 0$). In this case

$$\Phi = J_v \Delta p + IE \qquad (41)$$

The phenomenological equations corresponding to Equation 41

$$\begin{aligned} J_v &= L_{11}\Delta p + L_{12}E \\ I &= L_{12}\Delta p + L_{22}E \end{aligned} \qquad (42)$$

embrace the whole of electrokinetics. Indeed, symmetry relations have probably been studied more thoroughly in this system than in any other and were known long before the advent of nonequilibrium thermodynamics.

The conductance or resistance coefficients which appear in phenomenoloigcal equations are not always the most convenient quantities to measure; certain combinations are frequently more accessible by experiment. One of the sets of practical transport coefficients developed by Kedem and Katchalsky (17) is shown in the following flow equations, obtained by appropriate linear transformations of the basic phenomenological relations:

$$\begin{aligned} J_v &= L_p(\Delta p - \Delta\pi_i) - \sigma L_p \Delta\pi_s + \beta I \\ J_s &= \overline{c}_s(1-\sigma)J_v + \omega\Delta\pi_s + (\tau_1/\nu_1 z_1 F)I \\ I &= \kappa(\beta/L_p)J_v + \kappa(\tau_1/\nu_1 z_1 F)\Delta\mu_s^c + \kappa E \end{aligned} \qquad (43)$$

Equations 43 define the filtration coefficient L_p, the solute permeability ω, the electrical conductance κ, the transport number (of the ion to which the electrodes are not reversible) τ_1, the electroosmotic permeability β, and the reflection coefficient σ. (If the solute happens to be a nonelectrolyte or the system is open-circuited only the first two equations are required, with $I = 0$.) Since there are three degrees of freedom, the operations involved in measuring these coefficients always require the imposition of two restraints on the system. For example, the reflection coefficient, first introduced by Staverman (20), is given by

$$\sigma = \left(\frac{\Delta p - \Delta\pi_i}{\Delta\pi_s}\right)_{\substack{J_v=0 \\ I=0}} = 1 - \frac{1}{\overline{c}_s}\left(\frac{J_s}{J_v}\right)_{\substack{\Delta\pi_s=0 \\ I=0}} \qquad (44)$$

A systematic experimental study of the six NET conductance coefficients characterizing an ion-exchange membrane, as functions of solution concentration, has recently been described by Meares and co-workers (21, 22, 23, 24, 25).

Coupling Between Reactions and Flows. In bioenergetics one of our major concerns is the study of active transport—*i.e.*, transport which results as a consequence of a metabolic reaction. Kedem (26) first included chemical reaction explicitly in an array of interacting flows and forces describing membrane processes. Her formulation was based implicitly on the addition of a reaction term to the dissipation function appearing in Equation 38 to give

$$\Phi = J_w \Delta\mu_w + J_1 \Delta\tilde{\mu}_1 + J_2 \Delta\tilde{\mu}_2 + vA \tag{45}$$

The corresponding phenomenological equations were written as

$$\begin{aligned}
\Delta\mu_w &= R_{ww}J_w + R_{w1}J_1 + R_{w2}J_2 \\
\Delta\tilde{\mu}_1 &= R_{1w}J_w + R_{11}J_1 + R_{12}J_2 + R_{1r}v \\
\Delta\tilde{\mu}_2 &= R_{2w}J_w + R_{21}J_1 + R_{22}J_2 + R_{2r}v \\
A &= R_{r1}J_1 + R_{r2}J_2 + R_{rr}v
\end{aligned} \tag{46}$$

In Equations 46 nonzero values of the coefficients R_{1r} and R_{2r} indicate coupling between ionic flows and reaction; no direct coupling exists between the flow of water and reaction. Kedem defined the transport of an ion i to be active if the coefficient R_{ir} is different from zero. For example, in a short-circuited frog skin mounted between identical solutions, the flow of electric current coincides with the flow of sodium ions. This observation is consistent with the conclusion that $R_{2r} = R_{12} = 0$, while R_{1r} is nonzero.

There is an important difference between the coefficients R_{ir} and the other coefficients appearing in Equations 46, including R_{rr}. The latter are all scalar quantities; the former, however, are clearly vectors, since both v and A are scalars while $\Delta\tilde{\mu}_i$ and J_i are vectors. We must therefore ask ourselves the physical significance of a vectorial coupling coefficient.

The Curie–Prigogine Principle. To come to grips with the notion of a vectorial coupling coefficient, it is necessary to consider the nature of coupling between flows and forces of essentially different character. If a reaction within a membrane couples to a flow, as in the short-circuited frog skin mentioned above, intuitively the direction of the flow must be determined by a property of the membrane. If the membrane were completely isotropic— *i.e.*, its equilibrium properties were identical in all directions—it is not to be expected that such coupling could occur. This idea is embodied in the principle originally enunciated by Curie, which in its application to NET by Prigogine (1.) and later others (27), indicates that coupling between scalar and vector flows is impossible in isotropic media in the linear regime. However, in anisotropic media such coupling is no longer forbidden. The coupling coefficient must necessarily reflect the anistropy of the medium, and consequently will itself be vectorial.

Stationary State Coupling. If the number of restraints on a system in the steady state is changed, what happens? For example, take the system represented by Equations 46. If the maximum number of restraints are applied (all four forces being fixed) then the stationary state is fully defined since no more degrees of freedom are left. If no restraints at all are applied, the system will eventually reach equilibrium. Frequently, however, we impose an intermediate number of restraints. For such situations, Prigogine has shown that the entropy production assumes the minimal value compatible with the imposed restraints. Thus, if some of the forces are fixed, the remainder will reach values in the stationary state such that their conjugate flows become zero.

Stationary states of minimal entropy production can give rise to a special type of interdependence among flow processes, examples of which have already cropped up in the earlier discussion of chemical reactions. This interdependence has been termed by Prigogine stationary state coupling. Characteristically it manifests itself as a coupling between chemical reaction and diffusional flow, a direct consequence of the stationarity condition, which imposes a mutual linear dependence on some of the flows, in effect contracting the dissipation function to fewer terms. In a typical example discussed by Prigogine (*1*), which may be interpreted as a simplified model of a biological cell, a stationary nonequilibrium distribution of matter can arise, determined by the rate of a metabolic reaction within the cell. This case involves an open system such as region (0) of Figure 1a. The system, or cell, receives a component M from the environment and transforms it into a component N, which is then returned to the environment. It also receives another component O, which does not take part in any chemical reaction and is coupled frictionally to the flow of M only. To understand the consequences of stationarity in this system, it is sufficient to examine the dissipation function. If we apply our standard procedures, taking the view that the cell membrane is in a stationary state (since it has a much lower capacity than either the interior or the environment), we are led to the following dissipation function:

$$\Phi = J_M \Delta \mu_M + J_N \Delta \mu_N + J_O \Delta \mu_O + v A^{in} \qquad (47)$$

where the positive direction of flow is taken from environment to cell and A^{in} represents the affinity of the reaction within the cell. Considering the reactive components only, a condition for stationarity of the whole cell is

$$v = J_M = -J_N \qquad (48)$$

If we denote by A^{ex} the affinity of the reaction in the environment (where of course the requisite enzyme is absent), we have from Equation 18,

$$A^{ex} - A^{in} = \Delta \mu_M - \Delta \mu_N \qquad (49)$$

When the cell is in a stationary state the dissipation function therefore contracts to

$$\Phi = J_O \Delta \mu_O + v A^{ex} \qquad (50)$$

and since the only restraint that can be imposed is the fixing of A^{ex}, the result is that $J_O = 0$. On writing down the two phenomenological equations corresponding to the contracted dissipation function, $\Delta \mu_O$ is seen to be directly proportional to both v and A^{ex}.

In the above example, stationary state coupling occurs between the flow of component O and the reaction. In essence this is a two-compartment system, one compartment of which is inaccessible (except through the membrane) and contains an enzyme. The system is therefore basically unsymmetrical. We may infer that in order to obtain stationary state coupling between reaction and flow it is not necessary to have anisotropy as long as an appropriate unsymmetrical arrangement of isotropic elements is provided.

Active transport in cells such as muscle cells or red cells is not, however, considered to be associated with a mechanism of the above type. The coupling is clearly a property of the membrane itself, associated with enzymes which are an integral part of the membrane and attached either within it or to its surface. Nevertheless, in some cases we may have stationary state coupling on a molecular level.

Isotope Kinetics

Permeability Coefficients, Bulk Flow, and Single-File Diffusion. Radioactive isotopes have been used extensively to evaluate the permeability of biological membranes, with good reason. In order to determine permeability both concentration gradients and rates of flow must be measured. In other words, concentration changes on either side of the membrane have to be monitored. If the concentrations are low, or if the concentration gradients or flows are very small (they may be zero under physiological conditions), such measurements may be difficult to perform with reasonable accuracy. On the other hand, isotopes added in trace amounts, insignificant in comparison with the concentration of the parent substance, may present little difficulty with regard to concentration determinations. Thus a permeability may be calculated, for example, from measurements of tracer flow in the absence of net flow:

$$\omega_s{}^* = \left(\frac{J_s{}^*}{\Delta \pi_s{}^*}\right)_{J_V = 0,\ I = 0,\ J_s = 0} \tag{51}$$

The asterisk refers to the tracer species.

Now if, as is usually the case, we can ignore isotope effects, in other words we deal with isotopes differing so slightly in atomic weight that their kinetic and thermodynamic characteristics are to all intents and purposes identical, it may seem reasonable to expect that $\omega_s{}^* = \omega_s$, the permeability for net flow. This expectation, however, is subject to doubt, if only because the tracer and abundant species are present in very different concentrations and may have very different concentration gradients. Historically, the first discrepancy to come to the attention of physiologists arose in studies of the permeability of frog skin to water. This had different values depending on whether net flow induced by an osmotic pressure gradient or tracer flow in the absence of a net water flow was measured. An explanation of the discrepancy was put forward by Koefoed-Johnsen and Ussing (28), who suggested, in effect, that diffusion of tracer water might be coupled to flow of solvent water.

A closely related problem concerns the use of isotopes to study the nature of the forces which bring about the flows observed. Ussing (29) pointed out that if a substance moves as a consequence only of its own electrochemical potential gradient, then the flux ratio (here designated f) measured by adding two tracers, one to each side of the membrane, might be given by

$$RT \ln f = \Delta \tilde{\mu} \tag{52}$$

In the absence of electrical effects this simplifies to

$$f = c^{(1)}/c^{(2)} \tag{53}$$

Abnormality of the flux ratio, it was suggested, should indicate quantitatively the extent to which other forces, perhaps due to solvent drag or metabolism, might be operative. But the limitations of this approach quickly became apparent, for example, in the potassium flux measurements of Hodgkin and Keynes (30) in poisoned squid axons. A marked abnormality was found despite the absence of both solvent drag and active transport. Single-file diffusion was postulated as a possible mechanism for this.

The Flux Ratio and Isotope Interaction. These questions were re-examined by Kedem and Essig (31) by using the appropriate phenomenological equations. The equations for local flows within a homogeneous membrane are written down with the following convention: the subscript 0, sometimes omitted, refers to the total test substance under study, 1 refers to its abundant isotope, and 2 and 3 to tracer isotopes. The forces are local gradients of electrochemical potential:

$$-\frac{d\tilde{\mu}}{dx} = r_{00}J + \sum_{j=4}^{n} r_{0j}J_j \tag{54}$$

$$-\frac{d\tilde{\mu}_1}{dx} = r_{11}J_1 + r_{12}J_2 + r_{13}J_3 + \sum_{j=4}^{n} r_{1j}J_j$$

$$-\frac{d\tilde{\mu}_2}{dx} = r_{21}J_1 + r_{22}J_2 + r_{23}J_3 + \sum_{j=4}^{n} r_{2j}J_j \tag{55}$$

$$-\frac{d\tilde{\mu}_3}{dx} = r_{31}J_1 + r_{32}J_2 + r_{33}J_3 + \sum_{j=4}^{n} r_{3j}J_j$$

Here the flows J_j (where j is 4 or higher) include all species passing through the membrane other than the test species as well as any metabolic reactions that may be coupled to the flow of test species. The sum therefore represents the force contributed by all additional coupled processes. Equations 55 allow explicitly for the effects of *isotope interaction*, i.e., interaction of isotopes of the same chemical species, through the coupling coefficients r_{ik} ($i \neq k$; $i, k = 1, 2, 3$). Integrating Equation 54 across the membrane (of thickness Δx) in the steady state, from side (1) to side (2), leads to the result

$$JR = \Delta \tilde{\mu} - \int_0^{\Delta x} \sum_{j=4}^{n} r_{0j} J_j \, dx \tag{56}$$

where R, the resistance to net flow, is given by

$$R = \int_0^{\Delta x} r_{00} \, dx \tag{57}$$

For convenience, we denote the specific activity of a tracer species by ρ, e.g., $\rho_2 = c_2/c$. Considerations of thermodynamic and kinetic indistinguishability of the isotopes, as well as Onsager symmetry, are now invoked in order to arrive at the following important conclusions:

$$\begin{aligned} J_2 - \rho_2 J &= -\frac{RT}{r_{00} - r_{ik}} \frac{d\rho_2}{dx} \\ J_3 - \rho_3 J &= -\frac{RT}{r_{00} - r_{ik}} \frac{d\rho_3}{dx} \end{aligned} \tag{58}$$

where $r_{ik} = r_{ki} = r_{12} = r_{13} = r_{23}$. These equations are consistent with the requirement that in the absence of a gradient of specific activity through the membrane, the flow of tracer should be in a fixed ratio to the net flow. However, in the presence of a gradient of specific activity, tracer flow runs ahead of net flow, the relative rates depending only on the magnitude of the gradient and the resistance coefficient (despite the dependence of both flows on all forces).

Because of their simple form, Equations 58 may also be integrated across the membrane. The most straightforward case is to suppose that net flow is zero—i.e., only isotope exchange occurs. Then for a tracer species i,

$$(J_i R^*)_{J=0} = RT \Delta \rho_i \tag{59}$$

where R^*, the exchange resistance, is given by

$$R^* = \int_0^{\Delta x} (r_{00} - r_{ik}) \, dx \tag{60}$$

Comparing Equations 57 and 60, it is immediately apparent that the resistance to net flow will equal the exchange resistance only if isotope interaction is absent or can be neglected. Since on thermodynamic grounds r_{ik} may be of either sign, the exchange resistance may be either greater than or less than the resistance to net flow. Notice that for neutral solutes, or in the absence of an electrical field, we might write from Equation 56

$$\omega = \left(\frac{J}{\Delta\pi}\right)_{J_j=0} = \frac{1}{cR} \tag{61}$$

where j is 4 or greater and $\bar{c}$ represents the logarithmic mean concentration of test substance. On the other hand, if $\Delta c = 0$ we might write from Equation 59

$$\omega^* = \left(\frac{J_i}{\Delta\pi_i}\right)_{J=0} = \frac{1}{cR^*} \tag{62}$$

Thus the isotope permeability need not necessarily be equal to the permeability of the total test substance.

To determine the flux ratio, Equations 58 are integrated without setting the net flow to zero. This yields the widely used relation

$$\text{net flux} = \text{influx} - \text{outflux} \tag{63}$$

Thus if each tracer is added to only one compartment and $\rho_2^{(2)} \approx \rho_3^{(1)} \approx 0$, one obtains

$$J = \frac{J_2}{\rho_2^{(1)}} + \frac{J_3}{\rho_3^{(2)}} \tag{64}$$

where $J_2/\rho_2^{(1)}$ may be identified with the influx and $-J_3/\rho_3^{(2)}$ with the outflux. The flux ratio is defined as influx/outflux (29); for this quantity the integration of Equations 58 yields

$$JR^* = RT \ln f \tag{65}$$

It is clear from Equation 65 that measurements of influx and outflux may be used to calculate the exchange resistance but will yield the resistance to net flow only if isotope interaction is absent. Combining Equations 56 and 65 gives

$$RT \ln f = \frac{R^*}{R} \left(\Delta\tilde{\mu} - \int_0^{\Delta x} \sum_{j=4}^{n} r_{0j}J_j \, dx\right) \tag{66}$$

This equation incorporates the influence of active transport in the summation term. In the absence of isotope interaction $R^*/R = 1$ and the equation is analogous to the flux ratio equation of Ussing. In the presence of isotope interaction the factor R^*/R may be greater or less than one, depending on whether coupling is positive or negative.

The possible existence of isotope interaction obviously makes the interpretation of experiments utilizing tracers very much more complicated, and rather specific mechanisms must operate for it to occur. If frictional effects only are involved, one would scarcely expect to observe isotope interaction at physiological concentrations in nonselective membranes or in free aqueous solution since the probability of collisions with transfer of momentum is slight. This conclusion is borne out by an analysis of published data (32). Inside a biological membrane, however, circumstances may be very different. Gottlieb and Sollner published studies of highly selective collodion–matrix anion-exchange membranes activated with polyvinylmethylpyridinium bromide, in which they obtained findings consistent with an appreciable degree of negative isotope interaction (33). Membranes of this type were used by De Sousa et al. (34) to test the modified flux ratio equation in the absence of solvent drag or active transport, i.e., Equation 66 with the integral term zero. The quantities R and R^* were evaluated independently, R from the electrical resistance (using Equation 56, all J_j being zero), and R^* from the self-diffusion of ^{36}Cl (using Equation 59). The values of R^*/R were as low as 0.42. Tracer determinations of the flux ratio f were in excellent agreement with the predictions of Equation 66, while being abnormally low according to Equation 52.

In biological membranes such "abnormal" flux ratios are usually attributed to exchange diffusion by mobile carriers. A number of other plausible mechanisms have, however, been suggested (35, 36, 37). Thus, since mobile carriers can hardly traverse the synthetic membranes considered above, they may also not provide the mechanism for abnormal flux ratios in biological membranes.

Energy Conversion: General Considerations

Degree of Coupling. In the preceding sections, the idea that processes might be completely or incompletely coupled was taken as self-evident, although no measure of coupling was proposed. It will now be shown that the phenomenological description of two coupled flows leads directly to a thermodynamic definition of their degree of coupling (38). In the following discussion we consider a perfectly general system in which two processes are coupled and consequently in which energy conversion may occur. Many examples of energy converison by pairs of coupled processes are encountered in physiology and biochemistry, the most obvious ones being active transport and muscular contraction. We shall suppose that coupling takes place within a black box whose mechanism may not in general be understood. This black box usually constitutes the working element of the energy converter—e.g., the membrane in an active transport process. Since there are two processes, the dissipation function will consist of two terms. In general we can pick out one process which always gives rise to a positive term and hence may be regarded as the energy source or input. The other process generally gives rise to a negative term and constitutes the output. Since the output is usually more accessible than the input in biological systems, it is in a sense proximal to the observer, and we shall denote it by 1, while the input is denoted by 2. The dissipation function then takes the form

$$\Phi = \underbrace{J_1 X_1}_{-\text{ output power}} + \underbrace{J_2 X_2}_{\text{input power}} \qquad (67)$$

It is to be noted that the input flow J_2 always takes place in the spontaneous direction, *i.e.*, its direction is in accord with that of its conjugate force X_2. The output flow J_1, however, takes place *against* its conjugate force X_1 as a consequence of its coupling to J_2. Consequently the output term is negative, indicating that free energy is withdrawn from the working element into the surroundings. Thus part of the free energy expended by, and characteristic of, process 2 is converted into the form characteristic of process 1. Since the dissipation function can never be negative, output can never exceed input. It is intuitively clear that effective energy conversion requires tight coupling between the two processes.

The phenomenological equations of the system are given by Equations 22 or 25 (on deleting the last line and column in each set). In order to appreciate the fact that the degree of coupling emerges naturally from the phenomenological description, it is convenient to define a quantity Z as follows

$$Z = \sqrt{L_{11}/L_{22}} = \sqrt{R_{22}/R_{11}} \qquad (68)$$

The physical significance of Z will become clear shortly; for now it is sufficient to observe that Z has the dimenisons of J_1/J_2 or X_2/X_1. The ratio of flows can be written as a function of the ratio of forces in the following way:

$$\frac{(J_1/J_2)}{Z} = \left(\frac{L_{12}}{\sqrt{L_{11}L_{22}}} + Z(X_1/X_2)\right) / \left(1 + \frac{L_{12}}{\sqrt{L_{11}L_{22}}}\right) Z(X_1/X_2) \qquad (69)$$

The value of the quantity $L_{12}/\sqrt{L_{11}L_{22}}$, which we shall denote by q, is limited by the condition expressed in Equation 24:

$$-1 \leqslant q \leqslant 1 \qquad (q = L_{12}/\sqrt{L_{11}L_{22}}) \qquad (70)$$

This follows, of course, from the thermodynamic restriction that the dissipation function can never be negative. It is seen by examining Equation 69 that the nearer the absolute value of q is to unity, the slighter the dependence of J_1/J_2 on X_1/X_2. When q is zero, on the other hand, J_1/J_2 is proportional to X_1/X_2; the two processes are independent, and each flow is proportional to its conjugate force without any influence by the other force. Thus q is a measure of the degree of coupling, and a basis of comparison for different coupled systems. It is readily shown that q is also given by

$$q = -R_{12}/\sqrt{R_{11}R_{22}} \tag{71}$$

The significance of the sign of q is clearest if both flows are vectors. For example, the flow of a solute through a membrane may drag another solute along in the same direction ($q > 0$), or by an exchange process tend to push it back ($q < 0$). If q is positive, L_{12} is positive but R_{12} is negative, and vice versa.

The thermodynamic implications of $q = \pm 1$ are very important. In this case, according to Equations 69 and 70 we have

$$J_1/J_2 = \pm Z \tag{72}$$

(As the absolute value of q tends to unity the R's tend to $\pm \infty$, but the L's remain finite.) For such systems J_1 and J_2 are in a fixed ratio, and Z represents the stoichiometry of the processes. Since the flows are not linearly independent the dissipation function can be contracted to a single term—*i.e.*,

$$\Phi = J_2(X_2 \pm ZX_1) \quad (q = \pm 1) \tag{73}$$

If X_1 and X_2 can be varied independently, the composite force $X_2 \pm ZX_1$ can be made as small as desired. Hence Φ may become vanishingly small compared with input and output. This means that the process approaches reversibility as the rate tends to zero. An electrochemical cell, for example, can be completely coupled if there is one homogeneous electrolyte and no side reactions occur. The dissipation function is given by

$$\Phi = IE + vA \tag{74}$$

In the present notation $Z = nF$, where n is the number of electrons transferred per mole. Therefore, if coupling is indeed complete,

$$\Phi = v(A + nFE) \tag{75}$$

At reversible equilibrium, $\Phi = v = 0$ and $A = nF(-E)_{I=0}$. The quantity $(-E)_{I=0}$ is the emf of the cell *(38)*.

Energy Expenditure without Performance of Work. Two stationary states, in each of which the output term J_1X_1 is zero, are of particular significance; it should be borne in mind that in any stationary state the forces X_1 and X_2 are constant in time.

STATIC HEAD. If X_2, the input or driving force, is held constant but no restriction is placed on X_1, the system will eventually reach the state of minimal entropy production (when $J_1 = 0$ and X_1 thereafter remains constant). In this state we refer to the system as having developed a static head

(*38*). Examples of static head are an open-circuited fuel cell, the maintenance of concentration gradients across plant and animal cell membranes by active transport, and an isometric muscle contraction. In an incompletely coupled system, input energy must be expended to maintain static head, even though output is zero.

LEVEL FLOW. With external reservoirs of finite capacity, X_1 can be completely constant only if J_1 vanishes or is compensated by an equal flow through the external world. By suitable adjustment of this external flow one can clamp X_1 at any desired value, as for example in the customary measurement of the short-circuit current in biological systems. Instead of an electric current at zero potential gradient, any other force (*e.g.*, a pressure head) can be held at zero. In practice $X_1 = 0$ is usually achieved by using a suitable compensating device. We refer to this situation as level flow. Examples of level flow are a short-circuited fuel cell, the physiological transport of salt or water between isotonic solutions, and an unloaded muscle contraction. Clearly input energy must be expended to maintain level flow, even though output is zero.

If the driving force X_2 is either fixed or known to have the same value both at level flow and at static head, the measurement of q is straightforward. All that is required is to determine the input flow (or fuel consumption) in the two stationary states, since under these conditions

$$q^2 = 1 - \frac{J_2^{\text{static head}}}{J_2^{\text{level flow}}} \tag{76}$$

Effectiveness of Energy Utilization—Efficiency and Efficacy. The concept of efficiency has been interpreted in various ways by different workers. However, the efficiency of a system as an energy converter in some stationary state is clearly specified by the dissipation function, Equation 67. If efficiency is denoted by η, we have

$$\eta = \text{output/input} = -J_1 X_1 / J_2 X_2 \tag{77}$$

The condition for $J_1 X_1$ to constitute an output, *i.e.*, to become negative, while $J_2 X_2$ remains positive, is given in terms of the forces by

$$-1 \leqslant Z(X_1/X_2)/q \leqslant 0 \tag{78}$$

(The left-hand limit corresponds to static head, the right-hand limit to level flow.) Whenever this condition is satisfied, process 2 drives process 1, and

$$0 \leqslant \eta \leqslant 1 \tag{79}$$

The value of η may readily be obtained as a function of the force ratio and the degree of coupling:

$$\eta = -[q + Z(X_1/X_2)]/[q + 1/Z(X_1/X_2)] \tag{80}$$

Since efficiency is zero at static head and level flow, it must pass through a maximum at some intermediate value of the force ratio. This maximum value of the efficiency, η_{max}, depends only on the degree of coupling:

$$\eta_{max} = q^2/(1 + \sqrt{1-q^2})^2 \qquad (81)$$

Completely coupled systems can approach $\eta_{max} = 1$ (reversible equilibrium) as a limit. However, as pointed out earlier, the rate of the process under these conditions tends to zero.

The efficiency is obviously of interest when a system converts one form of energy into another. For example, energy is converted if a muscle raises a weight. Another example occurs in the loop of Henle, where sodium is transported against large differences of electrochemical potential at a rate which permits the osmotic water flow necessary to concentrate the urine. However, near static head and level flow, energy conversion becomes small and η approaches zero. The function of a biological system operating under these circumstances is clearly not energy conversion, and therefore the efficiency is of no significance.

Near static head the function of the system is apparently the maintenance of electrochemical potential difference, tension, or other appropriate force. In this case, a parameter of interest is the force developed per given rate of expenditure of metabolic energy, or force efficacy (*39*). Near level flow, the function of the system appears to be rapid transport of material. Here it is useful to consider the rate of transport per given rate of expenditure of metabolic energy, or flow efficacy (*39*).

Active Transport

Definitions. Of the several different identifying criteria which have been proposed for active transport, one that is frequently invoked is abnormality of the flux ratio, active transport being considered probable if $RT \ln f$ exceeds the electrochemical potential difference of the test species $\Delta\tilde{\mu}$. However, a glance at Equation 66 shows that this test may be misleading since other influences, including isotope interaction, may be operative. Rosenberg (*40*) considered the "demonstration of transport from a lower to a higher potential . . . the only certain criterion of active transport," but suggested also a broader definition: "movement of a substance which is influenced by other forces in addition to the chemical (or analogous) potential gradient." In phenomenological terms, for two-flow linear systems, Rosenberg's definition is equivalent to the definition of Kedem discussed earlier, *i.e.*, active transport of a species i occurs if the cross-coefficient R_{ir} is nonzero. The latter definition implies that if a system comprises several flows, a sufficient criterion for active transport is the existence of a nonzero $\Delta\tilde{\mu}$ for the test species when all flows have reached static head under continuing metabolism.

The energetics of active transport have been the subject of numerous studies from different points of view. Several of these divide the expenditure of metabolic energy into two parts: conversion into electroosmotic work and additional expenditure necessarily associated with a real transport process. The sum of these quantities is defined as work and is compared with estimates of the energy available. In principle, at least, the conversion into electroosmotic work may be evaluated by determining $-J\Delta\tilde{\mu}$, where J is the net rate of transport of the test species. On the other hand, the additional energy expenditure is not well defined. Ussing (41) and others regard this quantity, in the case of active sodium transport in frog skin, as the "work required to overcome the internal sodium resistance of the skin," evaluating it by means of the flux ratio. According to this interpretation, $RT \ln f$ represents the internal work done per mole of sodium transported. However, the flux ratio has serious interpretative shortcomings, and a precise evaluation of work (or an efficiency) which incorporates "internal work" cannot be made. Even if this were possible, a fundamental conceptual difficulty remains since the maintenance of static head across a membrane in the absence of leak should, in this view, require no expenditure of energy. This could only be the case if transport and metabolism were completely coupled.

An obvious example of complete coupling is an electrochemical cell without side reactions. Although electrophysiological data are customarily interpreted in terms of such a cell, the justification for doing so is doubtful. Ussing and Zerahn (42) evaluated the electromotive force of sodium transport, E_{Na}, considering it to represent the apparent driving force for active transport. The methods fall into two categories: (a) the determination of the static head quantity $(\Delta\tilde{\mu}_{Na})_{J_{Na}=0}$ by adjustment of the electrical potential difference or concentration difference across the membrane or by substitution of impermeant for permeant anions, and (b) the determination of the level flow (short circuit) flux ratio $(RT \ln f)_{\Delta\tilde{\mu}_{Na}=0}$. It follows from the work of Kedem and Essig (31), however, that these two quantities cannot be identical unless three conditions are satisfied, $viz.$, the absence of leaks, the absence of isotope interaction, and the identity of rates of metabolism at static head and level flow. Even if the first two conditions were satisfied, the third almost certainly is not since in the frog skin the rate of metabolism is dependent on $\Delta\tilde{\mu}_{Na}$ (43). Thus E_{Na}, however it is evaluated, is only an effective potential (as pointed out by Ussing). It is a complex quantity influenced by several parameters, including the affinity of the driving reaction, the degree of coupling between transport and metabolism, and the apparent stoichiometry of the transport process. The latter is the stoichiometry measured at level flow, which as can be seen from Equation 69 is just qZ.

NET Analysis. Essig and Caplan (39) treated the transport of a single cation driven by a single metabolic reaction and showed that a composite system comprising a pump, a series barrier, and a leak pathway may be described by linear equations if each element shows linearity. For simplicity, consider one cation, say Na^+, to be transported actively and one anion, say Cl^-, to be representative of all passively permeant anions. Two epithelial tissues which conform to this model in that they actively transport only sodium are frog skin and toad bladder (42, 44). Following our usual sign convention, the phenomenological equations for a membrane consisting of a pump and series barrier may be written

$$\Delta \tilde{\mu}_+ = R_+^a J_+^a + R_{+r}^a J_r$$
$$A = R_{+r}^a J_+^a + R_{rr}^a J_r \qquad (82)$$

where $\Delta \tilde{\mu}_+$ and J_+^a are the electrochemical potential difference across the membrane and the flow of the cation through the pump pathway, respectively, and R_+^a is the total series resistance of the pump pathway. A and J_r represent the affinity and velocity (per unit membrane area) of the metabolic reaction driving the pump. A parallel leak pathway must exist if there is to be continuing active cation transport *in vivo*. If passive flows are uncoupled to flows of other species,

$$\Delta \tilde{\mu}_+ = R_+^p J_+^p$$
$$\Delta \tilde{\mu}_- = R_-^p J_-^p \qquad (83)$$

Since $J_+ = J_+^a + J_+^p$, the membrane is characterized by linear equations. These would be of the same form as Equations 82, *sans* superscripts. The positive direction of cation flow is taken to be defined by level flow, which implies that the system is positively coupled.

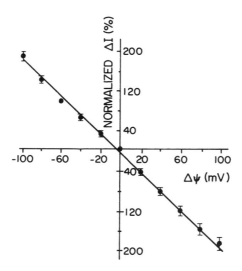

American Journal of Physiology

Figure 2. Mean normalized current-voltage relationship for toad bladder ($n=7$). *Values of* ΔI *in each hemibladder were expressed as fractions of the value of* ΔI *induced by changing* $\Delta \Psi$ *from 0 to* -60 *mV. Error bars represent* $\pm$ *1 SE (45).*

In recent years, studies in amphibian epithelia such as frog skins and toad bladders have demonstrated the remarkable linearity of behavior which these tissues manifest under various circumstances within the physiological range. Examples of this will be discussed below; however one manifestation, more familiar perhaps than the others, is ohmic behavior. Its significance has been disputed because of the undefined contribution of leak pathways. Recently Saito et al. (45) studied the current—voltage relationship in a group of toad bladders. They were able to show that much of the tissue conductance in this group was apparently by way of the active pathway ($\kappa^a/\kappa = 0.624 \pm 0.054$ SE, $n = 7$). Their results (see Figure 2) suggest strongly that active transport is a linear function of the electrical potential difference $\Delta\psi$ over the range examined. Similar results in frog skin have been reported by Biber and Sanders (46).

The Stoichiometric Ratio. In many studies of epithelial membranes, attempts have been made to determine the number of equivalents transported per mole of oxygen uptake (41, 47, 48). As we have seen, this quantity could have a unique value ($J_+/J_r = Z$) only if the system were completely coupled. Nevertheless, estimates of the stoichiometric ratio have commonly been made on the basis of an observed linear relationship between J_+ and J_r. If A is constant, linearity is a direct consequence of the linearity of the phenomenological equations, irrespective of the degree of coupling. Thus, although $(\partial J_+/\partial J_r)_A$ is identically equal to Z/q, (J_+/J_r) is constant only if q is unity. It is obvious that for effective functioning, the coupling in the active transport pathway must be rather tight and possibly complete. However the observable system is incompletely coupled, if only because of the leak pathway.

The apparent stoichiometric ratio at level flow, qZ, has been measured in frog skin by several workers (47, 48, 49). Vieira et al. (49) found that in standard glucose Ringer's solution both the short circuit current (reflecting sodium flow) and the associated oxygen consumption declined nonlinearly with time. The relationship between them, however, was linear, suggesting that the basal oxygen consumption was constant. For each skin, numerous experimental points were fitted by the best straight line. The intercept at zero short circuit current then gave the basal oxygen consumption and the slope gave the apparent stoichiometric ratio for a given skin. The values in ten skins varied significantly, ranging from 7.9 to 30.6 equivalents of sodium per mole of oxygen uptake. These results demonstrate clearly that there is no unique value appropriate for all skins.

The Affinity and Its Evaluation. If the affinity of some region of the metabolic chain coupled to active transport is constant, it can be evaluated. In principle, two independent methods exist, one being a static head measurement, the other a measurement at level flow. The first method depends on the relation

$$A = -F(\Delta\psi)_{J_+=0} / (\partial J_r/\partial J_+)_A \quad (\Delta c = 0) \tag{84}$$

the second on

$$A = (I)_{\Delta\psi=0}/\{\partial J_{\rm r}/\partial(\Delta\psi)\}_A \quad (\Delta c = 0) \tag{85}$$

Equation 85 has been used by Vieira et al. (43) and Saito et al. (50) to determine the apparent affinity driving the sodium pump in frog skin. Under specified conditions, linearity of J_r (rate of oxygen consumption) with $\Delta\psi$ was demonstrated over a range of as much as ± 160 mV in stable preparations. An example of this linearity is shown in Figure 3. The dependence of J_r on $\Delta\psi$ disappears entirely in the presence of ouabain, which blocks the sodium pump. The observation of linearity indicates that the phenomenological coefficients characterizing the system and the affinity of the metabolic

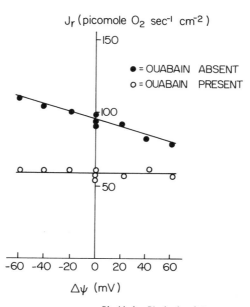

Biochimica Biophysica Acta

Figure 3. Dependence of the rate of oxygen consumption J_r on the electrical potential difference $\Delta\Psi$ in the aldosterone-treated frog skin; influence of ouabain (10^{-3}M) added 30 min before perturbation of $\Delta\Psi$ (50).

driving reaction must be invariant on perturbing $\Delta\psi$. Recently Danisi and Vieira have shown that in the short-circuited toad skin the rates of active sodium transport and oxygen consumption are linear functions of the chemical potential difference of sodium across the skin when changing the external sodium concentration (51).

Mode of Action of Substances That Influence Transport. The mechanisms by which various substances affect the rate of sodium transport in epithelia are of considerable interest. A case in point is the stimulatory effect of aldosterone. It has been suggested (52) that an aldosterone-induced protein may exert its effect at three possible sites: (1) at the apical entry step (permease theory), (2) directly on the Na K-ATPase (pump theory), or (3) on the oxidative pathway generating high energy intermediates to fuel the pump (metabolic theory). In studies of paired frog skins Saito et al. (50) found that the administration of aldosterone led to a significant increase in the apparent affinity as evaluated by Equation 85. The mean value in eight hormone-treated skins rose one day after administration from nearly 40 kcal/mole to nearly 100 kcal/mole. Depression of active transport and the associated metabolism with amiloride, which depresses permeability and is believed to act exclusively at the outer permeability barrier, also results in an increase in the apparent affinity. Evidently this was caused by a gradual rise in substrate level. These results suggest that in the case of aldosterone-induced stimulation energetic factors must be implicated.

Owen et al. (53) have recently reported comparative studies in frog skin of the effects of ouabain, a specific inhibitor of the sodium pump as mentioned earlier, and 2-deoxy-D-glucose, a competitive inhibitor of glucose metabolism. Both can be used to depress short-circuit current to approximately the same levels. The former agent affects only a phenomenological cross-coefficient, whereas the latter affects only the apparent affinity.

Oxidative Phosphorylation

Energetic Considerations. Thermodynamic treatments of oxidative phosphorylation are generally based on the determination of a stoichiometric ratio (P/O ratio), the ratio between the rates of phosphorylation and oxidation. The largest ratio obtainable experimentally, rounded off to an integer, is considered to represent the ideal stoichiometry of the system and may even be used in conjunction with other data to calculate the number of coupling sites (11). If n is the presumed stoichiometric ratio, the efficiency must be in accordance with the relation

$$-n\Delta G_{\text{phosphorylation}}/\Delta G_{\text{oxidation}} \leqslant 1 \qquad (86)$$

As we have seen above, however, this type of consideration is applicable only to fully coupled systems, and there is ample evidence that oxidative phosphorylation is not fully coupled (10). Incomplete coupling may be the consequence of side reactions or leaks (or both), but in any event respiration can certainly take place at appreciable rates without the occurrence of net phosphorylation. Stoichiometric ratios (as well as efficiencies) are therefore meaningless unless the state of the system is stipulated. Unfortunately, the parameters of Equation 86 are seldom determined simultaneously, and calculations tend to be based on extrapolated and estimated values that refer to different states.

Most of the models and theories proposed for mitochondrial oxidative phosphorylation accord with one of two major hypotheses: the chemical hypothesis or the chemiosmotic hypothesis. The former postulates direct chemical coupling—oxidation results in the formation of a high energy intermediate which is used to drive phosphorylation (11). The latter hypothesis (54, 55) suggests that oxidation is coupled to ejection of protons from the mitochondria. The resulting electrochemical potential gradient drives phosphorylation in the membrane by reversal of an ATPase-hydrogen ion pump. It has been pointed out by Rottenberg et al. that these mechanisms are not mutually exclusive (10).

There are two commonly used methods for determining the P/O ratio and the free energies of reaction (11). In carrying out these methods mitochondria are studied either in "state 3," i.e., during steady state phosphorylation after the addition of ADP, or in "state 4," i.e., during oxidation in the absence of phosphorylation owing to a limiting amount of ADP (56). State 4 is evidently static head, and thus a computation of efficiency in this state is meaningless. While such a computation is meaningful in state 3, mitochondria in state 3 rapidly phosphorylate to static head unless some means of clamping the phosphate potential is incorporated.

NET Analysis. It is possible in principle to envisage conditions in which the flow of hydrogen ion across the mitochondrial membrane may be maintained nonzero and constant for brief periods. For such a system it is reasonable to write

$$\Phi = J_P A_P + J_H \Delta \tilde{\mu}_H + J_O A_O \qquad (87)$$

where the subscripts P, H, and O refer to phosphorylation, net H⁺ flow, and substrate oxidation. Equation 87 is a perfectly adequate description of mitochondrial processes despite their complexity since we consider only steady states in which flows such as those of Ca^{2+}, Mg^{2+}, and H_2O have come to a halt. Net flow of H⁺ must be accompanied by an equivalent flow of anions and/or cations as a consequence of electroneutrality. If such flows occur through leak pathways, they need not be included since they are not coupled to the metabolic process.

Equation 87 is ambiguous as to whether the affinities are to be measured internally or externally. The same holds for the flows, which may be nonconservative (i.e., substances may be metabolized within the composite membrane). However, Caplan and Essig (57) have shown that the affinities and reaction rates may be evaluated externally if J_H is taken as the rate of decrease of H⁺ content inside the mitochondrion and $\Delta \tilde{\mu}_H$ is defined as $\tilde{\mu}_H^{in} - \tilde{\mu}_H^{ex}$. This is very convenient owing to the difficulty of determining internal activities. With this interpretation one may write the phenomenological equations:

$$J_P = L_P A_P + L_{PH} \Delta \tilde{\mu}_H + L_{PO} A_O \qquad (88)$$

$$J_H = L_{PH} A_P + L_H \Delta \tilde{\mu}_H + L_{OH} A_O \qquad (89)$$

$$J_O = L_{PO} A_P + L_{OH} \Delta \tilde{\mu}_H + L_O A_O \qquad (90)$$

This general formalism should be applicable near equilibrium whichever hypothesis is correct.

Chemical and Chemiosmotic Hypotheses. According to the chemical hypothesis, substrate oxidation produces a high energy intermediate which can drive either phosphorylation or outward H⁺ transport (and possibly cation transport). J_O is therefore positively coupled to both J_P and J_H. Since outward H⁺ transport may also result from the hydrolysis of ATP, J_P and J_H are negatively coupled. Each flow may be influenced by every force, so that all the coefficients are nonzero in the most general case. Partial uncoupling of oxidation and phosphorylation could result from breakdown of the high energy intermediate or by other side reactions or leaks.

According to the chemiosmotic hypothesis, J_P and J_O are coupled only through circulation of H⁺, rather than through a high energy intermediate. In this case, we can characterize the overall coefficients in terms of those of the two individual sites:

Site I:

$$J_H^I = L_H^I \Delta\tilde{\mu}_H + L_{OH}^I A_O \qquad (91)$$

$$J_O = L_{OH}^I \Delta\tilde{\mu}_H + L_O^I A_O \qquad (92)$$

Site II:

$$J_P = L_P^{II} A_P + L_{PH}^{II} \Delta\tilde{\mu}_H \qquad (93)$$

$$J_H^{II} = L_{PH}^{II} A_P + L_H^{II} \Delta\tilde{\mu}_H \qquad (94)$$

The net flow of H⁺ is given by

$$J_H = J_H^I + J_H^{II} \qquad (95)$$

and consequently the addition of Equations 91 and 94 gives

$$J_H = L_{PH}^{II} A_P + (L_H^I + L_H^{II}) \Delta\tilde{\mu}_H + L_{OH}^I A_O \qquad (96)$$

Equations 93, 96, and 92 correspond respectively to Equations 88, 89, and 90. The straight coefficients L_O and L_P and the cross coefficients L_{OH} and L_{PH} are simply those of the appropriate elemental sites, while L_H is the sum $L_H^I + L_H^{II}$. The important observation is that

$$L_{PO} = 0 \qquad \text{(chemiosmotic hypothesis)} \qquad (97)$$

i.e., if $\Delta\tilde{\mu}_H$ were maintained constant experimentally, phosphorylation and oxidation would be independent. In particular, if $\Delta\tilde{\mu}_H = 0$, uncoupling is complete, and $J_P = 0$ if $A_P = 0$, irrespective of J_O, while $J_O = 0$ if $A_O = 0$, irrespective of J_P.

This description of oxidative phosphorylation leads to the consideration of three degrees of coupling, *viz.* q_{PH}, q_{OH}, and q_{PO}. (In systems of more than two flows (58), degrees of coupling are defined in terms of resistance coefficients.) The condition of Equation 97 is synonymous with the relation

$$q_{PO} = -q_{PH}q_{OH} \quad \text{(chemiosmotic hypothesis)} \tag{98}$$

Since J_P and J_H are negatively coupled, q_{PH} is negative.

Experimental Application. There are clearly substantial difficulties in the experimental application of these phenomenological equations. It is possible, however, to vary and measure the external affinities A_O and A_P over a considerable range. Moreover, both the pH difference and the electrical potential difference may be evaluated and regulated in functioning mitochondria. Although it may be difficult to fix $\Delta\tilde{\mu}_H$ exactly at zero, it seems

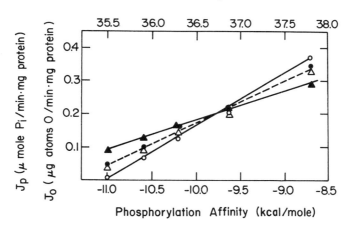

Figure 4. The rates of oxidation and phosphorylation as functions of the reaction affinities in rat liver mitochondria. ($\bigcirc$) J_P as a function of A_P (A_O constant); ($\bullet$) J_P as a function of A_O (A_P constant); ($\triangle$) J_O as a function of A_P; ($\blacktriangle$) J_O as a function of A_O (5).

possible to make it very small. Under these conditions, which approximate a level flow of H⁺, the P/O ratio tends to zero (59). At static head of H⁺, the phenomenological equations in terms of resistance coefficients are

$$\begin{aligned} A_P &= R_P J_P + R_{PO} J_O \\ A_O &= R_{PO} J_P + R_O J_O \end{aligned} \tag{99}$$

Rottenberg et al. (60) showed that, with the assumption of Onsager symmetry, all the coefficients of Equations 99 may be obtained experimentally.

In a recent study in rat liver mitochondria, Rottenberg (5) has verified the above assumptions of linearity and Onsager symmetry. There is a wide range over which the oxidation reaction is linear in both the oxidation and phosphorylation affinities, and similarly there is a range over which the phosphorylation rate is linear in both affinities. The dependences of phosphorylation on oxidation affinity and of oxidation on phosphorylation affinity are identical. These results are shown in Figure 4.

Muscular Contraction

The Hill Equation. A characteristic force–velocity relation for muscle which describes accurately the mechanical behavior of muscles of all types from many different species was discovered experimentally by A. V. Hill (*61*). It is generally written

$$(P + a)(V + b) = (P_0 + a)b = (V_m + b)a \tag{100}$$

where P is the tension and V the initial constant velocity in tetanic shortening; P_0 is the isometric tension, and V_m the maximal or unloaded velocity of contraction. The quantities a and b are purely mechanical constants (*62*). It follows that the ratios a/P_0 and b/V_m are equal; we shall denote them by θ. The most remarkable property of this relation, a property which sets it apart from all other proposed force–velocity relations, is its symmetry (*63, 64*). Normalizing both tension and velocity we obtain:

$$[(P/P_0) + \theta][(V/V_m) + \theta] = (1 + \theta)\theta \tag{101}$$

Equation 101 represents a family of hyperbolas cutting both the normalized tension and normalized velocity axes at unity. The curvature of any given hyperbola depends on θ, which was singled out by Hill (*61*) as an index of performance for the muscle. Typically θ has a value between 0.2 and 0.3 and the curvature is quite high.

The nonlinearity of the force–velocity relation suggests the possibility that a regulatory process is at work and that the muscle is able to adjust and match itself to its load. The existence of such a regulatory mechanism was proposed explicitly many years ago by Fenn (*65*). In phenomenological terms, regulation could be effected by either of two possible mechanisms. The phenomenological coefficients might be strong functions of tension or velocity; this would imply a fundamental nonlinearity of the system. On the other hand, the input (in particular, the affinity of the driving reaction) might be altered in response to changes in tension. An analysis of systems of the latter type was carried out by considering a linear energy converter in series with a regulator which modifies the input derived from a primary energy source (*66*). The output of the regulator is the input to the converter; nonlinearity is confined to the regulator. (The alternative view, assuming that regulation depends on a variable stoichiometry, has also been analyzed (*67*).) If the energy converter is incompletely coupled, operates between specified static head and level flow limits, and gives a unique adaptive response to any imposed load, then its behavior can always be described by the same general expression which may be regarded as the canonical regulator equation. In its simplest form, this expression reduces to the Hill force–velocity relation. This interpretation leads to a rather striking conclusion: the degree of coupling of the regulator is given by

$$q = 1/\sqrt{1 + \theta} \tag{102}$$

In this view, Equation 101 has only one adjustable parameter, the degree of coupling, which may be determined from the curvature of the force–velocity curve. Hill showed that θ is usually about 0.25 in frog sartorius (*61*). This gives a degree of coupling of about 90%, corresponding to a maximum efficiency of about 40%.

NET Analysis. The dissipation function for muscle is arrived at by a precise analog of the approach used earlier in the study of transport problems. The essential working element of the system is the myofibril, with an attached reservoir from which it derives its input. For small steady tetanic contractions, the fibril may be considered to be in a stationary state. This leads to the following form of the dissipation function (*66*)

$$\Phi = V(-P) + J_r A \tag{103}$$

where Hill's notation has been used for the output. It is convenient to write $-P$ as the output force (taking the direction of contraction as positive). This result implies that there is a unique reaction which is coupled to the mechanical process of contraction, and that this reaction is not coupled directly to any other reactions which may be occurring. The phenomenological coefficients corresponding to Equation 103 contain vectorial cross-coefficients. Consequently coupling must take place in an anisotropic medium in agreement with current ideas of structure (*68*). In this respect, the treatment is no different from that of Kedem (*26*) for active transport. Indeed muscular contraction may be considered a special case of this since, according to the sliding filament hypothesis, metabolic energy is expended to transport actin filaments relative to myosin filaments against mechanical tension.

Experimental Application. The NET approach requires simultaneous measurements of both chemical and mechanical parameters. The temporal resolution must be adequate if the rate determinations are to test the validity of the theory. Our knowledge of muscle energetics is based largely on data obtained from frog skeletal muscle. However, the speed of contraction of this muscle makes the required temporal resolution impossible to attain at present. On the other hand, if we use maximum velocity as an index of muscular speed, vascular smooth muscle (VSM) is two to three orders of magnitude slower than frog sartorius. In contrast to skeletal muscle preparations, VSM has no large pools of high-energy phosphate compounds, and consequently oxygen consumption rates in the steady state appear to be a valid measure of the rate of the driving reaction (*69*). Not only is a linear relation found between the rate of oxygen consumption and graded isometric

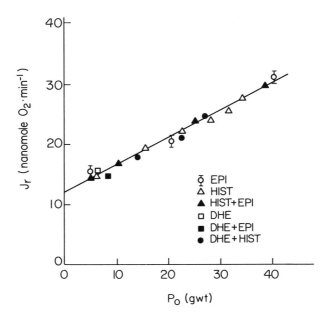

Journal of Mechanochemistry and Cell Motility
Figure 5. Oxygen consumption rate J_r vs. graded isometric tension P_o for bovine mesenteric vein in the presence of the adrenergic and non-adrenergic stimulants epinephrine (EPI) and histamine (HIST) respectively, as well as the adrenergic α-blocking agent dihydroergotamine (DHE). (▲) Stimulation with maximal concentrations of both HIST and EPI, and subsequent dilutions; (●) Maximal stimulation with HIST of DHE-blocked tissue, and subsequent dilutions. (70).

force in VSM, but the relation is independent of the specific pharmacological stimulants used. This is just what would be expected if the slope of this straight line reflected the magnitude of a phenomenological cross-coefficient R_{mc} (the mechano-chemical coupling coefficient) according to the equations

$$-P = R_m V + R_{mc} J_r$$
$$A = R_{mc} V + R_c J_r \quad (104)$$

These equations, of course, describe the essential contractile machinery only. Typical experimental results are shown in Figure 5.

Conclusions

We have been considering an analytical technique which has broad application to many systems, as for example in the recent studies of tracer flows by Li *et al.* (*71, 72*) and of energy coupling in active amino acid transport by Heinz *et al.* (*73, 74, 75*). However, the approach is strictly phenomenological in character. No attempt has been made here to relate the phenomenological description to explicit kinetic models and hence to function at the molecular level. An important class of membrane models for which this has in fact been accomplished, and which embraces coupled flows and forces including active transport, has been described by Hill and Kedem (*76*) and by Hill (*77*). The analysis makes use of a diagram method of great versatility and generality in the treatment of steady-state problems (*78, 79*). The diagrams represent states (points) connected by all possible first-order transitions (lines) between them; steady-state fluxes are associated with cycles (closed paths). Coupling arises when cycles share a common branch between two states. Near equilibrium, a calculation of the fluxes by the diagram method leads to conventional phenomenological equations in which the Onsager reciprocal relations are obeyed. The method has recently been used in an extensive theoretical treatment of the sliding filament model of striated muscle contraction (*80*).

The studies discussed above illustrate the primary thesis of this article: the tremendous ability of nonequilibrium thermodynamics to unify diverse phenomena. Whether or not the systems are truly linear over the entire range of biological interest is frequently beside the point—the provision of a single coherent logical framework is much more important. It should also be clear that the approach is rich in predictive value and can suggest novel experiments and points of view.

Acknowledgments

This undertaking was supported by the U.S. Public Health Service (Grant No. HL 14322 to the Harvard-M.I.T. Program in Health Sciences and Technology).

Literature Cited

1. Prigogine, I., "Introduction to Thermodynamics of Irreversible Processes," Wiley, New York, 1961.
2. Katchalsky, A., Curran, P. F., "Nonequilibrium Thermodynamics in Biophysics," Harvard Univ. Press, Cambridge, Mass., 1965.
3. Caplan, S.R., *Current Topics Bioenerg.*, (1971) **4**, 1.
4. Caplan, S. R., Mikulecky, D. C., In "Ion Exchange," J. A. Marinsky, Ed., Vol. 1, p. 1, Marcel Dekker, New York, 1966.
5. Rottenberg, H., *Biophys. J.* (1973) **13**, 503.
6. Hess, B., Brand, K., In "Control of Energy Metabolism," B. Chance *et al.*, Eds., p. 111, Academic Press, New York, 1965.
7. Klingenberg, M., In "Control of Energy Metabolism," B. Chance *et al.*, Eds., p. 149, Academic Press, New York, 1965.
8. Vernon, C. A., In "Size and Shape Changes of Contractile Polymers," A. Wasserman, Ed., p. 109, Pergamon Press, Oxford, 1960.
9. Blumenthal, R., Caplan, S.R., Kedem, O., *Biophys. J.* (1967) **7**, 735.
10. Rottenberg, H., Caplan, S. R., Essig, A., *Nature (London)* (1967) **216**, 610.
11. Slater, E. C., *Comp. Biochem.* (1966) **14**, 327.
12. Mikulecky, D. C., In "Transport Phenomena in Fluids," H. J. M. Hanley, Ed., p. 433, Marcel Dekker, New York, 1969.

13. Lakshminarayanaiah, N., "Transport Phenomena in Membranes," Academic Press, New York, 1969.
14. Läuger, P., *Angew. Chem., Int. Edn.* (1969) **8,** 42.
15. Kedem, O., Katchalsky, A., *Biochim. Biophys. Acta* (1958) **27,** 229.
16. Kedem, O., Katchalsky, A., *J. Gen. Physiol.* (1961) **45,** 143.
17. Kedem, O., Katchalsky, A., *Trans. Faraday Soc.* (1963) **59,** 1918.
18. Michaeli, I., Kedem, O., *Trans. Faraday Soc.* (1961) **57,** 1185.
19. Katchalsky, A., Kedem, O., *Biophys. J.* (1962) **2,** 53.
20. Staverman, A. J., *Rec. Trav. Chim.* (1951) **70,** 344.
21. Meares, P., Sutton, A. H., *J. Colloid Interface Sci.* (1968) **28,** 118.
22. Krämer, H., Meares, P., *Biophys. J.* (1969) **9,** 1006.
23. McHardy, W. J., Meares, P., Sutton, A. H., Thain, J. F., *J. Colloid Interface Sci.* (1969) **29,** 116.
24. Dawson, D. G., Meares, P., *J. Colloid Interface Sci.* (1970) **33,** 117.
25. Foley, T., Klinowsky, J., Meares, P., *Proc. Roy. Soc., Ser. A* (1974) **336,** 327.
26. Kedem, O., In "Membrane Transport and Metabolism," A. Kleinzeller and A. Kotyk, Eds., p. 87, Academic Press, New York, 1961.
27. Finlayson, B. A., Scriven, L. E., *Proc. Roy. Soc., Ser. A* (1969) **310,** 183.
28. Koefoed-Johnsen, V., Ussing, H. H., *Acta Physiol. Scand.* (1953) **28,** 60.
29. Ussing, H. H., *Acta Physiol. Scand.* (1949) **19,** 43.
30. Hodgkin, A. L., Keynes, R. D., *J. Physiol. (London)* (1955) **128,** 61.
31. Kedem, O., Essig, A., *J. Gen. Physiol.* (1965) **48,** 1047.
32. Curran, P. F., Taylor, A. E., Solomon, A. K., *Biophys. J.* (1967) **7,** 879.
33. Gottlieb, M. H., Sollner, K., *Biophys. J.* (1968) **8,** 515.
34. De Sousa, R. C., Li, J. H., Essig, A., *Nature* (1971) **231,** 44.
35. Essig, A., Kedem, O., Hill, T. L., *J. Theor. Biol.* (1966) **13,** 72.
36. Shean, G. M., Sollner, K., *Ann. N. Y. Acad. Sci.* (1966) **137,** 759.
37. Snell, F. M., Shulman, S., Spencer, R. P., Moos, C., "Biophysical Principles of Structure and Function," p. 320, Addison-Wesley, London, 1965.
38. Kedem, O., Caplan, S. R., *Trans. Faraday Soc.* (1965) **61,** 1897.
39. Essig, A., Caplan, S. R., *Biophys. J.* (1968) **8,** 1434.
40. Rosenberg, T., *Symp. Soc. Exp. Biol.* (1954) **8,** 27.
41. Ussing, H. H., In "The Alkali Metal Ions in Biology," H. H. Ussing *et al.* Eds., p. 1, Springer, Berlin, 1960.
42. Ussing, H. H., Zerahn, K., *Acta Physiol. Scand.* (1951) **23,** 110.
43. Vieira, F. L., Caplan, S. R., Essig, A., *J. Gen. Physiol.* (1972) **59,** 77.
44. Leaf, A., Anderson, J., Page, L. B., *J. Gen. Physiol.* (1958) **41,** 657.
45. Saito, T., Lief, P. D., Essig, A., *Am. J. Physiol.* (1974) **226,** 1265.
46. Biber, T. U. L., Sanders, M. L., *J. Gen. Physiol.* (1973) **61,** 529.
47. Zerahn, K., *Acta Physiol. Scand.* (1956) **36,** 300.
48. Leaf, A., Renshaw, A., *Biochem. J.* (1957) **65,** 82.
49. Vieira, F. L., Caplan, S. R., Essig, A., *J. Gen. Physiol.* (1972) **59,**60.
50. Saito, T., Essig, A., Caplan, S. R., *Biochim. Biophys. Acta* (1973) **318,** 371.
51. Danisi, G., Vieira, F. L., *J. Gen. Physiol.* (1974) **64,** 372.
52. Feldman, D., Funder, J. W., Edelman, I. S., *Am. J. Med.* (1972) **53,** 545.
53. Owen, A., Caplan, S. R., Essig, A., *Fed. Proc.* (1974) **33,** 215.
54. Mitchell, P., *Nature (London)* (1961) **191,** 144.
55. Mitchell, P., *Biol. Rev. Cambridge Phil. Soc.* (1966) **41,** 445.
56. Chance, B., Williams, G. R., *Advan. Enzymol.* (1956) **17,** 65.
57. Caplan, S. R., Essig, A., *Proc. Nat. Acad. Sci. U.S.* (1969) **64,** 211.
58. Caplan, S. R., *J. Theor. Biol.* (1966) **10,** 209 and **11,** 346.
59. Rottenberg, H., *Euro. J. Biochem.* (1970) **15,** 22.
60. Rottenberg, H., Caplan, S. R., Essig, A., In "Membranes and Ion Transport," E. E. Bittar, Ed., Vol. 1, p. 165, Wiley, New York, 1970.
61. Hill, A. V., *Proc. Roy. Soc., Ser. B* (1938) **126,** 136.
62. Hill, A. V., *Proc. Roy. Soc., Ser. B* (1964) **159,** 297.
63. Caplan, S. R., *Biophys. J.* (1968) **8,** 1146.

64. Caplan, S. R., *Biophys. J.* (1968) **8,** 1167.
65. Fenn, W. O., *J. Physiol. (London)* (1923) **58,** 175.
66. Caplan, S. R., *J. Theor. Biol.* (1966) **11,** 63.
67. Bornhorst, W. J., Minardi, J. E., *Biophys. J.* (1970) **10,** 137.
68. Huxley, H. E., *Science* (1969) **164,** 1356.
69. Paul, R. J., Peterson, J. W., Caplan, S. R., *Biochim. Biophys. Acta* (1973) **305,** 474.
70. Paul, R. J., Peterson, J. W., Caplan, S. R., *J. Mechanochem. Cell Motility* (1974) **3,** 19.
71. Li, J. H., De Sousa, R. C., Essig, A., *J. Membrane Biol.* (1974) **19,** 93.
72. Essig, A., Li, J. H., *J. Membrane Biol.* (1975) **20,** 341.
73. Heinz, E., In "Current Topics in Membranes and Transport," F. Bronner and A. Kleinzeller, Eds., Vol. 5, p. 137, Academic Press, New York, 1974.
74. Geck, P., Heinz, E., Pfeiffer, B., *Biochim. Biophys. Acta* (1974) **339,** 419.
75. Heinz, E., Geck, P., *Biochim. Biophys. Acta* (1974) **339,** 426.
76. Hill, T. L., Kedem, O., *J. Theoret. Biol.* (1966) **10,** 399.
77. Hill, T. L., "Thermodynamics for Chemists and Biologists," p. 141, Addison-Wesley, London, 1968.
78. Hill, T. L., *J. Theoret. Biol.* (1966) **10,** 442.
79. Hill, T. L., *Biochemistry* (1975) **14,** 2127.
80. Hill, T. L., In "Progress in Biophysics and Molecular Biology," A. J. V. Butler and D. Noble, Eds., Vol. 28, p. 267, Pergamon Press, Oxford, 1974.

Chapter

2

Biomembranes—Structure and Function

Donald F. H. Wallach, Director, Division of Radiobiology, Tufts-New England Medical Center, Boston, Mass. 02111

History, Morphology, Composition

History. The recognition of membranes as important elements of cell structure originated with the work of Nägeli (*1*). His microscopic observations showed that the surface layer of many plant cells is a self-sealing, viscous, semiliquid boundary, impermeable to numerous dyes. He proposed the term "plasma membrane" for this layer. Concurrently, Pringsheim (*2*) found that the limiting surface layers of cells are essential for their osmotic responsiveness. Soon thereafter, Ehrlich (*3*) identified the critical role of the cell surface in interactions between cells, as well as the responsiveness of individual cells to their physiologic environment, to drugs, and to poisons. Indeed, the concept of specific "surface receptors" originated with Ehrlich.

Historically, the next membrane system to be recognized was myelin. This lamellar membrane array is derived from the plasma membranes of axonal satellite cells. However, elucidation of myelin biogenesis—by electron microscopy (*4*)—came well after the overall architecture of myelin had been worked out by polarization optics and by x-ray diffraction (*5, 6, 7*).

The recognition of intracellular membranes rapidly followed the development of high resolution electron microscopy. In 1945 Porter observed that cells contain a labyrinth of membrane-bounded compartments, the endoplasmic reticulum (*8*). Shortly thereafter, Sjöstrand discovered that the outer segments of retinal photoreceptor cells consist of stacked, double-membrane discs (*9*). The Golgi complex, an extension of endoplasmic reticulum, consisting primarily of flattened membrane-enclosed cisternae was characterized almost concurrently. In 1952, Palade (*10*) and Sjöstrand (*11*) independently recognized that mitochondria are complex membranous structures. At the same time, Berthet and de Duve observed that lysosomes comprise defined membranous vesicles (*12*).

Some General Aspects of Membrane Morphology. The thickness of biomembranes lies near 100 A (10^{-6} cm). Such minute structures can be visualized only by electron microscopy. Thus, to obtain a 1-mm image requires an instrument magnification of $\sim 4 \times 10^4$ and subsequent 2.5-fold photographic enlargement. In addition, membranes per se do not absorb electrons to an unusual degree and cannot be discerned unless they first react with or are shadowed by heavy metals.

The two most important techniques for high resolution membrane electron microscopy measure electron transmission of stained thin sections or metal replicas. After reaction of fixed, dehydrated thin sections of tissue with heavy metals, cellular membranes appear as two electron-opaque layers, separated by a lucent zone. This is the "unit membrane" image (*13*), which was long interpreted to represent the structure proposed by Danielli and Davson (*14*)—*i.e.*, a lipid bilayer with protein bound to the two surfaces. We now know that the "unit membrane" appearance can arise from other molecular arrays. For example, the deposition of the electron-opaque metals used for contrast enhancement can occur exclusively at the polar ends of phosphatide molecules. This is the case with artificial bilayers made of saturated phospholipids (*15*). On the other hand, oxidizing heavy metal stains—*e.g.* osmium tetroxide—react covalently with unsaturated fatty acids, *i.e.*, in apolar lipid regions (*16*). Moreover, numerous proteins stain with heavy metals. Finally, most of the lipid can be extracted from glutaraldehyde-fixed myelin and mitochondria without affecting their appearance in stained thin sections (*17, 18*). This also indicates that much of the deposited metal is associated with membrane proteins.

The electron opacities responsible for a membrane image depend upon the stains and staining conditions used. Resolution depends on grain size which can be as little as 5–10 A under favorable conditions and would then allow recognition of individual protein molecules. However, the initial reaction with metal usually creates nucleation foci that induce additional stain deposition. This reduces resolution.

Proteins probably stain maximally at their polar amino acids; non-polar residues do not react significantly. However, high-resolution x-ray analyses of globular proteins show that polar amino acids lie at the surfaces of these macromolecules. The cores of such proteins are highly apolar. Metal-stained images of globular proteins should therefore exhibit an electron-luscent core and an electron-opaque perimeter. Globular lipid micelles should yield a similar appearance. Accordingly, staining techniques combined with high resolution electron microscopy may distinguish between polar and apolar domains but cannot establish *a priori* whether these are protein or lipid in nature. The "unit membrane" image thus merely indicates that biomembranes comprise an apolar core, bounded by two layers enriched in polar groups. The image could derive from a purely protein membrane with polar residues at the two surfaces. It can clearly also represent lipid bilayers and various lipid–protein combinations.

Freeze-fracture electron microscopy (*19*) reveals some of the structuring in the apolar cores of biomembranes. Here a piece of tissue or a cell suspension is first chilled to $\sim -180°C$ in the presence of a cryoprotectant (to avoid formation of large ice crystals). The frozen blocks are then split *in vacuo* with a cold knife. The resulting fracture planes tend to pass through predominantly apolar domains within membranes and to generate two membrane fracture faces. These are coated *in vacuo* with metal to produce coherent replicas, which are then examined by transmission electron microscopy. Optionally, material can be fixed, freed of soluble material and frozen. "Heating" *in vacuo* to $\sim -100°C$ ("etching") then sublimes off some of the ice and exposes true membrane surfaces.

Freeze-fracturing reveals several important structural features. Most informative, perhaps, is the fact that the cores of biomembranes (other than myelin) contain numerous "intramembranous particles" (*19*). These assuredly represent proteins penetrating into and/or through the membranes. Detailed analyses show considerable variation in the size and/or shape of intramembranous particles. This may mean that the particles comprise more than one biological entity and/or that a given molecule protrudes to varying degrees into the membrane core. In some membranes, the number, size, and distribution of intramembranous particles can change dramatically with temperature or pH.

The fracture faces of contacting epithelial cells commonly show hexagonally packed particle arrays of considerable size. These constitute "nexuses" or gap junctions and may correspond morphologically to the low resistance junctions inferred from electrical measurements (*20*).

An important conclusion about membrane structure can be drawn at this point: although the "unit-membrane" hypothesis has held sway for a long time and continues to be cited in textbooks as *the* structure of biomembranes, it is now thought to be wrong in at least one aspect—*i.e.*, in its localization of membrane proteins.

General Features of Membrane Composition. Nearly all known biomembranes contain protein, lipid, and carbohydrate; however, the gas vacuole membranes of halophile bacteria lack lipid. Membrane proteins constitute the major membrane macromolecules. Except for myelin they comprise 60–70% of the membrane mass. Membrane carbohydrate is associated primarily with glycoproteins and to a lesser extent with glycolipids.

LIPIDS. The biochemistry of membrane lipids has been reviewed often and well (*21, 22*). The lipids found in plasma membranes consist of three major classes: glycerophosphatides, sphingo- and glycolipids, and steroids. Most abundant in membranes are the glycerophosphatides, such as phosphatidyl choline, phosphatidyl serines, phosphatidyl ethanolamine and phosphatidyl inositol.

Glycerophosphatides comprise derivatives of glycerol-3-phosphoric acid. The hydrocarbon side chains are hydrophobic whereas the polar "head groups" containing the glycerol moiety, the phosphate group and the choline, serine, ethanolamine, or inositol, are water soluble. The phosphatides there-

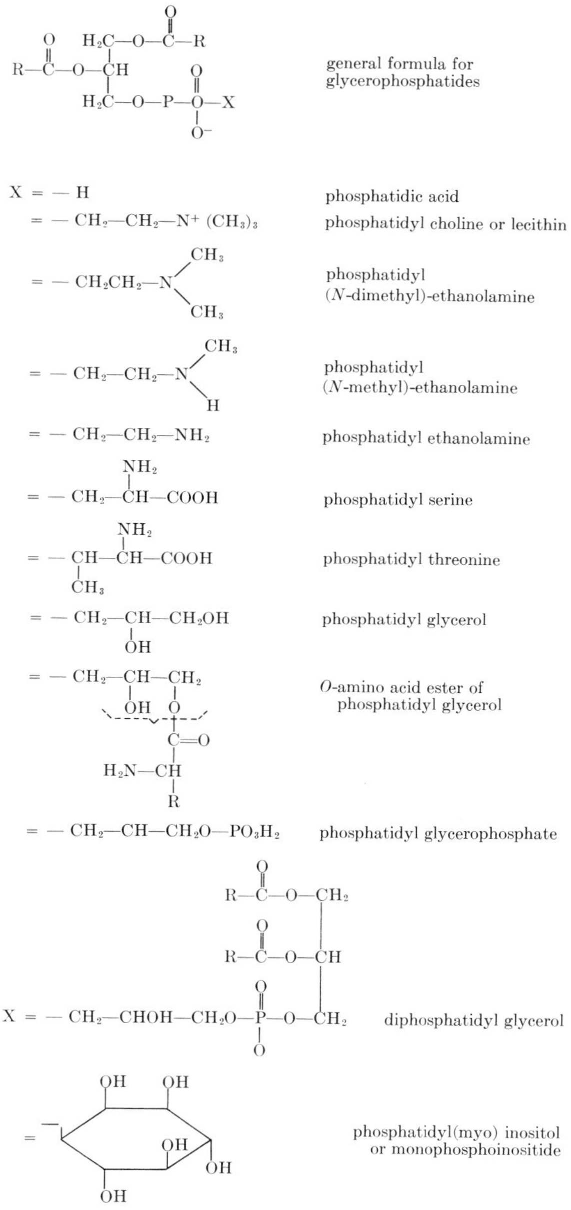

Figure 1. Glycerophosphatides

General formula for glycerophosphatides:

$$\begin{array}{c} H_2C-O-\overset{O}{\underset{\|}{C}}-R'' \\ R'-\overset{O}{\underset{\|}{C}}-O-CH \\ H_2C-O-\overset{}{\underset{}{P}}(=O)(-O^-)-O-X \end{array}$$

X =	Name
$-H$	phosphatidic acid
$-CH_2-CH_2-N^+(CH_3)_3$	phosphatidyl choline or lecithin
$-CH_2CH_2-N(CH_3)_2$	phosphatidyl (N-dimethyl)-ethanolamine
$-CH_2-CH_2-N(CH_3)(H)$	phosphatidyl (N-methyl)-ethanolamine
$-CH_2-CH_2-NH_2$	phosphatidyl ethanolamine
$-CH_2-CH(NH_2)-COOH$	phosphatidyl serine
$-CH(CH_3)-CH(NH_2)-COOH$	phosphatidyl threonine
$-CH_2-CH(OH)-CH_2OH$	phosphatidyl glycerol
$-CH_2-CH(OH)-CH_2-O-C(=O)-CH(NH_2)-R$	O-amino acid ester of phosphatidyl glycerol
$-CH_2-CH-CH_2O-PO_3H_2$	phosphatidyl glycerophosphate

For diphosphatidyl glycerol:

$$X = -CH_2-CHOH-CH_2O-\overset{O}{\underset{\|}{P}}(-O^-)-O-CH_2-CH(O-C(=O)R)-CH_2-O-C(=O)R$$

diphosphatidyl glycerol

X = inositol ring (with OH groups): phosphatidyl(myo) inositol or monophosphoinositide

R' = a saturated fatty acid commonly.
R'' = unsaturated commonly.
The chain length of R' and R² is usually 10–20. In inositol phosphatides X = $-C_6H_6(OH)_5$, but the 4-, and/or 5-positions of the inositol may also be phosphorylated.

Figure 1. Glycerophosphatides

fore orient at an oil–water interface in such a way that the side chains are in the oil and the polar groups are in the water. At physiologic pH the polar groups may lack a net charge as in phosphatidyl choline, be slightly anionic as in phosphatidyl ethanolamine, or bear a distinct net negative charge, as in phosphatidyl serine and in phosphatidyl inositol. Figure 1 shows the structures of the major glycerophosphatides.

Acetal phosphatides (plasmalogens) consist of glycerylphosphoryl-choline (or -ethanolamine, or -serine; X) with one esterified fatty acid and one fatty acid in enol–ether linkage, as shown below:

$$\begin{array}{c}
CH_2OCH{=}CHR \\
| \\
\overset{\displaystyle O}{\underset{\displaystyle \|}{RCH_2C}}{-}O{-}CH \\
| \\
\overset{\displaystyle}{}O \\
| \\
CH_2{-}O{-}\overset{\displaystyle}{\underset{\displaystyle \|}{P}}{-}O{-}X \\
O
\end{array}$$

The sphingolipids contain sphingosine (or dehydrosphingosine) with a fatty acid in amide linkage with the sphingosine. Fatty acid amides of sphingosine are known as ceramides (Figure 2). In sphingomyelin (Figure 2) a phosphate is in diester linkage between choline and the hydroxyl on the terminal carbon of ceramide. Sphingomyelin bears no net charge at neutral pH. Cerebrosides are glycosyl ceramides (Figure 2). The gangliosides are complex carbohydrate-containing derivatives of ceramide.

The third type of lipid observed in significant amounts in membranes are the steroids, particularly cholesterol (Figure 3). This compound does not have a charged polar group.

Biomembranes differ extensively in the proportions of protein and lipid, (Table I) as well as in lipid composition (Table II).

The unique composition of myelin, *i.e.*, its low protein content and the large proportions of cholesterol and glycolipids can be explained in terms of highly specialized function. This argument does not apply well to the differences between bacterial plasma membranes. However, diverse lipids may serve the same function if they participate suitably in the hydrophobic interactions which stabilize membranes. In this regard the structure of hemoglobins teaches some useful lessons (23). Although very similar functionally, the hemoglobins of various species show very large differences in amino acid composition. Only nine out of more than 140 residues are the same in all the monomers. However, in all hemoglobins 30 of the residues are always nonpolar and lie in the proteins' hydrophobic cores. Also, in all normally functioning hemoglobin molecules, the amino acid composition and sequence is

The structure of sphingomyelin is:

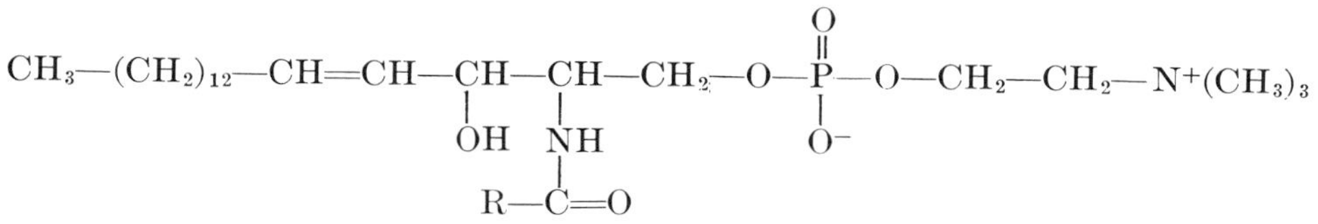

$$CH_3-(CH_2)_{12}-CH=CH-\underset{\underset{R-C=O}{NH}}{\overset{}{CH}}-\underset{OH}{\overset{}{CH}}-CH_2-O-\overset{\overset{O}{\|}}{\underset{\underset{O^-}{|}}{P}}-O-CH_2-CH_2-N^+(CH_3)_3$$

The general structure for glycolipids is:

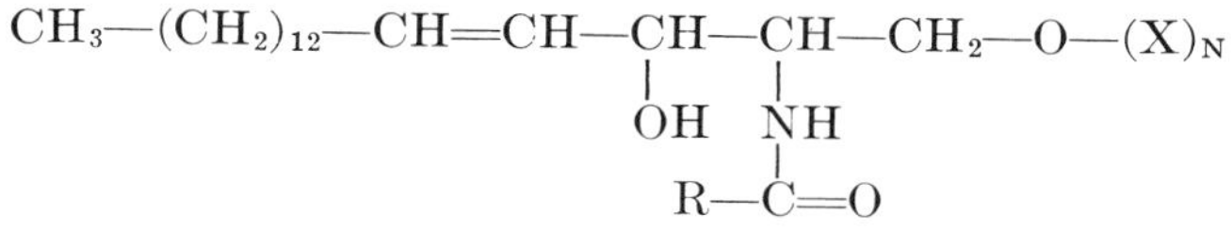

$$CH_3-(CH_2)_{12}-CH=CH-\underset{OH}{\overset{}{CH}}-\underset{\underset{R-C=O}{NH}}{\overset{}{CH}}-CH_2-O-(X)_N$$

In ceramide monohexosides or cerebrosides, X = galactose or glucose and $N = 1$. In ceramide oligohexosides $N = 1$, but there are equivalent molar amounts of sphingosine, hexose and fatty acid (R); the only nitrogen is in the sphingosine. In globosides (*e.g.*, from human erythrocytes) additional nitrogens occur in the sugar moiety, when X = (hexoses(s) + hexosamine(s)), or in mucolipids containing neuraminic acids (*e.g.*, in horse erythrocytes). Then X = (hexose(s) or hexosamine(s) + neuraminic acid).

Figure 2.   Spingo- and glycolipids

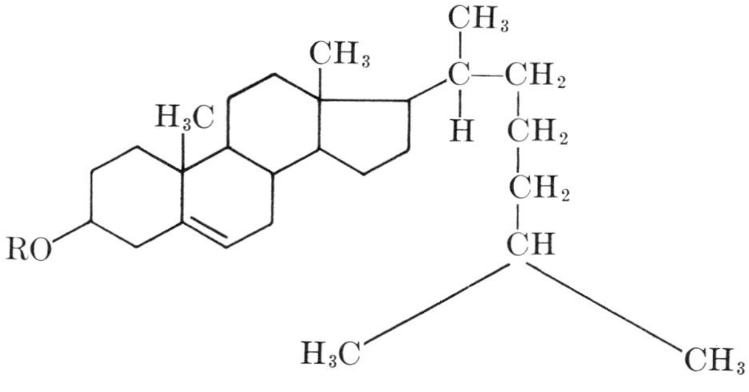

In cholesterol: R = H; in cholesterol esters: R = $-\overset{\overset{O}{\|}}{C}-R$

Figure 3.   Cholesterol and its esters

Table I. Protein, Phospholipid and Cholesterol Content of Some Biomembranes (16)

Membranes	Molar Ratio		
	Amino Acid	Phospholipid	Cholesterol
Erythrocyte	500	31.0	31.0
Myelin	264	111.0	75.0
Acholeplasma laidlawii	442	25.2	2.3
Bacillus Megaterium	520	23.0	0.0
Micrococcus lysodekticus	524	29.0	0.0
Streptococcus faecalis	441	31.0	0.0

Science

Table II. Lipid Composition of Some Animal and Microbial Membranes (wt. %) (16)

Lipid	Erythrocyte	Myelin	Mitochondria	Escherichia Coli	Bacillus Megaterium
Cholesterol	25	25	5	0	0
Phosphatidyl-ethanolamine	20	14	28	100	45
Phosphatidyl-serine	11	7	0	0	0
Phosphatidyl-choline	23	11	48	0	0
Phosphatidyl-inositol	2	0	8	0	0
Phosphatidyl-glycerol	0	0	1	0	0
Diphosphatidyl-glycerol	0	0	11	0	45
Sphingomyelin	18	6	0	0	0
Cerebroside	0	21	0	0	0
Cerebroside sulfate	0	4	0	0	0
Other	2	13	0	0	10

Science

such that the natively folded molecules form virtually indistinguishable secondary, tertiary, and quaternary structures.

The proportions of lipids in the membranes of a given cell type may vary strikingly from one species to the next. This appears to be caused by genetically determined differences in and among membrane proteins (24). Finally, the fatty acid composition of membrane phosphatides can change with the functional state of a cell, cell age, culture conditions, and so forth.

PROTEINS. *Introduction.* Membrane proteins remain among the more intractable biological entities. Indeed we even lack a clear consensus as to what constitutes a membrane protein. Numerous variables must be considered even for well-defined membrane domains, isolated by optimal methods from specific cells of defined genetic, developmental, and physiologic state. Membrane proteins are now often categorized into "integral" or "core" proteins and "extrinsic" proteins; this distinction is empirical and purely operational. Core proteins are those which remain water insoluble and associated with membrane lipids, unless drastic solubilization methods are used. Most would agree that the association of core proteins with their membranes involves hydrophobic forces. However, we lack assurance that what we define as a core protein may not exist in a guise that may obscure recognition of its membrane role, *i.e.*, in a soluble form, it might exist in a different configuration and/or functional capacity than when membrane associated. We know that well-defined, soluble proteins (*e.g.*, immunoglobulins) can serve as functionally important membrane proteins in lymphocytes; also, Langdon's recent data suggest that certain plasma lipoproteins can also serve as major membrane constituents of some species' erythrocytes (*25*).

Usually one can study membranes only after cell disruption and application of various subcellular fractionation procedures (*24*). A number of complications arise. For example, (a) one cannot equate membrane with organelle. Thus, in mitochondria and lysosomes most proteins are not membrane associated; (b) many membrane preparations are contaminated by absorbed proteins, *e.g.*, nucleohistones, or contain soluble proteins, trapped when membrane vesicles form during cell disruption; (c) most cells contain lysosomal cathepsins, and some membranes appear to bear endogenous peptidases. Both can introduce proteolytic artifact during the isolation of a membrane or fractionation of its proteins; (d) certain proteins can exhibit a high affinity for the membrane of intact cells but elute from the membranes during some isolation steps; (e) membrane topology is not homogeneous, and one surface domain may contain different proteins from the next; (f) cell variations arising from physiologic state, cell cycle, age, and so forth affect membrane protein composition.

Even a simple list of known membrane functions will contain hundreds of proteins. However, the contribution of each to the total membrane mass may be small. Thus, the number of ouabain-sensitive ATPase sites on human erythrocytes lies near 300 per cell (*26*)—a minute proportion of the total protein present. On the other hand, we know of instances where a biomembrane contains one predominant protein. For example, the membranes of some viruses possess only one protein (*27*), and rhodopsin accounts for nearly 90% of the protein mass in the membranes of retinal rod outer segments (*28*).

In general, the proteins of most membranes exhibit considerable diversity according to sodium dodecyl sulfate polyacrylamide gel electrophoresis (SDS PAGE), and their apparent molecular weights range from 15,000 to 300,000 daltons. However, within this diversity lie certain consistent patterns. For example, SDS PAGE of erythrocyte ghosts from different species, exhibit striking similarities (*29*). Also, lymphocyte plasma membranes reveal SDS PAGE patterns different from those of erythrocyte membranes but closely similar to each other (*30*). This leads one to suspect that proteins of a given membrane may possess certain structural homologies, which permit their interaction with the lipids of that membrane as well as its other protein constituents (*31*).

Amino Acid Composition and Sequence. Overall Hydrophobicity of Membrane Proteins. Apolar associations indubitably play a major role in the interactions of many proteins with their membranes. Studies on the solubilities of free amino acids in solvents of different polarity (*32, 33*) provide a reasonable assessment of protein hydrophobicity; each amino acid residue is assigned an experimentally determined "hydrophobicity" value, $H\phi$; these values range between 0.45 and 3.00 kcal per residue for Thr and Trp, respectively. Knowing amino acid composition of a protein and the residue hydrophobicities of its amino acids, one can compute its average hydrophobicity. The average hydrophobicities of soluble proteins range from 0.440 to 2.020 kcal per residue; with 50% of proteins $H\phi$ between 1.0 and 1.2 kcal per residue (*33*).

When one applies these stringent criteria, one cannot unambiguously distinguish membrane proteins as a class by virtue of hydrophobicity (Table III). Nevertheless, Capaldi and Vanderkooi (*34*) have devised an empirical

Table III. Average Protein Hydrophobicities, $H\phi$, of Some Membrane Preparations and Soluble Proteins[c]

Fraction	Average Hydrophobicity (kcal/Residue)
Erythrocyte ghosts[a]	1.068
Plasma membrane (Ehrlich ascites carcinoma)[a]	1.041
Smooth endoplasmic reticulum (Ehrlich ascites carcinoma)[a]	1.023
Smooth endoplasmic reticulum (liver)[a]	0.940
Myelin[a]	1.000
Glucagon[b]	0.930
Fibrinogen[b]	1.020
Hemoglobin (β chain)[b]	1.060
Serum albumin (bovine)[b]	1.120

[a] From Wallach and Winzler (*24*).
[b] From Bigelow (*33*).
[c] 50% of soluble proteins have $H\varphi$ values ranging between 1.0 and 1.2 kcal/residue.

scheme which allows a classification of membrane proteins into large multimeric proteins, tightly bound proteins, and loosely bound proteins according to overtime amino acid composition.

Amino acid analyses of whole membranes have not proved very informative, but subtle trends are detected when one classifies membrane proteins into categories that can be eluted by ionic manipulations and those requiring detergents or organic solvents for extraction. Ionically elutable proteins generally exhibit lower hydrophobicities than the more tightly bound species. For example, the erythrocyte protein "spectrin," *S. faecalis* ATPase, and bovine F_1 ATPase, all of which are eluted by ionic manipulations, exhibit $H\phi$ values of ≤ 1.10 kcal per residue while rhodopsin and bovine myelin proteolipid, which cannot thus be extracted, have hydrophobicities of ≤ 1.20 kcal per residue (Table IV). Interestingly, murein lipoprotein which is not easily eluted ionically exhibits an unusually low hydrophobicity (0.73 kcal per residue). Also, except for the last protein, both the "elutable" and "non-

Table IV. Average Hydrophobicities of Some Ionically Elutable and Non-elutable Membrane Protein[a]

Protein	Average Hydrophobicity (kcal/Residue)
Elutable	
Spectrin	0.960
S. Faecalis ATPase	1.10
Bovine F_1 ATPase	1.09
Non-elutable	
Rhodopsin	1.23
Myelin proteolipid	1.21
Murein lipoprotein	0.73

[a] From Wallach and Winzler (*24*); excluding Trp values.

elutable" proteins show hydrophobicities in the 0.9–1.2 kcal per residue range typical for most water-soluble proteins.

Amino Acid Distribution and Sequence. It has been proposed that membrane proteins lying at membrane surfaces might bind hydrophobically through apolar residues extending into apolar lipid regions. However, use of space-filling models shows that for a protein opposed to the surface of a phosphatide bilayer, even the largest apolar amino acid side chain, Trp, can only extend into the glycerol region of the layer; it cannot reach to the apolar fatty acid chains (*24*). Such side chain penetrations would also be energetically improbable. Therefore, apolar lipid–protein interactions, require that part or all of a membrane protein, rather than its apolar side chains, penetrate into or through the apolar membrane core (*24*).

A relatively simple model for the penetration of the protein into a membrane is suggested by the structure of melittin, a membrane-active polypeptide from bee venom. This is a generally hydrophobic protein (Hϕ = 1.24 kcal per residue). Also, nearly all of the first 20 residues from the amino terminal are highly apolar while the C-terminus contains a sequence of six polar and/or charged residues (Figure 4). This protein can intercalate its apolar residues among the relatively flexible fatty acid chains of a phosphatide bilayer in a random (*24*) or helical array.

```
 1         5                10
|Gly-Ile-Gly-Ala-Val-Leu-Lys-Val-Leu-Thr-Thr-Gly-Leu-Pro-

              15              20
              |Ala-Leu-Ile-Ser-Trp-Ile|-Lys-Arg-Lys-Arg-Glu-Glu
```

Figure 4. Amino acid sequence of melittin. Calculated hydrophobicity is 1.24. The boxed-in portion indicates the apolar segment.

The same structural principle appears in the major glycoprotein of human erythrocyte membranes. This protein spans the membrane thickness. Table V gives the amino acid composition of the intact glycoprotein, the soluble peptide segment released by proteolysis of intact erythrocytes and the insoluble peptide fraction remaining after proteolysis. The protease-accessible portion (bearing the sugar moiety) contains a large proportion of polar amino acids, giving it an average Hϕ of only 0.50 kcal per residue, compared with 0.93 kcal per residue for the whole protein. The very high average hydrophobicity of the residue (1.33 kcal per residue) is concordant with the concept that this hydrophobic portion "locks" the protein into the membrane (35).

Detailed studies indicate that the amino terminal segment, bearing the carbohydrate, lies extracellularly, while the apolar region occupies the hydrophobic membrane core, and a polar carboxyl terminal protrudes into the

Table V. Amino Acid Composition of the Major Human Erythrocyte Membrane Glycoprotein and Its Fragments

	Sialoglycoprotein[a]		
Amino Acid[b]	Intact	Soluble Peptide	Insoluble Peptide
Lys	3.5	4.3	2.0
His	3.8	4.9	3.8
Arg	4.1	3.3	3.9
Asp	6.0	7.6	2.4
Glu	10.0	4.7	7.0
Thr	13.8	23.8	6.2
Ser	13.6	23.8	6.0
Pro	6.5	4.0	4.5
1/2 Cys	0	0	0
Met	0	0	2.1
Gly	6.8	3.5	10.7
Ala	6.8	6.1	7.9
Val	7.7	4.9	8.1
Ile	4.5	2.8	14.3
Leu	4.5	1.6	11.6
Tyr	3.6	1.5	2.4
Phe	3.5	0	5.1
Trp	—	—	—

[a] From Winzler (44).
[b] Values are in residues per 100 residues. This protein contains 64% carbohydrate including most of the cell membrane's sialic acid in conjunction with galactose, mannose, fucose, N-acetyl-galactosamine, and N-acetyl-glucosamine. These comprise oligoheterosaccharides bound to the frequent threonine and serine residues usually *via* the N-acetyl-galactosamine.

cytoplasmic domain (35). The tentative amino acid sequence for the apolar segment, shown in Figure 5, shows an abundance of apolar residues. No ionizable residues occur for a stretch of 23 residues (11–35). However, two

```
     1                    5                        10                    15 16
Val-Gln-Leu-Ala-His-His-Phe-Ser-Glu-Ile-Glu-Ile-Thr-Leu-Ile- Gly-Phe-Gly
        (Pro)    (Pro)                                         (Ala)

        20                   25                    30 31                 35
-Val-Met-Ala-Gly-Val-Ile-Gly-Thr-Ile-Leu-Leu-Ile-Ser-Tyr-Gly-Ile-Arg-Arg

        40                       45 46                 50
-Leu-Ile-Lys-Lys-Ser-Pro-Ser-Asp-Val-Lys-Pro-Leu-Pro-Ser-Pro-
```

Biochemistry

Figure 5. Tentative amino acid sequence for the apolar portion of the major glycoproteins of human erythrocyte membranes. Parentheses indicate sequence uncertainties (35).

Glu residues (9, 11) and two Arg residues (35, 36) give the apolar segment two polar ends and an asymmetric charge distribution. As an α-helical array the 23-residue apolar segment could extend about 35 A, approximately one-half the width of many membranes.

Certain sequences of polar and apolar amino acids permit a polypeptide chain with an unremarkable amino acid composition to fold into secondary structures (e.g., α-helix) with opposite polar-apolar faces. Under such circumstances, tertiary structuring and/or quaternary associations will foster association of polar-with-polar and apolar-with-apolar faces. In water, minimal free energy obtains when the polar faces orient into the solvent and the apolar ones into the protein core (in analogy with the two halves of a phosphatide bilayer). Many well-documented examples of this architectural device exist, for example in hemoglobins. As noted above, the peptide chains of all vertebrate hemoglobins fold into virtually identical tertiary structures, in which polar residues are almost totally excluded from the interior of the molecules. Importantly, apolar residues occur at intervals of ∼ 3.6 residues along α-helical segments, giving each helix a hydrophobic face. This important principle of protein structure occurs in other soluble, partly helical proteins, whose structures have been elucidated by x-ray crystallography.

Theoretically, replacement of water with a low dielectric solvent—e.g., the hydrocarbon chains of phosphatides—should induce association of the polar interfaces at the interior of the folded polypeptide, giving it a hydrophobic perimeter. This would be analogous to the reversal of molecular orientation of amphipathic lipids, when transferred from an aqueous dispersion to an organic solvent. In the former, the molecules form arrays with the polar groups oriented away from each other into the solvent. In the latter, the hydrocarbon chains point away from each other into the solvent and the polar groups cluster in the interior. Accordingly, an appropriate combination of amino acid sequence, secondary, tertiary, and perhaps quaternary structures, might endow protein with the hydrophobic surfaces required for apolar membrane-protein associations, but without an unusually high average hydrophobicity and/or concentration of apolar residues in a given segment of a polypeptide chain. Clearly, a protein with a fully apolar perimeter, while suited for residence in a membrane's apolar core, cannot be considered ideally structured to deal with events involving the aqueous intra- and extracellular compartments. Such traffic would require (a) more or less cylindrical proteins, with (b) hydrophobic perimeters, (c) polar moieties projecting into aqueous spaces, and (d) possibly polar channels running through the protein assembly—analogous to the pore penetrating the hemoglobin tetramer.

Recent data on murein lipoprotein appear to provide an example of the above structural device in a biomembrane. The cell walls of *E. coli* comprise a rigid layer,stabilized by murein (peptidoglycan). This consists of polysaccharide chains crosslinked by short peptide segments. About 10^5 murein lipoprotein molecules per cell are linked to this net. The full amino acid sequence of this protein—as well as the murein and lipid attachment sites—has been published (36). As shown in Figure 6, the N-terminus consists of

Lipid
|
1   5   10   15
(Ser)-Ser-Asn-Ala-Lys-Ile-Asp-Glu-Leu-Ser-Ser-Asp-Val-Gln-Thr-Leu -

      20   25   30
Asn-Ala-Lys-Val-Asp- Glu-Leu-Ser-Asn-Asp-Val-Asn-Ala - Met-Arg-Ser-

      35
Asp-Val-Gln-Ala-Ala-Lys

European Journal of Biochemistry

Figure 6. N-terminal portions of murein lipoprotein. Apolar residues occur at every third or fourth position which in an α-helix, would give the structure a polar and an apolar face (36).

three very similar sequences. Starting with the third residue, apolar side chains occur at every third to fourth position in a consistently alternating set. In an α-helical conformation this would provide murein–lipoprotein with a polar and an apolar face, despite a very low hydrophobicity (Hϕ = 0.73 kcal/residue).

In summary, the requirements for apolar lipid–protein interactions can be met by high protein hydrophobocity, concentration of hydrophobic residues into one segment of the polypeptide chain, and through specialized sequences combined with appropriate chain folding. Sequence analyses of many more membrane proteins are required to determine which mechanism is prevalent and what structural homologies are most widely involved.

Geometric Disposition. The distribution of proteins across the membrane thickness has been repeatedly reviewed (37, 38), and there can be no doubt that most membranes exhibit a transverse asymmetry. This is apparent from x-ray diffraction studies (*e.g.*, on myelin), from membrane inversion studies from the differential susceptibilities of external and internal surfaces to proteases, phospholipases, and covalent labelling procedures, from the asymmetric images obtained by freeze-etch electron microscopy, from studies on the activities of membrane bound enzymes, and so forth. Such investigations not only indicate transverse asymmetry but also show that some proteins penetrate through the membrane. However, some major issues remain unresolved. For example, do the observed asymmetries reflect only differences of protein location, or can the reactivity of a given protein change with location? Also, protein disposition may not be static. Finally, there are indications that some proteins can turn over metabolically independently of other proteins in the same membrane (*39*).

The distribution of proteins parallel to the membrane plane is under dispute. As shown later, one of the favorite current membrane models suggests that membrane proteins are freely mobile in the membrane plane. On the other hand, substantial evidence indicates that the position of proteins parallel to the membrane plane is stabilized by some framework (*40*). This could be within the membrane or within the cytoplasm (cytoskeleton, microfilaments).

The emphasis of this section has been on the possible structural features which allow for the extensive apolar interactions between "intrinsic" membrane proteins and membrane lipids. One can expect that one or more of the architectural devices described participate in the membrane affinity for diverse enzymes, which are, of course, major membrane components and many of which cannot be dissociated from their membranes in functional form.

A very interesting example is the cytochrome b_5 isolated of rabbit liver endoplasmic reticulum (*41*). This enzyme consists of an enzymatically active portion, composed of 97 residues and a tail composed of ~ 40 residues. The enzymatically active moiety has a very low hydrophobicity ($H\phi_{Av} = 0.89$ kcal per residue) whereas the tail is highly hydrophobic ($H\phi = 1.7$ kcal per residue). It appears that the enzymatically active portion of the protein is anchored to the apolar portions of the membrane by its hydrophobic tail.

CARBOHYDRATES. Most cellular membranes bear greater or lesser amounts of diverse carbohydrates. These may be associated with mucopolysaccharide, such as exists in the glycocalyx which coats many cells' plasma membranes. As reviewed by Hughes (*42*) membrane carbohydrates are commonly also associated with membrane glycoprotein and glycolipid.

The carbohydrate moiety of membrane glycoproteins can usually be released by proteolysis. This has allowed considerable clarification of the oligosaccharide structure, which resembles that of soluble glycoproteins, as follows:

(1) In general the oligosaccharides are bound either *N*-glycosidically to Asn or *O*-glycosidically to a Ser or Thr hydroxyl.

(2) More than one heterosaccharide may be associated with one membrane or a given membrane protein.

(3) The sugar moieties comprise small heterosaccharides, usually with specific monosaccharide sequences.

(4) The major constituent monosaccharides are fucose, galactose, glucose, mannose, N-acetylgalactosamine, N-acetylglucosamine, and N-acetylneuraminic acid.

Two major membrane glycoproteins of very different character have been extensively studied. These are (1) the major glycoprotein of human erythrocytes and (2) bovine rhodopsin. The former contains about 64% carbohydrate by weight, bears most of the membrane's sugar and all of its N-acetylneuraminic acid. In its chemically extracted form it has a MW of 31,400, but SDS PAGE data suggest it may migrate as a dimer in that system. One cannot yet be certain that this substance is a single homogeneous molecule, particularly since it exhibits multiple biological function—e.g., binding of kidney-bean phytohemagglutinin, MN–antibody, and influenza virus. Probably these functions reflect the specificities of different heterosaccharides attached to a single polypeptide chain. Many of these heterosaccharids are tetrasaccharides comprising two molecules of N-acetylneuraminic acid, a galactose and an N-acetylgalactosamine linked O-glycosidically to Ser or Thr. Some oligosaccharides appear incomplete.

At least one of the heterosaccharides, N-glycosidically linked to the peptide chain, exhibits greater complexity. The following structure appears likely (24):

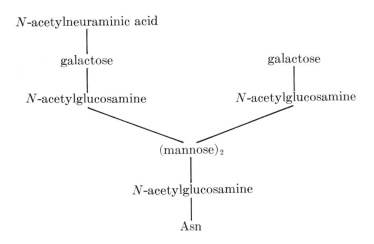

This heterosaccharide strongly binds kidney bean phytohemagglutinin.

Other carbohydrate side chains of this protein have also been characterized and attest to the complexity of this molecule or group of molecules. Of particular importance are the human blood group antigens A, B, H, and Lewis. These are attached not only to glycoproteins but also to glycolipids.

Rhodopsin provides a great contrast to the erythrocyte membrane glycoproteins. Only 4% of its mass is carbohydrate. Each 28,000-dalton peptide chain bears one heterosaccharide joined N-glycosidically *via* N-acetylglucosamine and Asn. The other sugars comprise two molecules of N-acetylglucosamine and three of mannose.

Some Major Membrane Functions

Compartmentalization and Organization of Functional Macromolecules. A critical function of biomembranes is the development and maintenance of electrochemical gradients between the compartments they separate. There are numerous important examples of this—e.g. the transport of ions, amino acids, sugars, and other metabolites, synthesis of ATP in mitochondria and chloroplasts, muscle contraction, axonal conduction, and synaptic transmission. Membranes also maintain certain enzymes in high local concentration. Also, in certain membranes, the membrane state creates a close topologic association between functional molecules, allowing for highly efficient linked processes. The respiratory chain of the mitochondria and associated phosphorylation processes are outstanding examples. Finally, membranes act as control points in regulating metabolic processes occurring in separate compartments. An important example is regulation of glycolysis and fatty acid synthesis (cytoplasmic) by citrate, which is generated within the mitochondria. The translocation of citrate requires a specific carrier located in the mitochondrial membranes.

The cytoplasmic space is highly segregated from the extracellular environment. However, extracellular material can be transferred intracellularly in bulk through pinocytosis and phagocytosis. In these processes, the plasma membrane invaginates and pinches off vesicles containing extracellular material. No direct continuity between intra- and extracellular compartments is involved. The reverse process also occurs in exocrine secretion, virus budding, and so forth. The cytoplasmic spaces of contacting cells often join through permeable junctions. These certainly allow passage of small molecules, but their real gating properties remain to be established.

Lysosomes are normally segregated from the cytoplasmic and other compartments but commonly fuse with phagocytic vesicles to form membrane enclosed spaces where the extracellular material is digested.

Mitochondria contain fluid compartments—one between the outer and inner mitochondrial membrane and another within the inner membranes. The inner space is particularly tightly segregated from the cytoplasm. Numerous transport systems mediating between these domains have been characterized.

The nuclear envelope separates the nucleoplasm from the cytoplasm. However, the nuclear membranes break down and re-form during the mitotic cycle. During interphase, the important nucleo-cytoplasmic traffic of RNA and protein proceeds through non-membranous, but highly structured gaps in the nuclear envelope, the nuclear pores.

The endoplasmic reticulum is a labyrinth of interconnecting cisternae and tubules. It appears to link to the space between the two membranes comprising the nuclear envelope and also connects to the Golgi cisternae. In some cells the endoplasmic reticulum connects to the extracellular space.

Permeability and Electrical Properties. Any meaningful concept of membrane structure must explain the permeabilities and electrical properties of biomembranes. The latter topics have been treated extensively (*44, 45, 46*). The electrical properties of natural and artificial membranes vary considerably (Table VI).

Table VI. Some Physical Properties of Natural and Model Membranes

	Natural Membrane	Black Lipid Membrane
Thickness (A)	50–120	68–73
Surface tension (dynes/cm)	0.03–3.0	0.5 –1.0
Water permeability (cm/sec $\times$ 10^3)	0.03–3.3	0.5 –1.0
Electrical capacitance (μF/cm^2)	0.5 –1.3	0.33–1.3
Electrical resistance (ohms/cm^2)	10^3–10^6	10^6–10^9
Breakdown potential (V)	0.1 –3.0	0.15–0.20

Experimentation on membrane permeabilities originated with Overton's studies on the permeation of various molecules into plant cells (*47*). He found that the passage of such entities through the cell periphery varied inversely with polarity. Moreover, the penetration rate of many, small polar molecules could be enhanced by modifying them to increase their solubility in apolar solvents. Accordingly, Overton concluded that the plasma membrane comprises an apolar permeability barrier. However, his studies also indicated that the passive penetration of molecules through plasma membranes varied critically with molecular size. His data and extensive information obtained since then indicate that (a) plasma membranes act as molecular sieves—*i.e.*, molecules below a certain size (80 daltons) permeate much more readily than larger entities; (b) for substances greater than 80 daltons, hydrophobic molecules penetrate less readily than apolar ones.

Many cells' plasma membranes allow free permeability to protons and K^+ and lesser permeability to Na^+. Some membranes are permeable to Cl^- as well or instead. Very large ion concentration gradients commonly exist across plasma membranes, and these can account for the fact that the interior of an animal cell is generally electronegative with respect to the exterior.

Measurements of the electrical capacitances and resistances of plasma membranes are consistent with the permeability studies. They indicate that plasma membranes act as insulating, low-dielectric barriers between the extra- and intracellular spaces. Some intracellular membranes probably act identically, but we lack detailed information on this.

The electrical resistance across plasma membranes can be estimated by measuring the potential difference between a microelectrode inserted into a cell and a similar extracellular electrode. The membrane resistivities of various cell types vary over a wide range. Also, the resistivities fluctuate with physiological state. For example, the resistance across the surface of *Drosophila* salivary gland cells lies near 10^4 ohms/cm². That across the wall of the squid giant axon is closer to 10^3 ohms/cm² during electrical impulse conduction (48).

Although the surface membranes of numerous cells exhibit very large resistances, contacting cells commonly allow the flow of ions (current) between them through low resistance junctions. These allow electrical coupling of long chains or large clusters of epithelial cells (49). The structures responsible for such ion flow appear to be the gap junctions or nexuses observed by electron microscopy (20). Mitochondrial membranes very likely have a high electrical resistance. Nuclear membranes show an area resistance of ~ 1 ohm/cm². This indicates that nuclear pores do not allow free ion flow.

Active Transport. The permeability properties of biomembranes are not limited to passive diffusion, and most membrane systems bear numerous, specific active transport systems (50, 51, 52, 53, 54). These comprise elaborate molecular assemblies which mediate the translocation of molecules against an electrochemical gradient. Numerous ions and biologically important organic entities, both large and small, are translocated actively. Some transport systems are coupled—*e.g.*, Na⁺ transport, K⁺ transport, and the translocation of certain amino acids (54). Membrane transport is not limited to the plasma membrane. For example, mitochondria and sarcoplasmic reticulum membranes (muscle endoplasmic reticulum) bear potent calcium transport systems. These maintain low intracellular Ca^{2+} levels.

Active transport utilizes energy derived from the hydrolysis of ATP and related high-energy compounds—*i.e.*, the membrane functions as an energy transducer. For the transport of Na⁺ and K⁺ through plasma membranes:

$$K_e^+ + Na_i^+ + ATP + H_2O \rightleftharpoons K_i^+ + Na_e^+ + ADP + P_i$$

K_e^+ and K_i^+ are the extracellular and intracellular potassium ions; Na_e^+ and Na_i^+ are the extracellular and intracellular sodium ions. Usually Na_i^+ is kept low relative to K_i^+.

In the case of calcium transport into sarcoplasmic reticulum:

$$Ca^{2+}{}_{cyt} + ATP + H_2O \rightleftharpoons Ca^{2+}{}_{sarc} + ADP + Pi$$

Here $Ca^{2+}{}_{cyt}$ and $Ca^{2+}{}_{sarc}$ refer to Ca^{2+} in the cytoplasmic and sarcoplasmic spaces respectively.

A somewhat detailed look into the active transport of Na⁺ and K⁺ across human erythrocyte membranes reveals certain important general characteristics of this crucial cation translocation system. [Different species demonstrate major, genetically determined differences in active erythrocyte membrane cation transport (55).] First, the process generally involves an exchange of cations across the membrane. Typically, the movement of external K^+ inward is coupled to the outward migration of intracellular Na^+, *i.e.*,

$$K_e^+ \rightleftharpoons Na_i^+$$

Other exchange mechanisms occur (*e.g.*, $K_e^+ \rightleftharpoons K_i^+$, $Na_e^+ \rightleftharpoons Na_i^+$), and in some cells, K^+ is pumped out and Na^+ in (*i.e.*, $Na_e^+ \rightleftharpoons K_i^+$).

Second, the cation transport system is regulated by the Na^+ and K^+ concentrations at the two membrane surfaces. Thus, in general, K_e^+ activates and Na_e^+ inhibits inward K^+ transport whereas K_i^+ independently inhibits and Na_i^+ independently activates. This illustrates the asymmetry of the system. This is also apparent from the fact that ATP hydrolysis proceeds at the cytoplasmic membrane surface, but cardiac glycosides—which specifically inhibit the alkali cation pump—react only at the external membrane surface.

Third, more than one Na^+ or K^+ is involved in the activation (inhibition) processes occurring at the two surfaces.

Fourth, ATP, or some high energy product of ATP metabolism is a necessary and adequate energy source for transport. Also, the transport system in erythrocytes and a Na^+-, K^+-dependent ATPase in erythrocyte membranes exhibit numerous common properties. Moreover, there is evidence (discussed later) that the process involves significant conformational changes in membrane proteins.

Specific transport requires some kind of discriminating "carrier" mechanism. Many hypotheses have been proposed; the most fascinating and embracing is the lattice concept of Changeux and associates (56) although it is speculative. This hypothesis rests on two critical features of biological membranes—*viz.*, membranes are infinite in two dimensions, and the *in situ* environment of a membrane is asymmetrical.

Changeux *et al.* (56) argue (a) membranes are isothermal lattices lying between two phases of different electrochemical potential and allowing a passive net flux of the permeating species; (b) the lattices contain equivalent "versatile ionophores," of which each bears at least two distinct, specific sites for a permeant ion, one on the inner face of the membrane, the other on the outer face. The permeant ion binds and crosses the membrane by a jump from one class of site to the other; (c) the binding and the permeability for the permeant ion change during the conformation transition of the versatile ionophore. The permeant ion thus controls its own translocation; (d) the ionophores interact cooperatively within the membrane lattice by structural coupling between nearest neighbors. This does not imply that the membrane lattices are continuous with each other; ionophore-bearing lattices could be organized in dispersed clusters provided that the ionophore density suffices for cooperation; (e) both sides of the membrane bear an "equilibration layer" in which the activity of the ion differs from those of both in the bulk and membrane phases. In such layers the ion concentration would depend on its rate of absorption on the membrane surface, on its rate of transport through the membrane, and on its rate of diffusion from the bulk phase.

Although developed to deal specifically with ion transport, the Changeux concept may be more generally applicable to membrane transport.

Information Translation. The plasma membranes of some animal cells bear machinery which translates external stimuli into intracellular information. This leads to modifications of metabolism, activation- or repression-specific synthetic pathways, and so forth. We will consider three examples of increasing complexity.

THE ADENYLATE CYCLASE SYSTEM: INTERACTIONS OF PLASMA MEMBRANES WITH HORMONES AND OTHER COMPOUNDS. Numerous hormones, and also neurotransmitters, etc., specifically control the function of their target cells. Such regulators combine with unique receptors on the target cells surfaces. This leads to the activation of membrane adenylate cyclase, which catalyzes the reaction:

$$ATP \rightarrow 3', 5'\text{-cyclic AMP (cAMP)} + \text{pyrophosphate}$$

cAMP acts as a diffusible, pleiotypic intracellular transmitter of the hormone effect—a "second messenger" (*57*).

A reasonable sequence of events is as follows. The extracellular agent (*e.g.*, hormone) changes the architecture of its membrane receptor in binding to it. This activates adenylate cyclase *via* a membrane transducer. Through its enzymatic action, the adenylate cyclase behaves as an amplifier. The operational sequence, therefore, is

$$\text{receptor} \rightarrow \text{transducer} \rightarrow \text{amplifier} \rightarrow \text{effector}.$$

The level of effector (cAMP) is further regulated by phosphodiesterase, which catalyzes the reaction cAMP $\rightarrow$ 5' AMP. The specificity of a hormone-mediated response thus lies in (a) the receptor and/or (b) the metabolic phenotype of the cell. The role of the cyclase is that of an amplifier and is not specific.

IMMUNE RECOGNITION. The immune response involves two highly specific recognition processes—one in the initiation of the response, the other in its execution. Plasma membranes appear to participate in both and lymphocytes are the principal cells involved.

Immunologic activation can be envisaged to proceed as follows. An immunogenic foreign substance—an antigen—enters an animal. When the antigen comes into contact with lymphocytes bearing receptors specific for the antigen, these are activated to proliferate, producing cell populations coded to react against the specific antigen. The descendants may be cells which produce antibody against the antigen, and/or they may be able to react directly with antigen. It is widely believed that the antigen receptors are membrane components, possibly specific immunoglobulins (*58*). In any event, the lymphocyte activation step quickly initiates significant changes in plasma membrane permeability (*59*), transport (*60*), and metabolism (*61*).

CELLULAR RECOGNITION. The cells of complex organisms participate in numerous surface-mediated interactions. During embryogenesis a number of differentiative processes are induced by contact of one cell type with another. On the other hand, when a stable differentiative stage is reached, like cells tend to associate with like. Motile cells constitute an exception and normal social interactions between cells break down in neoplasia.

How is cellular individuality expressed at cell surfaces and how is contact information translated into intracellular information? There is no satisfactory answer to the second question although cAMP has been repeatedly invoked. However, the matter of surface individuality has produced several interesting hypotheses. These fall into two categories: (a) chemical hypotheses and (b) coding concepts.

A number of chemical hypotheses have been proposed. The most elementary invoke the existence of specific bivalent proteins which link together cells bearing identical surface receptors. This is analogous to an antigen–antibody reaction with the antigens corresponding to the surface receptors (62). There is evidence that such mechanisms operate in very simple multicellular organisms.

Recently, Roseman (63) proposed an elegant enzyme–substrate model for cell recognition. He suggests that cell surfaces bear both certain enzymes and substrates (receptors) for these enzymes. More specifically, the substrates proposed are complex surface carbohydrates and the enzymes glycosyltransferases. Cell recognition is equated with the specific binding of a substrate by its enzyme and cell adhesion is mediated *via* enzyme–substrate complexes. In support of this concept are observations that glycosyltransferases, as well as complex carbohydrates, occur on cell surfaces. Roseman further elaborates his hypothesis in suggesting a metabolic mechanism for the modification of intercellular adhesion and recognition. He suggests that the glycosyltransferases might, under appropriate conditions, be supplied with appropriate sugar nucleotides, synthesized intracellularly. They would then add these to the existing receptors (substrates) converting them into products. Since enzymes bind poorly to products, there would be loss of adhesion and recognition. One major objection to the enzyme substrate concept is that a cell with one type of enzyme can attach (or modify) a different cell that happens to bear the enzyme substrate.

Very recently, Najjar (64) has proposed another elegant chemical hypothesis. This has its roots in the patterns of subunit associations in polymeric proteins. Such (*e.g.*, the interaction between hemoglobin chains) are extremely specific, can be very tight, and could be modified by subtle biochemical mechanisms. Najjar suggests that the surfaces of single, like cells bear the monomers of a multimeric protein. Contact between the cells allows formation of the multimer. The model provides for cell recognition and for cell attachment.

An important weakness in the chemical theories is the fact that in vertebrates the number of cells making specific connections are too large to be individually coded by specific proteins. This point is particularly striking when one considers the organization of the nervous system. The DNA required to allow coding by specific proteins would vastly exceed the quantity of DNA in a fertilized egg. Accordingly, Changeux and associates (56) suggest that a gene-conserving coding machinery must be involved in tissue organization. They propose that (1) the stability of intercellular association derives from the affinity of their surfaces; and (2) interaction specificity arises from "coding units" made up of distinct protomer classes, each coding unit corresponding to a given combination of membrane subunits in the membrane plane. Accordingly, the affinity between cells will depend upon protomer content and/or disposition of the surface membrane. A protomer could reside in a predetermined, fixed position in the membrane plane (ordered code) or be randomly distributed (statistical code). An ordered code requires different protomers, and coding units must be precisely placed; it is difficult to envisage such machinery. In contrast, a statistical code only requires structural homology of membrane proteins, allowing regular lattice structure. Then cell–cell affinity would depend on the overall protomer composition of the participating lattices and would yield a virtually unlimited number of different coding units.

Some Physical Approaches to Membrane Structure

X-Ray Diffraction Studies. Low angle x-ray diffraction analyses have significantly elucidated the organization of multilamellar membrane arrays, both natural and synthetic. This topic is well reviewed by Shipley (65). The presence of lipid bilayers in multilamellar systems is readily identified. However, major difficulties arise when one attempts to locate proteins with similar electrodensity distributions as might occur in lipid bilayers and/or micelles. This is a particular problem in biomembranes other than myelin, in which protein accounts for more than 65% of the membrane mass. Indeed, the apolar regions of proteins are not likely to exhibit electron densities very different from those in the hydrocarbon cores of phosphatide bilayers. Work on certain polypeptides supports this conclusion.

MYELIN. Of all membranes, myelin has been most thoroughly investigated by x-ray methods. These show that nerve myelin consists of concentric lipoprotein layers wrapped round the nerve axon. Each myelin lamella contains a double bilayer of lipid. The myelin proteins appear concentrated at the polar regions of the lipid. The double bilayers are asymmetric because of differences in the protein at their inner and outer surfaces. The most refined study (66) indicates that the cholesterol in each bilayer is asymmetrically distributed. The core of each bilayer is considerably disordered because of flexible hydrocarbon chains. Packing becomes more rigid as one moves out from the center, owing to the presence of cholesterol and stiff hydrocarbon chains. The proteins appear associated with the hydrated polar groups of the phosphatides. The concentration of amino acid residues is highest near the lipid surface. The possibility that some proteins extends across the lipid bilayer is not excluded.

RETINAL ROD OUTER SEGMENTS. X-ray analyses have yielded a rather detailed picture of this membrane system (67). The data suggest that the photoreceptor membranes have a lipid bilayer structure with the lipid head groups about 40 A apart. This means that the hydrocarbon core of the bilayers is very narrow. A significant proportion of the membrane protein (predominantly rhodopsin) penetrates into or through the lipid bilayer.

ERYTHROCYTE MEMBRANES. X-ray analyses have been recently extended to membranes which are not natively in a multilamellar array. One such approach is to stack the membranes centrifugally. Analyses of erythrocyte ghosts prepared in this fashion indicate a membrane thickness of $\sim$ 100 A, with polar material concentrated at the surfaces. Recently Wilkins *et al.* (*68*) have obtained x-ray diffraction patterns from suspensions of hemoglobin-free erythrocyte ghosts. The data are interpreted to indicate extensive but not necessarily continuous regions of phospholipid bilayers in the membranes. However the membrane proteins could not be specifically localized.

ACHOLEPLASMA LAIDLAWII. Dispersions of the plasma membranes of *Acholeplasma laidlawii* yield patterns analogous to those obtained with suspensions of erythrocyte ghosts (*69*). There is evidence of extensive regions of lipid bilayer, but the membrane proteins remain to be located.

HALOBACTERIUM HALOBIUM. The purple membranes of this microorganism yield x-ray diffraction patterns which are again consistent with the presence of lipid bilayer (*70*). However, they also point to a significant proportion of the membrane protein which lies at the membrane core.

CONCLUSION. All biomembranes studied show x-ray features consistent with the presence of phospholipid bilayers. The presence of lipid in other arrays cannot be excluded. The localization of membrane protein remains rather vague, but most recent experimentation indicates that some proportion of membrane protein penetrates into or through the bilayer regions.

Spectroscopic Approaches. Membrane biologists increasingly employ spectroscopic techniques to elucidate membrane structure. Two strategies are used: (a) analysis of signals emanating from intrinsic membrane components (by infrared spectroscopy, optical activity measurements, nuclear magnetic resonance) and (b) utilization of molecular probes, either fluorescent or paramagnetic. The area of membrane spectroscopy is expanding very rapidly, and several reviews have been published recently (*24, 71*). However, despite very extensive effort and expenditure, spectroscopic techniques are only now beginning to yield significant returns.

OPTICAL ACTIVITY. *Polypeptides and Soluble Proteins.* The secondary peptide structures (conformations) in membrane proteins have been evaluated by optical activity measurements [optical rotatory dispersion (ORD) and circular dichroism (CD)]. The optical activity of peptide bonds arises from (a) $\pi^0-\pi^-$ electronic transitions near 190 nm and (b) $n-\pi$ transitions near 220 nm. The contribution of both types of transition depends upon the spatial relationships of the individual peptide bonds and α-helical, β-, and "unordered" conformations yield distinct ORD and CD spectra. This has allowed the conformational analysis of some soluble polypeptides and proteins.

One initially hoped that the contributions of α-helical, β-structured, and unordered segments of globular proteins would sum, permitting conformational analyses by comparing their ORD and CD spectra with those of reference polypeptides. These expectations have faded because:

(a) Numerous proteins contain significant proportions of β-conformations; unknown proteins must, thus, be analyzed in terms of at least three known conformations. This is difficult since seven optically active chromophores are concerned.

(b) In α-helices, optical activity increases with the number of consecutive, helically arrayed residues up to 20; it also diminishes when the helices are distorted. Both situations are common in globular proteins.

(c) Optical activity is side-chain dependent.

(d) "Unordered" segments in proteins, if irregular, are usually fixed and structured; they vary from protein to protein and are non-uniformly distributed elements of diverse size and structure. They differ from the "random" states of polypeptides.

(e) Many peptide linkages of globular proteins reside in an apolar, highly polarizable microenvironments. This tends to reduce the rotational strengths of the optically active bands and shift these to higher wavelengths.

(f) Conformations other than the conventional right-handed α-helical, β, and "unordered" structures may exist in proteins. For example, while the left-handed α-helix is intrinsically less stable than the right-handed helix structure, extrinsic forces may favor it in certain situations (*e.g.*, 72).

The difficulties encountered in the conformational analysis of soluble globular proteins apply also to membrane proteins but may be further compounded by artifacts arising from light scattering and absorption statistics (73). Also, except for some viral envelopes, membranes lack any one, predominant protein. Optical activity measurements, therefore, can at best yield only the average conformational proportions in membranes. However, these might mirror structural homologies among the proteins of a given membrane or diverse membranes.

Membranes. When compared with the spectra of polypeptide standards, the CD and ORD spectra of nearly all membrane types studied so far show shapes close to that obtained with pure, right-handed α-helix, low intensity, and displacement of the entire spectrum to-longer wavelengths than observed with α-helix. With mitochondria and their fragments, these deviations are extreme, probably because of the presence of α-structured peptide (24).

Some of these anomalies arise from light-scattering artifacts. However, as pointed out in Ref. 74, most features of membrane optical activity do not arise from the particulate nature of the membrane and one can explain the major features of membrane optical activity as follows:

(a) Globular and membrane proteins tend to give low helical signals because of short helical segments, distortions from perfect helicity and apolar environments.

(b) In native proteins, peptide linkages not in α-helical or β-structured disposition comprise multiple, irregular, but non-random arrays, each yielding a spectrum other than that of unordered polypeptides. The summed contribution of such segments will be diffuse, allowing a helical shape and underestimating the proportion of unordered peptide.

Therefore, membrane proteins may be more helical and contain more unordered peptide than indicated by their optical activity. Also, the lesser CD amplitude of one membrane type, compared with another, could reflect differences in overall "helicity" or in the proportion of long helical segments. Membrane helicity cannot, therefore, be quantified unambiguously by optical activity measurements at present.

Despite many obstacles, optical activity measurements have yielded the important information that most membranes contain more than 30% of their peptide linkages in a right-handed α-helical conformation. This is greater than what is found in most globular proteins and may at that be an underestimate.

INFRARED (IR) SPECTROSCOPY. IR measurements in some regions of amide absorption can prove useful in the conformational analysis of polypeptides and proteins (24). The amide absorptions of polypeptides arise principally from in-plane C=O, C—N, and C—H stretching, OCN bending, CNH bending, and out-of-plane C—N twisting, and, in polypeptides, C=O and N—H bending. Because of their similarity to the absorption bands of secondary amides, the major peptide bands in polypeptides are termed "Amide" bands.

Polypeptides and Proteins. The amide absorption bands of polypeptides are sensitive to the folding of the peptide backbone and to the hydrogen bonding with other peptide linkages, *i.e.*, to conformation or 2° structure. Polypeptides in α-helical and/or unordered conformations exhibit IR spectra very different from those of β-structures, particularly in the Amide I region.

Studies of proteins whose detailed structures are known by x-ray crystallography have demonstrated that it is valid to use Amide I absorption in conformational analyses (24) in that the presence in proteins of β-structures together with α-helical and/or unordered conformations can readily be distinguished. For example, lysozyme, which has about 10% of its peptide linkages in the anti-parallel β-conformation (24) exhibits a 1650 cm^{-1} peak with a shoulder at 1632 cm^{-1}.

Membranes. The IR spectra of cellular membranes have been correlated with those of synthetic polypeptides and proteins with known conformation, to yield information concerning the secondary structure of membrane proteins and their possible conformational transitions.

The Amide I region of erythrocyte membranes lyophilized from 7 mM phosphate buffer, pH 7.4, shows maximal absorption at 1652 cm^{-1}, indicative of protein in the α-helical and/or unordered conformation; no band is observed at 1630–1640 cm^{-1}, implying little β-structure. The data indicate that proteins of erythrocytes membranes are mixtures of α-helical and unordered conformations. However, quantification of the exact proportions of these structures has not been achieved as yet. Conformational transition caused by drying cannot be excluded, but the Amide I spectra of erythrocyte ghosts freshly suspended in D_2O closely resemble those of dried films.

IR studies on purified plasma membranes and endoplasmic reticulum fragments of Ehrlich ascites carcinoma, dried at neutral pH, also yield no evidence of β-structured peptide, but the spectra of membranes dried from acid solvents, indicate some transition to the β-structure. Myelin—where most of the membrane protein probably lies at the surface of lipid bilayers— also demonstrates an Amide I band near 1655 cm^{-1} with no unusual absorption near 1630 cm^{-1}.

In contrast, the IR spectra of dried *Micrococcus lysodekticus* membranes (75) exhibit a shoulder near 1635 cm^{-1}, indicating presence of some peptide in the β-conformation. The membranes of *Acholeplasma laidlawii* also show IR evidence of β-structured peptide (76), and their optical activity is best interpreted in these terms. Finally, according to IR criteria, isolated adipocyte plasma membranes contain an appreciable proportion of peptide in the antiparallel β-conformation (77).

IR spectroscopy indicates that the proteins of conventionally isolated rat liver mitochondrial membranes contain a significant proportion of peptide in the antiparallel β-conformation (78); the exact proportion depends upon the metabolic state. Elution of soluble, "matrical" mitochondrial proteins by osmotic lysis and comparison of outer and inner mitochondrial membranes indicates that β-structured peptide is associated with the inner membranes.

FUNCTIONALLY RELATED CHANGES IN MEMBRANE PROTEIN CONFORMATION. *Optical Activity Measurements.* Few studies have been conducted to monitor possible functionally relevant changes in membrane optical activity. This is in large part because many metabolic substrates, cofactors, inhibitors, and so forth, absorb strongly in the region of peptide absorption and cannot be used satisfactorily at their biologically active levels. Nevertheless, Masotti *et al.* (79) were able to record the effects of ATP hydrolysis on the CD spectrum of microsomal membrane vesicles, derived from dog gastric mucosa cells. These membranes bear an ATPase which is insensitive to Na$^+$ or K$^+$ but is stimulated by HCO$_3^-$ and inhibited by SCN$^-$; it is thought involved in acid secretion. Untreated vesicles exhibit nearly equivalent CD intensities near 223 and 208 nm. Addition of 1 mM ATP shifts the 223-nm band to 224 nm, lowers CD intensity, and shifts the crossover from 202 nm to 207 nm. Mg–AMP is without effect. The authors suggest the possibility of an $\alpha \to \beta$ transformation but did not test this possibility by infrared spectroscopy.

In a different direction, Sonenberg and associates have demonstrated significant changes in the CD of erythrocyte ghosts (80) and liver bile front membranes (81), in response to physiologic levels of growth hormone. The effects were species specific. They were also accompanied by alterations in the intrinsic fluorescence of the membrane proteins. Thus, the CD spectra of liver bile fronts obtained from hypophysectomized rats showed greater CD in the region of peptide absorption than those from normal animals or rats treated with bovine growth hormone. Studies on the effects of growth hormone on previously isolated liver bile fronts support these findings (82). Sonification of liver bile fronts, as well as treatment with phospholipase A, destroyed the characteristic CD response to growth hormone, altered the fluorescence properties of the membranes, and inactivated their Na^+–K^+ sensitive ATPase.

These data, as well as those of Masotti *et al.*, cannot be easily defined in terms of the structure changes involved. However, they support IR evidence showing that membrane proteins can undergo significant alterations in secondary structure in response to biologic stimuli.

IR Studies. IR analyses show that addition of ATP and Mg^{2+} to (0.5–1.0 mM) to erythrocyte membranes induces a shift to the β-conformation, away from an α-helical or unordered structure (*83*). The spectral change appears to be a consequence of ATP utilization. The conformational transition is prominent within 2 min after ATP addition when the reaction is carried out at room temperature and stopped rapidly with liquid nitrogen prior to lyophilization for IR examination. The transition is also observed when the metabolic reaction and IR measurements are done in D_2O.

The inner membranes of rat liver mitochondria exhibit large differences in the Amide I region of the IR spectra, depending upon the metabolic state of the mitochondria. Induction of electron transport by the addition of succinate (0.1 mM) in the presence of oxygen, shifts the protein conformation towards the antiparallel β-structure, as indicated by the increase in absorption around 1635 and 1685 cm^{-1}. This can be prevented or reversed by inhibiting electron transport or by inducing phosphorylation. The shift towards the antiparallel β-structure becomes extreme when electron transport is uncoupled from oxidative phosphorylation (*78*).

In both erythrocyte ghosts and mitochondria, the metabolically induced spectral changes observed in the Amide I region are considerable; however,

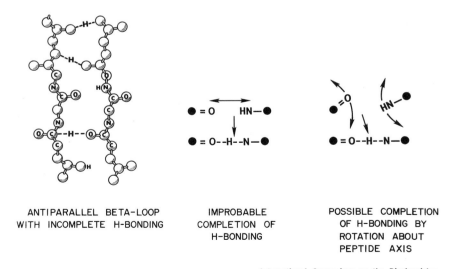

International Symposium on the Biochemistry and Biophysics of Mitochondrial Membranes

Figure 7. Possible basis for the changes in signal observed by IR spectroscopy of biomembranes under certain metabolic conditions (84)

they might arise without a large rearrangement of the peptide backbone (*84*). One may thus be dealing with peptide arrays such as exist in several soluble globular proteins—*e.g.*, carboxy-peptidase A— where there is extensive antiparallel folding, but less than maximal β-structured H-bonding. In such cases, a small change in the orientation of the peptide groups by rotations about certain bonds in the backbone would favor more extensive antiparallel-β H-bonding; this would be reflected in the IR spectra (Figure 7).

NUCLEAR MAGNETIC RESONANCE (NMR). Many atomic nuclei of immediate or potential interest exhibit a net nuclear spin. When placed in a constant magnetic field, such nuclei orient their magnetic moments either parallel or antiparallel to the spin axis, producing two energy levels.

Considering ^{1}H nuclei as an example, a population of protons represents a Boltzmann distribution between the two energy levels. Because of the small energies involved, the proportion of nuclei in the low-energy population exceeds that of the high energy group by only 10^6 nuclei per mole. When one aligns protons in an imposed magnetic field, H_o and exposes them to electromagnetic radiation of the appropriate frequency ν_o, they are promoted from the lower-energy level ($-\mu H_o$) to the higher energy state ($+\mu H_o$) when

$$h\nu_o = 2\mu H_o$$

where μ is the magnetic moment of the proton and $h=$ Plank's constant. Since the protons in the relaxed state are in a slight excess, electromagnetic radiation of appropriate frequency produces a measurable absorption of energy.

Different NMR-active nuclei absorb at different resonance frequencies. Therefore, molecules containing several such nuclei, exposed to electromagnetic radiation of varying frequency, yield a number of frequency absorption maxima. NMR data can be expressed either in frequency units ν_o, or field units, H_o. It is simpler technically to operate at a constant applied frequency (*e.g.*, 40, 60, 100, 120, and 220 MHz) and vary the magnetic field, H_o. Then the resonant absorption is expressed in terms of the magnetic field (in gauss). Biological experimentation has concentrated on the study of ^{1}H-nuclei (protons) because of their signal intensities and their prevalence in biomolecules. Considerable effort is now also directed into studies of ^{13}C resonance and ^{19}F resonance. ^{19}F substitution for ^{1}H in hydrocarbon chains introduces only minor structural perturbations.

The field experienced by a particular nucleus depends not only upon the applied field, H_o, but also upon the extent (positive or negative) to which the local environment (other nuclei, orbital electrons) modifies the applied field. The various nuclei in molecules thus sense a field H different from the applied as follows

$$H = H_o (1 - \sigma')$$

where σ' is constant reflecting a nucleus' microenvironment. σ' differs for nuclei of atoms participating in different chemical bonds. The resonance displacement caused by such effects is called the chemical shift. Diverse chemical configurations yield different values of σ'.

Membranes. NMR studies on natural and synthetic membranes are multiplying and have been recently reviewed (*24, 85, 86*). In the domain of biomembranes, studies of erythrocyte ghosts and their components have proved particularly informative. For example (*87*), suspensions of erythrocyte lipid extracts (phosphatides + cholesterol) in D_2O produce several well defined, high-resolution NMR signals arising from

    (a) —$N(CH_3)_3$ protons of choline,
    (b) —CH=CH— protons of unsaturated hydrocarbon chains,
    (c) CH_2 protons of hydrocarbon chains,
    (d) CH_3 protons.

However, erythrocyte membranes exhibit only feeble chain signals although head group signals are very prominent. Dissociation of protein from lipid with sodium deoxycholate brings out the chain signals. Protein denaturation by heating above 80°C, treatment with trifluoroacetic acid or sodium dodecyl sulfate does the same but also frees resonances arising from aromatic amino acid residues. In contrast, phospholipase C eliminates the choline signal, but does not detectably influence the hydrocarbon signals. The results are most readily explained by steric restraints of lipid chain mobility, arising from apolar interactions with membrane proteins. ^{13}C NMR, deuteron NMR and ^{19}F NMR generally support the data obtained with ^1H NMR.

A different but productive application of NMR has been the use of ^1H NMR "probes" (*88, 89, 90*). This approach rests on the fact that numerous membrane active agents, *e.g.*, anaesthetics, aliphatic, and aromatic alcohols, vitamin A, below a critical concentration increase the osmotic stability of erythrocytes but disrupt the membranes above this concentration. Several of these substances, *e.g.*, xylocaine and benzyl alcohol (BeOH), exhibit distinctive NMR bands, whose width and intensities differ in the free and membrane-bound states. This allows deductions about the fluidities of the probe environments, the number of binding sites, and the nature of the molecular events occurring during certain membrane perturbations.

Metcalfe and associates have used the aromatic protons of BeOH to report on perturbations produced by this agent and/or other anaesthetics. The aromatic resonance band of BeOH narrows progressively and reversibly until the critical lytic concentration is reached. Then the band broadens abruptly and irreversibly, indicating a sharp restriction in the mobility of the probe. Independent studies indicate that new protein binding sites for BeOH appear during membrane lysis. It is possible that in the intact membrane the lipid masks protein sites which become exposed when the lipid dissociates from the protein at high BeOH levels. It is also possible that high, local BeOH levels denature membrane proteins.

The characteristic "labilization–stabilization" NMR pattern found in going from prelytic to lytic levels of BeOH also obtains when BeOH is used at prelytic levels—as a "reporter molecule"—to monitor changes wrought by other membrane-active substances.

Spectroscopic Probes. INTRODUCTION. Some small molecules reflect the polarity, polarizability, viscosity, and other aspects of their microenvironment in their spectroscopic properties. Such probes might be fluorescent, paramagnetic, optically absorbing, or optically active. Many such probes have been bound to biological macromolecules or inserted into macromolecular systems in efforts to determine the nature of binding sites and to uncover the relations between structure and function. Ideally, a molecular membrane probe should

(a) be non-perturbing,
(b) be stable in its membrane location,
(c) reflect the characteristics of the binding site interpretably,
(d) be restricted to a unique, defined site.

The probe approach is very popular in membrane biology. However, probes very likely perturb their microenvironment and cannot often be precisely located (24).

ELECTRON SPIN RESONANCE (ESR) AND SPIN LABELS. The spinning charge of an electron induces a magnetic field. When electrons are paired as in most chemical bonds, their spins are opposite, and their magnetic moments cancel. However, molecules such as nitroxides, $\rangle N \rightarrow O$, bear solitary electrons, whose unpaired spins generate measurable magnetic moments; these can be covalently coupled to appropriate carrier molecules as "spin-labels." Spin-label probes inform about the number of probes bound, their molecular motion, their orientation in an external magnetic field, the polarity of their binding sites, and the interaction between various labeled molecules and between the spin labels and nearby fluorophores. The use of spin probes in membrane studies has been well reviewed recently (24, 91). A whole armamentarium of spin-labeled reagents exists, including steroids, fatty acids, phosphatides, and a variety of compounds reacting covalently with -SH, -NH$_2$, and other groups. Many spin-labels suitable for membrane studies are now available, but their application has not yet reached maturity because of major uncertainties concerning the location of the spin-label probes, the biological significance of the probe signals, and the possibility that the probes might seriously perturb membranes (24). Nevertheless, spin-label approaches have extracted important information, particularly concerning the state of lipids in model systems (24, 91).

When incorporated in phosphatide, or phosphatide–cholesterol bilayers, fatty acid labels, steroid labels, and phosphatide labels orient predictably (91). However, comparisons of such molecules spread at air–water interfaces, with the native lipids, reveals large discrepancies (92, 93). Other model studies show that lipid spin probes tend to generate their own fluid microenvironment (94); also experimentation on erythrocyte membranes indicates that several lipoidal spin probes can perturb membrane structure reversibly and irreversibly (95). Finally, there is increasing evidence that many lipoidal spin probes also bind to membrane proteins (96, 97). I will return to this point in the section on membrane models.

An interesting approach, applying spin-label probes to biomembranes has been published by Metcalfe and et al. (98, 99). They have used the probes to elucidate the action of erythrocyte membrane perturbants such as BeOH. With dodecyl dimethyl tempoyl ammonium:

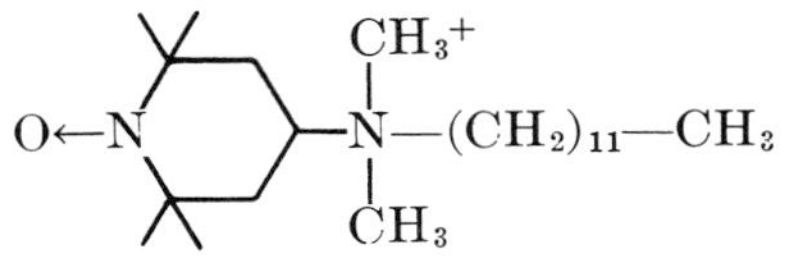

$$O \leftarrow N \underset{}{\bigcirc} N^+(CH_3)_2 - (CH_2)_{11} - CH_3$$

increasing concentrations of BeOH progressively lower the viscosity at the binding sites; however, no new binding sites appear, even at lytic concentrations of BeOH. Also, the label is similarly bound by the separated lipid and protein components of the membrane, and its environment is similarly "fluidized" by BeOH in both cases.

A steroid nitroxide:

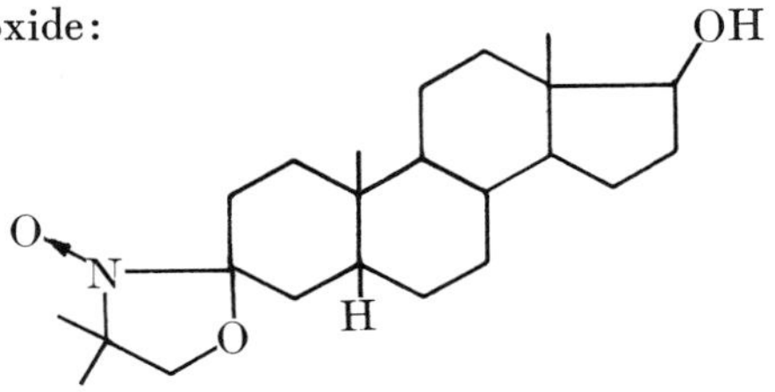

behaves differently. Increasing, non-lytic, levels of BeOH progressively sharpen the spectrum of the steroid nitroxide, again suggesting a "fluidizing" effect. However, at lytic concentrations of the alcohol, the signals from the steroid spin label change, reflecting the appearance of new, immobilized binding sites. These are attributed to membrane protein, since BeOH "fluidizes" the environment of the spin label in lipid multilayers.

Metcalfe makes a very important generalization. He argues that even primarily apolar probes will bind to both protein and lipid components of membranes because of the common characteristics of the chemical forces involved in the interactions with either component.

The spin probe technique clearly represents a very potent approach to the evaluation of membrane structure. This is evident from the reviews in (*24, 91*). However, the interpretation of results will require more caution than generally exercised heretofore (*24*).

FLUORESCENT PROBES. These substances provide high sensitivity and yield information about the polarity and viscosity in the vicinity of the probe, the flexibility of the binding site, and the proximity of the probe to other fluorophores. The measurements necessary to exploit fully the potential of fluorescent probes include determination of their excitation and emission spectra, their quantum yields (*i.e.*, quanta emitted/quantum absorbed), the lifetimes of their excited states, and the polarization of their fluorescence. The application of fluorescence probes to membrane studies has been well reviewed recently (*100*).

The most widely used fluorescent probe in biomembrane studies is 1-anilino-8-naphthalene sulfonate (ANS):

This tends to concentrate into the apolar membrane regions, probably within membrane proteins or at protein–lipid interfaces. However, ANS and related substances cannot be localized well enough to give very precise information about the structure of membranes and possible dynamic variations therein. Accordingly, there has been considerable effort to devise more specific probes —e.g., phosphatides with covalently bound fluorescent residues. Unfortunately, x-ray diffraction analyses of model phosphatide bilayers, containing ANS and other fluorescent probes, show that these molecules severely alter their microenvironment (*101*).

Some experiments with ANS have yielded interesting biological information. The work of Tasaki *et al.* is particularly impressive. They have used probes such as ANS to monitor possible structural changes associated with electrical excitation of axons, developing extremely elegant techniques to realize their goal. They find that the nerves from lobsters, crabs, and squid, when exposed to ANS and irradiated at 365 nm, emit typical ANS fluorescence. After nerve stimulation, the fluorescence increases upon arrival of the nerve impulse at the site of optical irradiation. The fluorescence increment ranges from 5×10^{-5} to 4×10^{-4} times the intensity before stimulation. Importantly, the probes which change fluorescence during the passage of the nerve impulse maintain a constant polarization angle relative to the nerve axis (*102*).

The small size of the signal indicates that the excitable membrane is only a minute portion of the total structure observed. Accordingly, the fluorescence change in the functional regions of the axonal plasma membrane may be much greater than those observed. Indeed Tasaki *et al.* suggest major conformational changes of the macromolecules in the nerve membrane during nerve conduction.

In another direction, Kasai *et al.* find a high polarization of ANS bound to the membranes of the electric organ of *Electrophorus electricus* (*103*). Polarization is further independent of solvent viscosity, indicating that ANS is strongly immobilized by the membrane and that it is sequestered deeply within the membrane phase. 5-Dimethyl-amino-1-naphthalene sulfonyl chloride (DNS), coupled covalently to the proteins of the membrane fragments exhibits similar fluorescence behavior. However, solvent separation of the membrane lipids and proteins, as well as the effects of detergents, indicate that disruption of membrane organization lowers polarization and makes it responsive to solvent characteristics. The high polarization, insensitive to solvent viscosity, requires association of lipids and proteins into an organized membrane structure.

Membrane Models

The membrane literature contains a profusion of membrane models. Most of these address membrane structure rather than function. In all probability, no single model can apply to all membranes, all parts of a given membrane or all states of a membrane system.

Structural Models. THE PAUCIMOLECULAR MODEL (1935). The "paucimolecular" membrane model of Danielli and Davson (14) was the first, specific molecular–biological hypothesis (Figure 8). It proposed that biomembranes consist of bilayers of phosphatides (and associated lipids). The polar lipid headgroups lie at the membrane surfaces, and their hydrocarbon chains constitute the apolar core. Membrane proteins were thought to localize at the polar surfaces interacting with the lipids primarily through ionic and/or hydrogen bonds. As pointed out elsewhere (24), the evidence on which the model was based is consistent with many other interpretations. Moreover, most of the current experimentation concerning this model relies on

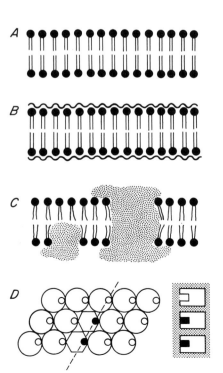

Figure 8. Schematic of some major membrane models

A. Phospholipid Bilayer. This was first proposed as part of the "Paucimolecular Model" of Danielli and Davson (13). Some forms of this lipid array very likely exist in many biomembranes. Model membranes of this type can be readily manufactured. In many membranes the lipid composition is complex but can still be accommodated in this basic array.

B. Phospholipid-bilayer with Surface-adsorbed Protein. This scheme was proposed by Danielli and Davson (13) to account for the properties of many membranes and to locate the membrane protein (~~). This model fits many of the properties of myelin.

C. Lipid–globular–protein Mosaic Model. Here membrane proteins (stippled) are postulated to penetrate through the lipid bilayer. The latter is anomalously structured around the apolar perimeter of the penetrating proteins (55, 106, 107, 108, 109). In the fluid lipid–globular–protein mosaic model, the penetrating proteins are thought to be freely mobile in the plane of the membrane, depending on the viscosity of the membrane lipid (112).

D. Cooperative-lattice Model (56, 118, 119, 120). Membranes are envisaged to contain extended, two-dimensional multimolecular aggregates, i.e., lattices. On the left a section of a hypothetical lattice is viewed face on. On the right is a sagittal section along the dashed line. Open small circles indicate unliganded binding sites. Full circles indicate occupied binding sites. Two structural states are indicated (see text for details).

inference and correlation. Nevertheless, it appears highly probable that some form of phospholipid bilayer exists in most biomembranes. Certainly, membrane phosphatides (± other membrane lipids) can spontaneously form such structures, and recent x-ray diffraction data convincingly demonstrate the existence of bilayer structures in a variety of biomembranes. However, the Danielli–Davson hypothesis and related models do not adequately describe the disposition of membrane proteins.

MOSAIC MODELS. *The Hypothesis of Parpart and Ballentine.* Although Danielli and Davson presented the first detailed membrane model in 1935, a lipid–protein mosaic structure had been proposed by Nathanson (*104*) as early as 1904. However, the first detailed mosaic hypothesis was presented by Parpart and Ballentine in 1952. The authors developed this model to account for the state of proteins in erythrocyte membranes. They proposed that the membrane consists of a protein continuum. As diagrammed in their Figure 3 (*105*), the protein penetrates the membrane core. The protein continuum is interrupted by cylindrical channels 50A in diameter with hydrated walls. The channels are filled with cylindrical lipid plugs. The head groups of the amphipathic lipids orient toward the hydrated walls of the channels, and their hydrocarbon chains form an apolar core. The authors propose two types of lipid binding—strong polar associations between phosphatide headgroups and protein and weak apolar association of cholesterol and some other lipids with the apolar residues of the strongly bound phosphatides.

Present information lends little support for the details of the Parpart–Ballentine model. However, there is increasing evidence that significant proportions of membrane protein penetrate into or through the membrane core. Parpart and Ballentine (*105*) do not envisage apolar lipid–protein associations, but these play a major role in modern mosaic hypotheses.

Lipid–Globular Protein Mosaic Models (LGPM). Modern mosaic models have been proposed by Wallach and associates (*55, 106, 107*) and Singer and associates (*108, 109*). Both groups conclude that some proteins penetrate into or through the membrane core. Both suggest that the penetrating portions of these proteins bear apolar residues which form hydrophobic associations with hydrocarbon chains of membrane lipids. Wallach and associates (*55, 106, 107*) describe the LGPM as follows. Membranes contain subunit domains, assembled as a tangentially mobile patchwork, penetrated to varying depths by both protein and lipid. Contact between protein subunits and between protein and lipid occurs in both polar and apolar regions of the membrane. Although in a bilayer array, the lipid adjacent to the protein has a composition and organization determined by the surrounding protein, the "ordering" influence of the protein falling off with distance.

Membrane peptide lies on both of the membrane surfaces and also within the apolar core of the membrane. Surface-located peptide may be coiled irregularly, but other conformations may exist. Penetrating protein segments may be predominantly helical rods, each with opposite hydrophobic and hydrophilic faces, analogous to the H-helix of hemoglobin packed to form subunit assemblies, whose perimeters bear apolar aminoacid residues. Helicity is not an essential requirement; unordered or β-structured peptide segments with suitable amino acid sequence could also provide the apolar perimeters required of penetrating protein segments. Clearly, high hydrophobicity is also not a requirement and lipid protein interactions are not exclusively hydrophobic.

The apolar amino acid residues, which make up the perimeters of the subunits, comprise binding sites for the hydrocarbon residues of tightly retained membrane lipids. Polar lipid head groups can also participate in polar associations with surface-located protein side chains. The association of tightly bound lipids with membrane proteins is analogous to the apolar heme–protein interactions in hemoglobin. Lipid, lying more distant from the protein, *e.g.*, cholesterol is bound less tightly and less specifically. The axes of the subunits lie normal or nearly normal to the membrane surfaces. The distribution of polar amino acids in the protein is conjectured to be such that their side chains lie at the membrane surfaces or cluster around the central axis of each subunit or both, possibly forming polar channels penetrating through the membrane. This is analogous to what occurs in hemoglobin and provides a concept of what membrane "pores" may be. The permeability of the hypothetical channels would be highly sensitive to the conformational state of the protein.

The listed mosaic hypotheses, like all other membrane models, must be viewed with some reserve. However, one prediction of the model of Wallach *et al.* has been borne out: proteins in several membranes are surrounded by shells of lipid in apolar association with the proteins (*96, 97, 110, 111*). Moreover, in at least one case (*97*) these lipid layers can be perturbed by subtle protein modifications.

The Fluid Lipid–Globular Protein Mosaic Model (FLGPM). This hypothesis, first proposed by Singer and Nicolson (*112*) is closely related to the LGPM. It suggests that the matrix of membranes is not stabilized by protein–protein interactions (whether within the membrane or through a cytoskeleton). Instead, membrane lipid is proposed to form the membrane continuum. Since membrane phosphatides tend to have fluid hydrocarbon chains at physiologic temperatures, the continuum would be fluid. Thus, membrane lipids and proteins could undergo free translational diffusion in the membrane plane. Except for short range interactions, the protein distribution parallel to the membrane plane would be random.

The FLGPM represents a synthesis of information showing that membrane lipids can be expected to exhibit fluid properties at physiological temperatures and macroscopic experiments demonstrating considerable mobility of certain surface components (*e.g.*, immunoglobulins on lymphocytes, some histocompatibility antigens on fibroblasts). This matter of mobility has been critically reviewed by Cherry (*113*) who shows that the situation is far more complicated than realized initially. Indubitably, some membranes contain regions of fluid lipid (*114*) and some proteins in some membranes can exhibit rapid rotational and translational motion (*113*).

However, proteins in many membranes show remarkably little rotational mobility (*102*). Also recent spectroscopic data (*113, 114*) and electromicroscopic evidence (*115, 116, 117*) document that the mobility of proteins in many membranes is severely constrained. The FLGPM can thus not be simply generalized, and many phenomena ascribed to plasma membrane "fluidity" very likely represented movement of molecules *on* rather than *within* the membrane.

The Cooperative Lattice Model. Most membrane model derive from structural features and static properties. In contrast, concepts of Changeux and associates (*56, 118, 119*), further extended by Wyman (*120*) focus on membrane functions and the interaction of functional components. Their dynamic model builds on the behavior of regulatory enzymes because these molecules and membranes share certain crucial properties:

Both associate with specific, regulatory, structure- and function-determining "ligands," which alter the properties of their receptors without involving active site in catalytic or covalent reactions. Both behave cooperatively; that is their physical state and biologic function depend upon critical levels of regulatory substances: this produces their transduction and amplifying characteristics.

Regulatory enzymes and membranes differ in their symmetry properties. In the first, stimulation generally requires the stereospecific reaction between enzyme and activator. In membranes, recognition may occasionally proceed in an analogous fashion, but it may also occur on a larger scale [as envisioned by Ehrlich and Weiss (*3, 62*)]. Specific recognition, as a membrane process, may derive from a unique structural component. It could also involve a specific, topologic subunit array within the membrane plane. Biological responsiveness may, thus, depend upon structural asymmetries within the membrane.

Membranes constitute plastic, condensed, non-covalently bonded arrays of diverse proteins, plus their associated lipids, separating two distinct aqueous phases by an apolar one. They differ from other macromolecular assemblies in that they comprise virtually unlimited, two-dimensional multimolecular aggregates. Moreover, while soluble proteins are ordinarily encompassed by a symmetrical solvent environment, the membrane forms an unlimited barrier separating two physicochemically and metabolically dissimilar phases. It is asymmetrical because its environment is asymmetrical.

The cooperative model suggest that each membrane subunit can exist in at least two structural states, R and S, and that all R $\rightleftarrows$ S transitions depend upon the state of near neighbors. The functional subunits could be distributed irregularly over the membrane surface as small, localized, oligomeric clusters or form part of a lattice with homologous associations between neighbors. Only the second case can explain both the graded and all-or-none responses observed in membranes.

The cooperative lattice hypothesis, while stimulating, is perhaps too general and will be extremely difficult to test. Nevertheless, one requirement of the hypothesis—the capacity of membrane subunits to change structurally in response to external stimuli—has been met (*41, 42, 43, 44*).

Literature Cited

1. Naegli, C., Cramer, C., *Pflanzenphysiol. Untersuchungen* (1855) Heft 1, Zurich, F. Schultess.
2. Pringsheim, 1854, quoted by H. W. Smith, *Circulation* (1962) **26,** 987.
3. Ehrlich, P., *in* Paul Ehrlich: *Gesammelte Arbeiten* Vol. II, p. 178, Springer Verlag, Heidelberg, 1957.

4. Green, B. B., *Exp. Cell. Res.* (1954) **7**, 558.
5. Schmitt, F. O., Bear, R. S., Palmer, K. J., *J. Cell. Comp. Physiol.* (1941) **18**, 31.
6. Fernandez-Moran, H., Finean, J. B., *J. Biochem. Biophys. Cytol.* (1957) **3**, 725.
7. Finean, J. B., *Progr. Biophys. Molec. Biol.* (1966) **16**, 145.
8. Porter, K. R., Claude, A., Fulham, E. F., *J. Exp. Med.* (1945) **81**, 233.
9. Sjöstrand, F. S., *J. Appl. Physiol.* (1948) **19**, 1188.
10. Palade, G. E., *Anat. Rec.* (1952) **114**, 427.
11. Sjoestrand, F. S., *Nature* (1953) **171**, 30.
12. Berthet, F., de Duue, C., *Biochem. J.* (1951) **50**, 174.
13. Robertson, J. D., *Biochem. Soc. Symp.* (1959) **16**, 3.
14. Danielli, J. F., Davson, H., *J. Cell. Comp. Physiol.* (1935) **5**, 495.
15. Stoeckenius, W., *J. Biophys. Biochem. Cytol.* (1959) **17**, 443.
16. Korn, E. D., *Science* (1966) **153**, 1491.
17. Fleischer, S., Fleischer, B., Stoeckenius, W., *J. Cell. Biol.* (1967) **32**, 193.
18. Napolitano, L., LeBaron, F., Scaletti, J., *J. Cell. Biol.* (1967) **34**, 817.
19. Weinstein, R., *in* "The Red Blood Cell," D. M. Surgenor, Ed., 2nd ed., p. 213, Academic Press, New York, 1974.
20. McNutt, S. N., Weinstein, R. S., *Progress Biophys. Molec. Biol.* (1973) **26**, 47.
21. van Deenen, L. L. M., *Progr. Chem. Fats Other Lipids* (1965) **8**, pt. 1.
22. Davenport, J. B., *in* "Biochemistry and Methodology of Lipids," Wiley, New York, 1971.
23. Perutz, M. F., Kendrew, J. D., Watson, H. C., *J. Molec. Biol.* (1965) **13**, 669.
24. Wallach, D. F. H., Winzler, R., "Evolving Strategies and Tactics in Biomembrane Research," Springer Verlag, New York, 1974.
25. Langdon, R. G., *Biochim. Biophys. Acta* (1974) **342**, 213.
26. Dunham, P. B., Hoffman, J. F., *Proc. Natl. Acad. Sci. U.S.* (1970) **66**, 939.
27. Straus, J. H., Jr., Burge, B. W., Darnell, J. E., Jr., *J. Mol. Biol.* (1970) **47**, 437.
28. Dryer, W. J., Papermaster, D. S., Kühn, H., *Ann. N.Y. Acad. Sci.* (1972) **195**, 61.
29. Knüfermann, H., Schmidt-Ullrich, R., Ferber, E., Fischer, H., Wallach, D. F. H., *in* "Erythrocytes, Thrombocytes, Leukocytes," E. Gerlach, K. Moser, E. Deutsch, Wilmanns, Eds., p. 12, G. Thieme, Stuttgart, 1973.
30. Schmidt-Ullrich, R., Ferber, E., Knüfermann, H., Fischer, H., Wallach, D. F. H., *Biochim. Biophys. Acta* (1974) **332**, 175.
31. Wallach, D. F. H., Gordon, A. S., *in* "Regulatory Functions of Biological Membranes," J. Jarnefelt, Ed., p. 87, Elsevier, Amsterdam, 1968.
32. Tanford, C., *J. Amer. Chem. Soc.* (1962) **84**, 4240.
33. Bigelow, C. B., *J. Theoret. Biol.* (1967) **16**, 187.
34. Capaldi, R. A. C., Vanderkooi, G. V., *Proc. Natl. Acad. Sci. U.S.* (1972) **69**, 930.
35. Jackson, R. L., Segrest, J. P., Kahane, I., Marchesi, V. T., *Biochemistry* (1973) **12**, 3131.
36. Braun, V., Bosch, V., *Eur. J. Biochem.* (1972) **28**, 51.
37. Wallach, D. F. H., *Biochim. Biophys. Acta* (1971) **265**, 61.
38. Zwaal, R. F. A., Roelofson, B., Colley, C. M., *Biochim. Biophys. Acta* (1973) **300**, 159.
39. Schmidt-Ullrich, R., Wallach, D. F. H., Ferber, E., *Biochim. Biophys. Acta* (1974) **356**, 300.
40. Wunderlich, F., Wallach, D. F. H., Speth, V., Fischer, H., *Biochim. Biophys. Acta* (1974) **373**, 34.
41. Strittmatter, P., Rogers, M. J., Spatz, L., *J. Biol. Chem.* (1972) **247**, 7188.
42. Hughes, R. C., *Progress Biophys. Mol. Biol.* (1973) **26**, 191.

43. Winzler, R. J., in: "Red Cell Membrane, Structure and Function," G. A. Jamieson, T. Z. Greenwalt, Eds., p. 157, J. S. Lippincott, Philadelphia, 1969.
44. Cole, K. S., in: "Physical Principles of Biological Membranes," F. S. Snell, J. Wolken, G. J. Iverson, J. Lam, Eds., p. 1, Gordon and Breach, New York, 1970.
45. "Membranes," G. Eisenman, Ed., Vol. 1, Marcel Dekker, New York, 1972.
46. "Biophysics and Physiology of Excitable Membranes," W. J. Adelman, Ed., Van Nostrand, New York, 1971.
47. Overton, E., *Vierteljahrschr. Naturforsch Ges. Zurich* (1899) **44**, 88.
48. Cole, K. S., *Cold Spring Harbor Symp. Quant. Biol.* (1940) **8**, 110.
49. Furshpan, E. I., Potter, D. D., in: "Current Topics in Developmental Biology," Vol. 3, p. 95, Academic Press, New York, 1968.
50. Kotyk, A., *Biochim. Biophys. Acta* (1973) **300**, 183.
51. Lieb, W. R., Stein, W. D., *Biochim. Biophys. Acta* (1972) **265**, 187.
52. Hope, A. B., "Ion Transport and Membranes," Butterworths, London, 1971.
53. Bolis, L., Katchalsky, A., Keynes, R. D., Loewenstein, W. R., Pethica, B., Eds., "Permeability and Function of Biological Membranes," North Holland, Amsterdam, 1970.
54. Christensen, H. N., Cespedes, C. de, Handlogten, M. E., Ronquist, R., *Biochim. Biophys. Acta* (1973) **300**, 487.
55. Wallach, D. F. H., "The Plasma Membrane," p. 15, Springer Verlag, New York, 1972.
56. Changeux, J. P., Blumenthal, R., Kasai, M., Podleski, T., in: "Molecular Properties of Drug Receptors," R. Porter, M. O'Connor, Eds., p. 197, J. & A. Churchill, London, 1970.
57. Birnbaumer, L., *Biochim. Biophys. Acta* (1973) **300**, 129.
58. Jerne, N. K., *Europ. J. Immunol.* (1971) **1**, 1.
59. Hulser, D. F., Peters, J. H., *Eur. J. Immunol.* (1971) **1**, 494.
60. van den Berg, K. J., Betel, I., *Exp. Cell Res.* (1973) **76**, 63.
61. Resch, K., Gelfand, E. W., Hansen, K., Ferber, E., *Eur. J. Immunol.* (1972) **2**, 589.
62. Weiss, P., *Proc. Nat. Acad. Sci. U.S.* (1960) **46**, 993.
63. Roseman, S., *Chem. Phys. Lipids* (1970) **5**, 720.
64. Najjar, V., in: "Biological Membranes," D. Chapman, D. F. H. Wallach, Eds., Vol. 3, Academic Press, London, in press.
65. Shipley, G., in: "Biological Membranes," D. Chapman, D. F. H. Wallach, Eds., Vol. 2, p. 1, Academic Press, London.
66. Caspar, D. L. D., Kirschner, D. A., *Nature* (1971) **231**, 46.
67. Blaurock, A. E., Wilkins, M. H. F., *Nature* (1972) **236**, 313.
68. Wilkins, M. H. F., Blaurock, A. E., Engelman, D. M., *Nature* (1971) **230**, 72.
69. Engelman, D. M., *J. Molec. Biol.* (1970) **47**, 115.
70. Blaurock, A. E., Stoeckenius, W., *Nature* (1971) **233**, 152.
71. "Membrane Molecular Biology," C. F. Fox, A. D. Keith, Eds., Sinauer Associates, Stamford, Conn., 1972.
72. Malcolm, B. R., *Biopolymers* (1970) **9**, 911.
73. Holzwarth, G., in: "Membrane Molecular Biology," C. F. Fox, A. D. Keith, Eds., p. 228, Sinauer Associates, Stamford, Conn., 1972.
74. Wallach, D. F. H., Lowe, D., Bertland, A. V., *Proc. Natl. Acad. Sci. U.S.* (1973) **70**, 3235.
75. Green, D. H., Salton, M. R. J., *Biochim. Biophys. Acta* (1970) **211**, 139.
76. Choules, G. L., Bjorklund, R. F., *Biochemistry* (1970) **9**, 4750.
77. Avruch, J., Wallach, D. F. H., *Biochim. Biophys. Acta* (1971) **241**, 249.
78. Graham, J. M., Wallach, D. F. H., *Biochim. Biophys. Acta* (1969) **193**, 225.
79. Masotti, L., Urry, D., Krivacic, J. R., Long, M. M., *Biochim. Biophys. Acta* (1972) **255**, 420.
80. Sonenberg, M., *Biochim. Biophys. Res. Commun.* (1969) **36**, 450.

81. Rubin, M. S., Swislocki, N. I., Sonenberg, M., *Arch. Biochem. Biophys.* (1973) **157**, 243.
82. Rubin, M. S., Swislocki, N. I., Sonenberg, M., *Arch. Biochem. Biophys.* (1973) **157**, 252.
83. Graham, J. M., Wallach, D. F. H., *Biochim. Biophys. Acta* (1971) **241**, 180.
84. Wallach, D. F. H., Graham, J. M., *in:* "International Symposium on the Biochemistry and Biophysics of Mitochondrial Membranes," F. Azzone, N. Siliprandi, Eds., p. 231, Academic Press, New York, 1972.
85. Chapman, D., Dodd, G. H., *in:* "Structure and Function of Biological Membranes," L. I. Rothfield, Ed., p. 28, Academic Press, New York, 1971.
86. Chapman, D., *in:* "Biological Membranes," D. Chapman, D. F. H. Wallach, Eds., Vol. 2, p. 91, Academic Press, London, 1973.
87. Chapman, D., Kamat, V. B., de Gier, J., Penckett, S. A., *J. Molec. Biol.* (1968) **31**, 101.
88. Metcalfe, J. C., Burgen, A. S. U., *Nature* (1968) **220**, 587.
89. Colley, C. M. C., Metcalfe, S. M., Turner, B., Burgen, A. S. U., *Biochim. Biophys. Acta* (1971) **233**.
90. Metcalfe, J. C., Metcalfe, S. M., Engelman, D. M., *Biochim. Biophys. Acta* (1971) **241**, 412.
91. Keith, A. D., Sharnoff, M., Cohn, G. E., *Biochim. Biophys. Acta* (1973) **300**, 379.
92. Cadenhead, D. A. C., Katti, S. S., *Biochim. Biophys. Acta* (1971) **241**, 709.
93. Tinoco, J., Ghosh, D., Keith, A. D., *Biochim. Biophys. Acta* (1972) **274**, 279.
94. Melhorn, R., Snipes, W., Keith, A. D., *Biophys. J.* (1973) **13**, 1223.
95. Bieri, V., Wallach, D. F. H., Lin, P. S., *Proc. Natl. Acad. Sci. U.S.* (1974) **71**, 4797.
96. Jost, P. C., Griffith, O. H., Capaldi, R. A., Vanderkooi, G., *Proc. Natl. Acad. Sci. U.S.* (1973) **70**, 480.
97. Wallach, D. F. H., Verma, S., Weidekamm, E., Bieri, V., *Biochim. Biophys. Acta* (1974) **356**, 68.
98. Hubbell, W. L., Metcalfe, J. C., Metcalfe, S. M., McConnell, H. M., *Biochim. Biophys. Acta* (1970) **219**, 415.
99. Metcalfe, J. C., *in:* "The Dynamic Structure of Membranes," D. F. H. Wallach, H. Fischer, Eds., p. 202, Springer-Verlag, Heidelberg, 1971.
100. Radda, G. R., Vanderkooi, J. M., *Biochim. Biophys. Acta* (1972) **265**, 509.
101. Lesslauer, W. L., Cain, J. E., Blasie, J. K., *Proc. Natl. Acad. Sci. U.S.* (1972) **69**, 1499.
102. Tasaki, I., Watanabe, A., Hallett, M., *J. Membr. Biol.* (1972) **8**, 109.
103. Kasai, M., Changeux, J. P., Monnerie, L., *Biochem. Biophys. Res. Commun.* (1969) **36**, 420.
104. Nathenson, A., *Jahrb. wiss Botan.* (1904) **39**, 607.
105. Parpart, A. K., Ballentine, R., *in:* "Trends in Physiology and Biochemistry," E. S. G. Barron, Ed., p. 135, Academic Press, New York, 1952.
106. Wallach, D. F. H., Zahler, P., *Proc. Natl. Acad. Sci. U.S.* (1966) **56**, 1552.
107. Wallach, D. F. H., Gordon, A. S., *Fed. Proc.* (1968) **27**, 1263.
108. Lenard, Singer, S. J., *Proc. Natl. Acad. Sci. U.S.* (1966) **56**, 1828.
109. Singer, S. J., *in:* "Structure and Function of Biological Membranes," L. I. Rothfield, Ed., p. 146, Academic Press, New York, 1971.
110. Träuble, H., Overath, P., *Biochim. Biophys. Acta* (1973) **307**, 491.
111. Stier, A., Sackmann, *Biochim. Biophys. Acta* (1973) **311**, 400.
112. Singer, S. J., Nicolson, G. R., *Science* (1972) **175**, 720.
113. Cherry, R. D., *in:* "Biological Membranes," D. Chapman, D. F. H. Wallach, Eds., Academic Press, London, 1975, Vol. 3, in press.
114. Wallach, D. F. H., Bieri, V., Verma, S. P., Schmidt-Ullrich, R., *Ann. N.Y. Acad. Sci.* (1975), in press.
115. Wunderlich, F., Batz, W., Speth, V., Wallach, D. F. H., *J. Cell Biol.* (1974) **61**, 633.

116. Bhakdi, S., Speth, V., Knüfermann, H., Wallach, D. F. H., *Biochim. Biophys. Acta* (1974) **356,** 300.
117. Wunderlich, F., Wallach, D. F. H., Speth, V., Fischer, H., *Biochim. Biophys. Acta* (1974) **373,** 34.
118. Changeux, J. P., Thiery, J., *in:* "Regulatory Functions of Biological Membranes," J. Jarnefelt, Ed., p. 115, Elsevier, Amsterdam, 1968.
119. Changeux, J. P., *in:* "Symmetry and Function of Biological Systems at the Macromolecular Level," A. Engström, B. Strandberg, Eds., p. 235, Wiley-Interscience, New York, 1969.
120. Wyman, J., *in:* "Symmetry and Function of Biological Systems at the Macromolecular Level," A. Engström, B. Strandberg, Eds., p. 267, Wiley-Interscience, New York, 1969.

Supported by grants No. CA13061, CA13052, CA12178 from the United States Public Health Service, grant No. 6B32123 from National Science Foundation and Award PRA-78 from the American Cancer Society (DFHW).

Chapter

3

Membrane Transport

Joseph D. Robinson, Department of Pharmacology, State University of New York, Upstate Medical Center, Syracuse, N. Y. 13210

A MAJOR FUNCTION OF biological membranes is that of a barrier: to prevent the loss of valuable constituents from the cell and to hinder the random entry of possibly noxious substances from the outside. The lipoidal plasma membrane covering the cell surface separates effectively two aqueous phases and their (largely) polar solutes. Recent studies support formulations of the cell membrane as a bimolecular lipid sheet; the original Danielli-Davson model is modified to include proteins inserted in and through the lipid bilayer as well as on its surface. The highly non-polar membrane interior (containing fatty acid chains of the phospholipids and hydrophobic amino acids of the intercalated membrane proteins) thus severely hinders passage of water-soluble non-lipoidal substances into or from the cell. However, it is also vital for the cell to have access to the exterior (to extract nutrients from the environment and to dispose of cellular wastes). This is the subject of transport. The barrier is modified in certain characteristic ways to permit passage across the membrane.

For highly lipid-soluble substances, of course, there is only a relatively insignificant barrier; such non-polar molecules can readily cross the lipoidal membrane by simple diffusion, "dissolving" in the non-polar phase. For polar substances, however, such a route should be forbidden. Nevertheless, many small polar molecules seem to cross the membrane by simple diffusion, and to account for this diffusion it has been proposed that the lipoidal barrier is penetrated by pores—polar channels through which hydrophilic molecules sufficiently small can pass. Thus, water, certain ions, and small polar organic molecules can equilibrate across the membrane. The numbers and diameters of the pores afford certain selectivity. Consequently, the valuable macromolecules of the cell can still be retained although the barrier may be somewhat leaky.

A second transport category—facilitated diffusion—permits both a finer discrimination than one based on size alone and a more rapid equilibration between cell and environment. This process, also known as carrier-mediated transport, reflects specific interactions between the solute transported and a membrane element and results in certain kinetic properties, notably saturable transport (the system has a maximal velocity, V_{max}). A modification of this system permits active transport. By linking the facilitated diffusion mechanism to cell metabolism it is then possible to transport substances against their concentration gradients. Instead of merely speeding equilibration, active transport systems can attain net accumulations of a substance.

How the membrane is able to transport substances across its lipoidal interior, what structures and forces are involved, and how metabolic energy is coupled to achieve active transport are clearly questions of fundamental importance. This chapter is concerned with efforts to provide biochemical answers to these questions. First, the general characteristics of the various transport processes are described since they serve as the bases for recognition, identification, and are the phenomena that must ultimately be accounted for. Next, methods of examining such systems are outlined. Finally, to exemplify the processes and approaches, a detailed account of the biochemical properties of two active transport systems (the Na/K–ATPase of the sodium/potassium pump and the Ca–ATPase of the sarcoplasmic reticulum calcium pump) is presented (*1–8*).

Transport Processes

Simple Diffusion. FICK'S LAW OF DIFFUSION. Simple diffusion is a process of equilibration arising from random thermal motion. It is described by Fick's law of diffusion:

$$\frac{dn}{dt} = -DA \frac{dc}{dx}$$

where dn/dt is the rate of diffusion across an area A, dc/dx is the concentration gradient being dissipated, and D is a diffusion coefficient—a function of the substance diffusing, the solvent, and other system properties, *e.g.*, temperature. [Instead of concentrations all formulas should deal with thermodynamic activities; however, since the appropriate corrections for biological systems are often uncertain, the concentration of the freely available solute is generally used.] The rate of diffusion is thus directly proportional to the concentration gradient dc/dx.

Assuming that the concentration gradient across the membrane is linear, the gradient may be represented as $(C_1 - C_2)/x$, where C_1 and C_2 are the concentrations on either side of a membrane of thickness x. For convenience the membrane thickness is often combined with the diffusion coefficient across a unit area to give a permeability coefficient, P, so that

$$\frac{dn}{dt} = -P (C_1 - C_2)$$

In addition, the rate of transport across a membrane is frequently termed a flux, J.

Simple diffusion thus represents an intuitively apparent equilibration process in which thermal agitation dissipates a concentration gradient. It can apply to the passage of lipoidal molecules diffusing across the membrane and the movement of small molecules through the membrane pores. In this latter case, however, complicating circumstances can arise (2, 3, 8, 9, 10).

WATER FLOW AND SOLVENT DRAG. In the absence of a net driving force, movement of water across the membrane follows the kinetics of simple diffusion and is termed the diffusive flow of water. Addition of a net driving force (hydrostatic or osmotic pressure) frequently produces a rate of flow faster than can be attributed to diffusion alone. When the pores in the membrane are large relative to the diameter of the water molecules, a pressure gradient produces a bulk flow of water. That is, water flows as a bulk phase limited by frictional resistance with the walls of the pore, not under the constraints of molecular diffusion. Flow then approximates that described by Poiseuille's law of the capillary (10, 11).

When a bulk flow of water occurs, dissolved solutes may then be carried across the membrane more rapidly than they would otherwise diffuse. This phenomenon is called solvent drag; the flowing solvent sweeps solutes along with it. Bulk flow through a porous membrane resulting from a hydrostatic pressure gradient is frequently termed filtration. There is then a sieving effect: the molecules carried in the bulk flow are selected by the pore size; those with smaller diameters are carried by the flow (solvent drag), but those larger are rejected.

SEMIPERMEABLE MEMBRANES. An ideal semipermeable membrane is one that is permeable to solvent but not to solutes. When such a membrane separates two compartments—one containing pure water and the other water plus solute—there is an osmotic flow of water through the semipermeable membrane into the compartment containing solute. This osmotic flow can be halted by applying a pressure—the osmotic pressure Π—to the solute side. For dilute solutions Π is given by the van't Hoff relationship:

$$\Pi = c\,RT$$

where c is the concentration of solute, R the gas constant, and T the absolute temperature. The driving force for the osmotic flow can be considered in terms of the chemical potential of water in the two compartments, with flow from pure water (high potential) to solution (lower potential).

Actual membranes are not semipermeable in the ideal sense since they allow the passage of certain solutes as well as water. The deviation from the ideal semipermeable membrane—the degree of leakiness of the membrane—must be expressed relative to a given solute. This is often stated in terms of the reflection coefficient σ (2, 10, 12), the extent to which a given membrane is permeable to the solute in question, and is the ratio of the observed osmotic pressure of a given solute and semipermeable membrane to the theoretical osmotic pressure. Thus, when the given membrane is impermeable to a particular solute, the reflection coefficient of that solute is 1.0; for successively greater permeabilities σ tends toward zero.

A further consequence of semipermeable membranes separating solutions arises in situations where some charged solutes (but not others) can cross the membrane; an impermeable charged solute on one side of the membrane will influence the equilibrium distribution of the permeable ions. This distribution is described by the Gibbs-Donnan equilibrium: the product of the concentratons of the permeable ions on one side of the membrane equals the product of the concentrations of the permeable ions on the other side. For a system with one permeable cation and anion, C^+ and C^-:

$$C^+_1\ C^-_1\ =\ C^+_2\ C^-_2$$

where the subscripts refer to the side of the membrane. Since on each side of the membrane the sum of all the cations must equal the sum of all the anions (regardless of whether they are permeable or not) the equilibrium distribution of the ions can be calculated from the simultaneous equations. At equilibrium there is an asymmetric distribution of permeable ions and also of the number of particles so that an osmotic pressure differential results (Figure 1).

A major consequence of systems with nondiffusable ions is the creation at equilibrium of this pressure differential between the two compartments. Since mammalian cell membranes cannot withstand significant pressure gradients, Gibbs-Donnan equilibrium distributions cannot be tolerated. Instead,

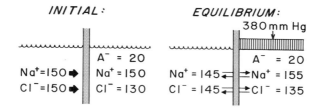

Figure 1. Gibbs-Donnan equilibrium

The properties of the Gibbs-Donnan equilibrium distribution are illustrated for a simple case in which a membrane impermeable to the univalent anion A^- separates two NaCl solutions and is permeable to the latter ions. Although in the initial conditions shown there are an equivalent number of particles on each side of the membrane, the Na^+ concentrations are equal, and electrical neutrality exists, there will be a movement of both Na^+ and Cl^- across the membrane to the compartment containing the impermeable anion A^- until the product of $[Na] \times [Cl]$ on one side equals that on the other, electrical neutrality being preserved. The equilibrium distribution of ions results in a concentration of osmotically active solutes on one side of the membrane, and to prevent an osmotic flow of water, a pressure of 0.5 atm must be exerted on that compartment to maintain its original volume.

these cells maintain a non-equilibrium steady state distribution of ions such that essentially no pressure differential exists (*13, 14*). Probably the major rationale behind active transport systems for ions has been the need to regulate cell pressures (and volumes) by maintaining a steady state ion distribution against the equilibrium distribution (*15*). Nevertheless, although the distribution of the cell's total ion population is not at equilibrium, individual species of ions may be distributed passively according to the Gibbs-Donnan equilibrium (with other ions kept at non-equilibrium levels by active transport). For example, in erythrocytes where Na^+, K^+, and Ca^{2+} levels are maintained at non-equilibrium values by active transport systems, Cl^- appears to be passively distributed according to the Gibbs-Donnan description (*16*).

MEMBRANE POTENTIALS. When there is an asymmetric distribution of ions, there is an electrical potential across the membrane separating the two compartments, a diffusion potential in terms of the ions that can permeate the membrane. This membrane potential, E_m, is described by the Nernst equation:

$$E_m = \frac{RT}{ZF} \ln \frac{C_1}{C_2}$$

where Z is the charge on the ion and F is Faraday's constant. For a single monovalent cation asymmetrically distributed, the potential, in millivolts, is equal to $59 \log C_1/C_2$ at $25°C$.

For biological membranes the reality is more complex. Not only are there several species of permeant ions which contribute to the membrane potential, but they appear to contribute to differing degrees. This latter complication may be simplified by assuming different relative permeabilities for the various ion species in the framework of Goldman's constant field approximation (*17*) as presented by Hodgin and Katz (*18*):

$$E_m = \frac{-RT}{F} \ln \frac{P_K [K^+_1] + P_{Na} [Na^+_1] + P_{Cl} [Cl^-_2]}{P_K [K^+_2] + P_{Na} [Na^+_2] + P_{Cl} [Cl^-_1]}$$

where P_K, P_{Na}, and P_{Cl} are the relative membrane permeabilities of K^+, Na^+, and Cl^-. The resting membrane potentials are conventionally written with respect to the exterior at zero potential; the cell interior is thus usually negative. Here subscript 1 refers to concentrations inside the cell and 2 to those outside.

UNSTIRRED LAYERS. A final complexity should be noted; there exists over all membranes an "unstirred layer" of solvent in which mixing is far slower than in the bulk phase (*19*). Despite the vigor with which the bulk phase is agitated the unstirred layer remains adjacent to the membrane, perhaps as thick as 10^{-1}–10^{-2} mm (*20, 21*). Consequently, kinetic studies of membrane transport can be distorted by permeation through this layer, which may represent for some solutes a greater barrier than the membrane itself (*22, 23*). Furthermore, experimental alterations from a steady state may lag because of equilibration times between the bulk phase and the unstirred layer.

Facilitated Diffusion. GENERAL PROPERTIES. Facilitated diffusion differs from simple diffusion both in mechanism and in kinetics, but like simple diffusion it is an equilibrating process driven by thermal agitation. In facilitated diffusion the transported substance interacts reversibly with a specific membrane element that permits its movement across the membrane. Although originally depicted in terms of mobile carriers that shuttled substances through the lipoidal barrier, the molecular mechanisms underlying facilitated diffusion remain unclear, but the distinguishing characteristics are generally apparent. As the concentration of the transported substance increases, the rate of transport increases hyperbolically (Figure 2), approaching asymptotically a maximal rate, V_{max} (or T_{max} for transport maximum). This kinetic pattern suggests a carrier mechanism in which the transported substance, S, must first combine with a membrane element, C, to form a complex that facilitates diffusion:

$$S + C \rightleftarrows SC \rightarrow \text{transport}$$

There is thus a parallel with enzyme catalysis, in which substrate combines with enzyme to form the enzyme–substrate complex that proceeds to product formation. For facilitated diffusion an equation of the Michaelis-Menten form can thus be written:

$$v = \frac{V_{max}}{1 + \frac{K_m}{S}}$$

where v and V_{max} are observed and maximal rates of transport, and K_m the concentration of substance for half-maximal transport. Like the K_m for enzyme catalysis, this K_m reflects the dissociation constant for the carrier–substrate complex but need not be identical to it (*e.g.*, if the rate of transport approaches the rate of binding or dissociation of the complex). Linear transformations of this equation are generally more convenient (Figure 2).

Like enzyme kinetics, this formulation deals with initial velocities. More complex equations are available for time dependence of transport (integrated rate equations), for net transport when bidirectional systems are operating, and for pump-and-leak systems where active transport opposes a simple diffusion leak (*1, 2, 5*).

SPECIFICITY AND COMPETITION. Selection by the carrier enables the cell to discriminate between candidates for transport beyond criteria of polarity, charge, and size that regulate simple diffusion. Various transport systems display widely differing degrees of specificity, but the specificity is, of course, never absolute. Thus competition for the transport system can occur between structurally similar molecules. The kinetic pattern for competitive inhibition is identical to that for enzymatic competition (Figure 2). Non-competitive inhibition of facilitated diffusion can also occur, and inhibitors can be reversible and irreversible; the usual enzymological criteria may be applied to these situations as well.

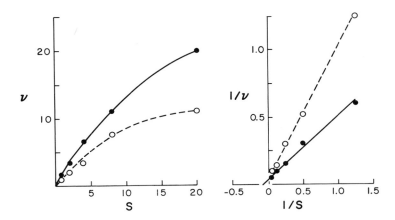

Figure 2. Kinetics of facilitated diffusion with a competitive inhibitor

In the experiments shown the transport of a glucose analog across the intestine is shown in the absence and presence of a competitive inhibitor, phlorizin. Rings of everted hamster intestine were incubated in the concentrations shown (mM) of ^{3}H-labeled glucose analog, 1,5-anhydro-D-glucitol, in the absence (●) or presence (○) of 0.4 μM phlorizin. The velocity of transport was estimated from accumulations of radioactivity after brief incubations at 37°C. Data are plotted in terms of the velocity of transport (in arbitrary units) against substrate concentration (left), and in double-reciprocal form (right). (Redrawn from Ref. 299.)

COUNTER TRANSPORT. Under certain conditions addition of one substance that can be transported by a facilitated diffusion system causes a net flow in the opposite direction of a different substance that also uses the same transport system. This process is termed counter transport or flow driven by counter flow, and it has been explained in terms of mobile carriers capable of moving solutes in either direction across the membrane (1, 2, 5). In such a system one solute, A, moves equally in either direction at equilibrium. However, addition of a second solute, B, to one compartment (e.g., the outside) at sufficiently high concentration produces competition with A for entry. Thus there is an influx of B and a lessened entry of A. On the inner surface B is absent initially; consequently, A continues to move outward as before. Thus, more A leaves than enters the cell, and a concentration gradient is created for substance A. As substance B accumulates within the cell, however, the efflux of A is retarded by competition. Ultimately both A and B are equally distributed. The net counterflow of A proceeds only as long as there is a driving force of B resulting from its asymmetric distribution.

MODELS FOR FACILITATED DIFFUSION. Counter transport is perhaps most significant for the constraints it places on models for facilitated diffusion and has been frequently presented as a strong argument for mobile carriers (1, 2). The recognition of ionophores (e.g., valinomycin, an antibiotic that forms a lipid-solution complex with K^+ and thus accelerates its equilibration across lipoidal membranes) provided concrete examples for such a mechanism. Nevertheless, firm evidence for the natural occurrence of such a carrier in physiological systems is lacking. Although transmembrane channels would not be expected to couple influx of B with efflux of A, variants of such systems may be formulated to accommodate these processes. For example, Lieb and Stein (5, 24) propose coupled alternating conformational changes in tetrameric transport units spanning the membrane so that influx of A by one subunit is symmetrically tied to efflux of B by another subunit (Figure 3). Similarly, oscillating-pore mechanisms (25) may be coupled to achieve the same result (Figure 3).

Active Transport. GENERAL PROPERTIES. Active transport is defined as transport against a concentration gradient (or for charged solutes against an electrochemical gradient) that is energized by cell metabolism. Because it shares the same kinetic properties with facilitated diffusion (*e.g.*, saturation kinetics, competitive inhibition), it is usually regarded as a modification of this process with an energy input coupled to drive the system. However, in mechanistic terms, since the molecular processes underlying neither process are known, this plausible assumption is not demonstrable.

The coupling of metabolic energy to the transport system may pass through the common currency of cellular potential energy stores (*e.g.*, ATP) or be more directly linked to ongoing metabolic processes (*e.g.*, utilization of the energized state generated during electron transport). In what may be considered as another variant of active transport, cellular metabolic energy

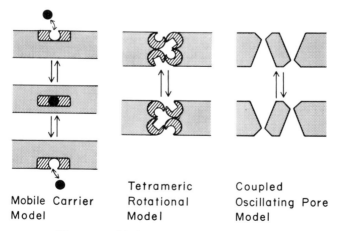

Figure 3. Models for facilitated diffusion

A mobile carrier in the membrane (left) can account for both the saturation kinetics seen in facilitated diffusion and the process of counter transport. Alternatively, coupled systems (center and right) could also accomplish counter transport, one channel mediating influx and the other efflux, with either rotational (5) or oscillating-pore (25) mechanisms.

is stored by the active transport of another solute, and the dissipation of the concentration gradient of that solute is then coupled to energize active transport, moving either in the same or opposite direction as the driving solute. This latter process is often termed cotransport. For example, the concentration gradient for Na^+ across the cell membrane—itself established by active transport of Na^+ from the cell—may be used for the active transport of sugars into the cell: the influx of Na^+ down its electrochemical gradient can energize an influx of sugar against its gradient (26, but see Ref. 27).

ELECTROGENIC PUMPS. In addition to creating concentration gradients across membranes some active transport systems can directly create an electrical potential by separating charge; such a mechanism is termed an electrogenic pump (28). The sodium/potassium pump of the cell membrane apparently transports three Na^+ outward for each two K^+ inward (29) and thus generates a potential across the membrane (negative inside). In the steady state, of course, the Na^+ current outward produced by the pump is balanced by an equal leak of cation inward. Although such electrogenic pumps have been demonstrated, they generally seem to contribute far less to the total membrane potential than do the ionic asymmetries (28).

MODELS FOR ACTIVE TRANSPORT. To adapt facilitated diffusion models to active transport, two modifications are often made: (a) an energy input is added at the binding site to propel the solutes across the membrane, and (b) the binding site is also specified as undergoing cyclical changes in affinity sufficient to select the solute in media of low concentration and then discharge it into media of high concentration (see Figure 3). For the latter purposes the solute itself may be chemically modified. Models of mobile carrier mechanisms thus must be severely altered to incorporate these considerations.

Alternatively, rotational elements have been proposed (30, 31). To span the full width of the membrane (70 A) enormous motions are necessary so that models frequently link smaller rotational elements in series (32) or postulate narrower barriers, with free diffusion of solutes to the thinned region of the membrane. However, in all cases a major problem is rotation of a superficial and thus hydrophilic binding element into the hydrophobic membrane interior.

Oscillating-pore mechanisms (25) also suffer from the long distances to be traversed; thus, they also often resort to an *ad hoc* thinning of the membrane. Addition of an energy source to this model provides a sort of microperistalsis that can produce both the necessary translation and change in affinity (33, 34). The magnitude of conformational changes required and the energy barriers confronted seem perhaps less formidable in this proposal.

Finally, although transport systems are often categorized into the three phenomenological classes—simple diffusion, facilitated diffusion, and active transport—differences in mechanism may not always be sharply drawn. In the absence of an energy input active transport systems may equilibrate gradients by facilitated diffusion (35, 36), and there may be a relationship between active transport sites and channels for simple diffusion or "leaks" (37).

Thermodynamic Descriptions. A more thorough and precise treatment of these transport processes is possible within the framework of irreversible thermodynamics. Unlike classical thermodynamics which deals with reversible processes (specifically those proceeding infinitely slowly so that an infinitesimal change in the opposite direction can reverse the process), irreversible thermodynamics deals with phenomena (*e.g.*, the dissipation of concentration gradients) that are inherently not reversible but associated with significant changes in entropy (*see* Chapter 1; for discussions of the irreversible thermodynamics of transport processes *see* Refs. *2, 38, 39*).

Approaches

Studies of Transport Processes. REQUIREMENTS. In contrast to most biochemical approaches which traditionally emphasize isolation and purification of components for individual scrutiny, the examination of transport phenomena obviously requires an intact membrane separating two accessible compartments, at least one of which should be modifiable. Given the complexity of such systems, various approaches can categorize transport systems in the terms of the preceding section and quantitate the transport parameters.

Kinetics should differentiate between simple and facilitated diffusion; in the latter case it should be possible to demonstrate not only saturation kinetics (a V_{max}) but also specificity, competition, and perhaps counter transport. To distinguish between facilitated diffusion and active transport the criteria are (a) transport against an electrochemical gradient, and (b) a link to cellular metabolism.

Criterion a ought to be straightforward, but two major complications may arise. To define the gradient, true electrochemical activities should be specified; obviously solutes bound or sequestered within subcellular compartments must not be included with freely available solutes in calculating the gradient. It is also essential in quantitating solute transport that metabolites of the solute not be included in the calculations inadvertently. Such errors can readily occur in studying fluxes of labeled organic solutes where chemical identity is only assumed. Since in some circumstances the rate of metabolism is quite rapid, non-metabolizable analogs of the solute are sometimes used instead (*e.g.*, 3-*O*-methyl glucose for glucose).

To establish criterion b, dependence on cellular metabolism, demonstrations of the actual energy source are most convincing. Showing a decrease in transport rate with depletion of energy stores (*e.g.*, induced by anoxia or general metabolic poisons) is less persuasive because of the diverse consequences of such manipulations. Temperature dependence of transport does not discriminate between active and passive systems since both simple and facilitated diffusion in some systems have equally high temperature coefficients (*2, 40, 41*).

For the kinetic determinations unidirectional fluxes under steady state or equilibrium conditions can be measured by tracer techniques. Under conditions of net transport (accumulation of a solute) unidirectional fluxes can be measured by tracer techniques or by chemical methods. The equations presented above are for initial velocities of unidirectional fluxes; more complex formulas are available (2) for bidirectional transport and for prolonged time courses (rate equations integrated with respect to time). Simultaneous flux measurements in opposite directions may be made by tracer techniques using different isotopes (*e.g.*, ^{22}Na and ^{24}Na).

Where charged solutes are transported the net fluxes can often be followed in terms of the resultant electrical potential or current. This approach is more convenient in macroscopic systems—*e.g.*, Na$^+$ transport across epithelial sheets of toad urinary bladder: here a "short circuit current" may be determined as the applied current necessary to abolish the potential difference across the epithelium normally resulting from ion transport (42, 43).

Since it is generally assumed that water itself is not transported in biological systems by facilitated diffusion or active transport (9, 44), water movements must imply a flux of solute (in the absence of other driving forces, *e.g.*, pressure gradients). Consequently, solute movements can be monitored by osmotic shrinkage and swelling of cells, either by direct determinations of water content or volume, or in certain cases by changes in turbidity of cell suspensions reflecting the volume changes (9, 44, 45).

AVAILABLE SYSTEMS. Ideally, transport should be studied across the membrane of an individual cell, and many of the early investigations of transport processes were on large plant cells and marine oocytes (46, 47). Giant axons of the squid—diameters on the order of 0.5 mm—may be freed of the bulk of their cytoplasm and perfused internally; such preparations continue to display transport processes (*e.g.*, active transport when supplied with energy sources and the variable permeability sequences of the action potential) (48, 49).

For smaller cells transport must be studied in populations. Dispersed cells (*e.g.*, free-living microorganisms or cells grown in tissue culture) minimize topological problems of access and diffusion pathways. By contrast, studies with tissue slices are complicated by cellular heterogeneity, hindered extracellular diffusion, and obvious damage at the cut surfaces.

Mammalian erythrocytes are extensively used in transport studies (45) in part because of their ready availability and uniformity but largely because of certain major advantages. These cells contain no internal compartments (*e.g.*, mitochondria) which can complicate analyses of solute accumulation. In addition, by the technique of reversible hemolysis (45, 50) erythrocyte membrane permeability may be increased transiently not only to empty the cells of their constituents (forming ghosts) but also to permit loading with solutes that would not ordinarily penetrate the membrane (*e.g.*, ATP). These resealed ghosts may then be used to examine the effects on transport of changes in internal constituents. Erythrocytes, however, suffer from several disadvantages; notably, the rates of permeability and active transport for many solutes (*e.g.*, Na$^+$, K$^+$) are extremely slow (45), corresponding to the low density of transport systems in the membrane.

With certain subcellular systems transport processes across the organelle membranes may be measured, and mitochondrial and chloroplast transport are extensively studied (*see* Chapters 4 and 5). Cellular membrane systems that exist *in vivo* as sheets or tubular processes may on homogenization seal off to form vesicles; thus, the endoplasmic reticulum forms microsomes, and nerve endings form synaptosomes (*51, 52*). Vesicles formed from bacterial cell membranes are extremely useful (*53*).

Transport across epithelial layers plays an important role in many physiological processes (*e.g.*, sugar absorption by the intestine); thus, it is often appropriate to examine transport across an entire epithelium. Moreover, manipulative advantages of working with a gross membrane that can be placed between two accessible chambers are apparent. Among the obvious difficulties is the problem of transport across various routes, representing different cell types, layers, and intercellular channels. However, some preparations (*e.g.*, the toad urinary bladder) have an epithelium of nearly a single layer with minimal extraneous tissue, and containing chiefly (but not entirely) a single cell type (*54*). Given an epithelium a single cell thick there may still remain the task of determining whether one of the two cell membranes representing the two faces of the epithelium determines the transport characteristics (*55, 56*).

In epithelial sheets an additional complexity may arise from topological properties. In some epithelia (*e.g.*, the gall bladder) (*57*) the transporting cells lie in a sheet with their membranes closely approximated near one face of the sheet but separated at the opposite face so that wedge-shaped channels lie between neighboring cells. Solute actively transported into the apical portion of the wedge-shaped space produces a local osmotic flow of water from the cell; this process of active transport and corresponding osmotic flow along the channel sums to produce a bulk flow of solution, ultimately emptying at the broad end of the channel. By this mechanism a flow of water can occur across the epithelial sheet driven by the active transport of solute.

A more general model has been proposed by Curran (*58*) for two membranes of differing permeability arranged in series. Active transport of solute into the space between the two membranes can result in an osmotic flow of water across the tighter membrane impermeable to solute, and a bulk flow across the more porous membrane. There can thus be a net water flow even in the presence of an opposing osmotic gradient across the system.

At the opposite extreme of complexity are artificial membrane systems. Bimolecular lipid films (called black lipid films from their optional interference properties) may be formed over small apertures (*ca.* 1 mm diameter) in septa separating two chambers, by painting solutions of certain lipids across the aperture (*59, 60*). Transport can then be examined by isotopic or electrical techniques as functions of the solute and of membrane composition. Ionophores—organic molecules of natural and synthetic origin that accelerate the equilibration of ions across lipoidal barriers—are studied extensively in such systems (*59, 60, 61*). These films are also used as matrices in attempts to position isolated components so that the actual transport phenomena may be examined, as yet with limited success (*62, 63*).

Certain lipid mixtures when dried, swollen in salt solutions, and then dialyzed form suspensions containing multilayed bimolecular sheets in roughly spherical form (myelin figures or liposomes) (*41, 64*). These preparations offer greater bulk and ease of preparation in comparison with black lipid films, but the multilaminate nature complicates interpretation. Various treatments of these liposomes (*e.g.* sonic irradiation) can produce vesicular structures bound by a single bilayer; these are used in studies on permeability (*65*) and also in certain physical approaches to bilayer properties (*66, 67*).

Biochemistry of Transport. IDENTIFICATION OF COMPONENTS. A first step in the biochemical approach to transport is a chemical description of the components. In the process of isolation, however, several problems arise, notably the loss of transport phenomena for identifying the components and the possible destruction of organized complexes (*e.g.*, disaggregation of subunits forming a transport channel). Nevertheless, considerable progress has been made.

Approaches to isolation and identification depend somewhat on the nature of the system and the preconceptions about its characteristics. For facilitated diffusion solute recognition and translocation mechanisms are presumed, plus for active transport a coupled energy input. On the other hand, attempts to identify components of a simple diffusion channel are hindered by the absence of specific properties that could be postulated for the membrane proteins forming the channel. In certain systems, however, approaches even to simple diffusion are suggested; for example, the inward flow of Na^+ associated with the nerve action potential can be blocked by tetrodotoxin, and some progress was made toward isolating components of this Na^+ channel, identified by their capacity to bind ^{3}H-tetrodoxotin (*68, 69*).

Specificity in facilitated diffusion and active transport systems implies recognition sites, and identification of such sites may be approached from demonstrable affinities. Following the isolation by Pardee (*70, 71*) of a crystalline protein from bacteria with both a high affinity for sulfate and a convincing relationship to the sulfate transport process, a host of proteins binding a variety of transported solutes were described (*6*). Although for many of these proteins there is indirect evidence of their involvement in transport processes, demonstrations of high affinity need not imply participation in a transport process. Binding sites on proteins may serve as storage sites (*72*) or as regulatory (or catalytic) sites in other systems; comparison of affinities and competitive characteristics with the transport system can help establish specificity. With bacterial transport the use of mutants and inducible systems has been particularly valuable in this respect (*see* below).

Components of transport systems may be labeled for identification by using specific reagents (affinity labeling), reflecting the binding properties of the system as in the approaches above (*68, 69*). Alternatively, nonspecific reagents may be used to label components in the presence of specific modifiers (*e.g.*, substrates or inhibitors of the system that interact with the components to favor or hinder labeling). Fox and Kennedy (73, 74) used this latter technique to isolate the M-protein, a membrane-bound component of the galactoside transport system of *E. coli*. The basic approach was to react crude membrane preparations first with unlabeled *N*-ethylmaleimide (NEM)—a general reagent binding covalently to sulfhydryl groups—in the presence of a substrate analog, thiodigalactoside. The protective agent was then removed and the membrane preparation next reacted with radioactive NEM; the M-protein could then be identified by this label. The specificity of this approach was greatly increased by using membranes from two cultures of *E. coli*, identical except that one had been induced to synthesize the galactoside system and the other had not (and thus lacked the transport system). Membranes from both cultures were first incubated with unlabeled NEM in the presence of thiodigalactoside, and after removal of the protective agents the membranes from induced cells were incubated with ^{14}C-NEM and those from the uninduced cells with ^{3}H-NEM. Equal amounts of each were mixed together and the M-protein isolated by following the ^{14}C to H^3 ratio (73).

Enzymatic processes associated with active transport are useful tags for identification as well as aspects to be investigated themselves. On occasion a complex of enzymatic properties may suggest the presence of a transport system. Roseman (75, 76), while examining bacterial sugar phosphorylation pathways, discovered the phosphotransferase system that transfers the phosphate of phosphoenolpyruvate to various sugars; one of the components is membrane bound, and Roseman proposed that the system is associated with bacterial sugar transport. Subsequent investigations demonstrated (a) a cytoplasmic protein (Enzyme 1) that transfers the phosphate of phosphoenolpyruvate to one histidine of a second soluble protein (HPr); (b) the phosphorylated HPr then reacts with a membrane-bound complex (Enzyme II) to phosphorylate the sugar. Enzyme II is actually a family of sugar-specific entities which can be fractionated into two protein components (one of which is specific for certain sugars) plus lipid. The sugar apparently is transported simultaneously with its phosphorylation.

Controlled and defined perturbations of the system can greatly assist labeling and identification. In addition to using specific substrates, cofactors, and inhibitors alluded to above, genetic variants have proved to be extremely valuable (6, 74, 76). Mutants lacking a specific transport element are obvious controls in identification and in distinguishing between specific and nonspecific binding (77). Moreover, in complexes such as the phosphotransferase system mutants lacking certain elements verify the involvement of that component with transport and can lead to specific predictions about its role. For example, loss of one sugar-specific element of the Enzyme II family should still allow sugars using other elements to be transported (76). Another virtue of studying bacteria is the ready induction of many transport systems (6, 73). Thus, the concentration of transport elements may be varied for correlation with extracted components.

Reconstitution to form a functioning unit is the definitive demonstration of the participation of isolated components in a transport system. Clearly the intricacies of such systems may make reconstitution difficult, and the use of irreversible inhibitors for labeling would prevent it. Nevertheless, some success was achieved in both bacterial systems (6, 78) and certain mammalian preparations. The reconstitution of the sarcoplasmic reticulum ATPase into Ca-transporting vesicles is described in the last section.

DETERMINING MOLECULAR MECHANISMS. The ultimate aim in studying the biochemistry of facilitated diffusion and active transport is to describe the process in molecular terms. Unfortunately, little progress has been made toward this goal, although a number of associated events have been determined (e.g., enzymatic reaction sequences) that, like the phenomenological characteristics, can limit possible molecular models. Membrane proteins are commonly assumed to be the essential elements, although the lipids of their environment can strongly modify their properties. The usual techniques of enzymology and protein chemistry applied here suffer at least the same frustrations as in other applications.

Standard enzymological approaches to active transport center on delineating the reaction sequence and its sites of interaction with the transported solute; these include measurements of reaction kinetics, thermodynamic properties, substrate and inhibitor effects, and binding properties. Effects of various perturbations on these properties may then be revealing. Specific chemical modifications of the active unit (79, 80, 81) or the surrounding lipid matrix (82, 83, 84) can be undertaken. Mutations may be helpful, just as the study of hemoglobin variants contributed to the understanding of oxygen binding processes.

However, as attempts to describe detailed reaction mechanisms for enzymes in the absence of precise structural information often has in retrospect seemed premature, so attempts to describe how membrane elements accomplish solute translocation in the absence of such structural information seems unencouraging. For this reason various physical techniques have been attempted to provide details of the three-dimensional structure of the components. X-ray crystallography, which should solve the problem, appears inapplicable (85, 86). Low-resolution diffraction can describe the gross organization of the membrane elements, but to define conformational changes postulated to effect transport more detail is required. For high-resolution diffraction studies it is necessary that the unit (e.g., enzyme) have a distinct conformation and be regularly arrayed (i.e., be crystalline). Native membranes do not provide such a repetitive array, and even if the units were so arranged in the plane of the membrane, the membranes would have to be stacked in register. If isolated membrane units could be crystallized, then the problems of low natural abundance and repetitive array would be solved; unfortunately, the prospect for crystallizing such lipoprotein complexes is not now encouraging.

Nuclear magnetic resonance (NMR), probably the next most powerful approach, can in principle define the three-dimensional structure of proteins in non-crystalline arrays. ^{13}C-Fourier transform NMR techniques are potentially powerful (87), but with current methodologies this approach would require selective enrichment with ^{13}C amino acids, so that resonances of the specific residues would stand out from the mass of resonances at the natural abundance of ^{13}C (1.1%). The necessity for selective enrichment—from considerations of cost and feasibility—limits this approach to transport systems of microorganisms.

Electron spin resonance (ESR) techniques provide considerable information about motional freedom of lipids in membranes (*88, 89*). Spin labels attached to proteins while reporting on conformational changes in the membrane proteins as yet have not led to distinguishing precisely what changes are occurring (*90*).

Of the various optical techniques fluorescence measurements appear most promising, especially nanosecond emission kinetic studies. With such approaches distance relationships can be approximated (*91*), and with the appropriate fluorescent probes could quantitate specific conformational changes.

The outlook for descriptions akin to those of hemoglobin or lysozyme or chymotrypsin is thus somewhat pessimistic; nevertheless, available techniques should supply considerable information now generally lacking. Although mechanistic descriptions of transport processes in terms of defined conformational changes of specific amino acid residues may not lie in the immediate future, current approaches should achieve certain less ambitious but still important goals, such as cataloging transport sites as a function of the enzymatic reaction progress and distinguishing between, say, mobile carriers and oscillating pores in certain transport systems.

Studies on Active Transport Systems

The Na/K Pump and the Na/K–ATPase.

CHARACTERISTICS OF THE NA/K PUMP. Most animal cells maintain high potassium and low sodium concentrations in their cytoplasm, against concentration gradients and dependent on metabolic energy; thus, an active transport system can be inferred —the Na/K pump. Studies on the transport processes for these ions have been pursued in various systems but most extensively in mammalian erythrocytes and in squid giant axons where accessibility to both faces of the membrane permits greater experimental opportunities. In these systems the salient characteristics of the Na/K pump are: (a) ATP is the energy source, (b) the efflux of Na is coupled to the influx of K, and (c) cardioactive steroids (*e.g.*, ouabain) inhibit the pump at sites on the external membrane surface, serving as reagents both to manipulate transport activity and to identify it (*92, 93, 94*). Detailed examinations of these characteristics provide considerable insight into possible molecular mechanisms of the transport process.

ATP is the physiological energy source for the pump, as demonstrated by following its consumption and by selective replacement (*49, 94, 95*). Other nucleotide triphosphates—substituted for ATP—do not produce measurable rates of active transport (*49, 95, 96*) although they may not be totally ineffective. ATP is hydrolyzed to ADP and P_i; conversely, ATP can be synthesized from ADP and P_i by allowing the pump to run backward, dissipating a transmembrane ion gradient (*97, 98*). To demonstrate this, Garrahan and Glynn (*97*) loaded erythrocytes—by reversible hemolysis—with ATP, ADP, $^{32}P_i$, and K and then incubated these cells in a high Na medium; the downhill fluxes of Na and K (opposite to the direction in which they are pumped) led to an incorporation of ^{32}P into ATP that could be blocked by ouabain. This incorporation of ^{32}P into ATP was favored by steep gradients of Na and K and by a low ratio of ATP to ADP and P_i. Later experiments demonstrated a net synthesis of ATP (*99*).

Under experimental conditions approximating the normal physiological state the erythrocyte pump transports three Na outward and two K inward for each molecule of ATP hydrolyzed (*100, 101*). A similar stoichiometry between Na and K was found in squid axon preparations in the presence of normal internal Na concentrations; nevertheless, the coupling ratio may vary to accommodate marked changes in ionic milieu and perhaps other stresses on the pump mechanism (*94*).

With a coupling ratio of three Na ejected for each two K taken up the pump is electrogenic unless there is also a coupled transport of a compensating cation (*e.g.*, H^+) or anion. No indication of such compensatory coupling is apparent, and considerable evidence for electrogenic properties has been amassed (*28*). Thomas (*102*) showed that in a snail neuron injecting 25 pequiv of Na caused a hyperpolarization of 20 mV, and that the response was abolished by inhibiting the Na/K pump with ouabain. Moreover, he found that the pump current and the rate of Na extrusion were proportional to the elevation in internal Na concentration, and that (in keeping with the assumed stoichiometry) two-thirds of the Na extruded was coupled to the transport of a compensating ion (presumably K) while one-third produced the electrogenic effect.

Although the net outward transport of Na depends on the presence of extracellular K (or certain analogs), in the absence of K the pump catalyzes a Na/Na exchange—a facilitated diffusion in which internal Na is exchanged with that in the medium on a one-to-one basis (*35*). Energy is not required for this process, but adenine nucleotides are. Studies in which the levels of nucleotides are controlled over the course of the experiment indicate that ADP is the required form. Glynn *et al.* (*96*) proposed that this ADP-requiring Na/Na exchange represents oscillations over a segment of the total reaction sequence involved in Na/K transport. In the absence of Na the pump similarly catalyzes a K/K exchange dependent on adenine nucleotides, which also may represent reversible traffic over a segment of the reaction sequence (*96*). Both the Na/Na and K/K exchange reactions can be blocked by ouabain, which is a major criterion for identifying the activities with the Na/K pump. A recent report (*103*) describes an uncoupled efflux of Na into Na- and K-free media that requires ATP and is inhibited by ouabain; the Na flux is small compared with the Na/Na exchange and Na/K pump rates.

The rate of transport increases sigmoidally with the K concentration of the medium rather than hyperbolically (*104, 105*). This kinetic response was interpreted as representing a requirement for occupancy of multiple sites to effect transport, in accord with the stoichiometry of the pump. The specific kinetic formulation is, however, uncertain; there might be two independent K sites, perhaps of differing affinity, both of which must be filled for transport to occur, or there might be cooperative allosteric interactions between the sites which could produce a similar response (104, 105, 106, 107). In the absence of extracellular Na (replaced with choline) the apparent affinity for K increases and the sigmoidal response seemingly disappears; nevertheless, the inflection point may merely be shifted to lower K concentrations where (in the presence of K leaks from the cells) accurate measurements of velocity vs. K concentration are difficult (94, 104). The kinetic response to Na at the inner membrane surface would also be expected to be sigmoidal, but available data do not demonstrate this (49).

Extracellular Na apparently competes with K at the K sites, and intracellular K competes with Na at the Na sites (94, 104, 108). On the other

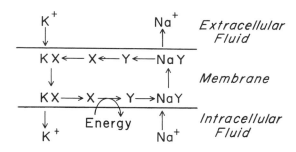

Journal of Physiology (London)

Figure 4. Shaw model for coupled Na and K transport

A carrier (X) transports Na outward; its specificity is changed (to Y) and it transports K inward, thereupon reverting to its original form (X). Metabolic energy drives the system to effect the changes in affinity and perhaps also to move the carriers (112).

hand, the apparent affinity of the extracellular K sites is unaffected by intracellular Na, and that of the intracellular Na sites unaffected by extracellular K (109, 110, 111), observations of particular importance to formulations of the pump mechanism (see below).

Most models for the Na/K pump are based on the proposal of Shaw (112) in which a carrier alternately binds and transports Na and then K (Figure 4). Thus there is only one class of cation sites that alternates in specificity. Such a model readily accounts for the coupled transport of Na and K as well as the Na/Na and K/K exchanges; the stoichiometry of three Na for two K is less easily accommodated.

Models featuring separate coexisting sites for Na and K are also proposed (109, 110, 111, 113, 114, 115), and could function with either a simultaneous or sequential binding (and transport) of Na and K; in a sequential scheme the second cation could bind before the first is released, in contrast to the Shaw model.

Arguments to differentiate between these models were presented recently in terms of the pump kinetics cited above; the lack of effect of Na at its internal sites on the apparent affinity of the external K sites, and vice-versa (*109, 110, 111*). Considered in terms of a bisubstrate reaction these data do not fit the Shaw model, which is equivalent to the ping-pong mechanism in enzyme kinetics (*116*). In a ping-pong bisubstrate reaction the first substrate binds and product is released before the second substrate binds. Consequently, double-reciprocal plots of velocity *vs.* one substrate, at various fixed concentrations of the other substrate, result in a family of parallel lines. By contrast, the data from erythrocytes plotted in this form result in a family of lines intersecting on the abscissa, consistent with bisubstrate reactions only under certain limiting conditions. The standard kinetic formulation for a bisubstrate reaction (*117*) is:

$$\begin{array}{ccc} E + S_1 & \overset{K_1}{\rightleftarrows} & ES_1 \\ + & & + \\ S_2 & & S_2 \\ K_2 \updownarrow & & K_4 \updownarrow \\ ES_2 + S_1 & \overset{K_3}{\rightleftarrows} & ES_1 S_2 \longrightarrow \text{Product(s)} \end{array}$$

where K_1, K_2, K_3, and K_4 are the dissociation constants for the indicated processes, and

$$v = \frac{V_{max}}{1 + \frac{K_3}{S_1} + \frac{K_4}{S_2} + \frac{K_1 K_4}{S_1 S_2}}$$

With this formula the family of double reciprocal plots of v against S_1, at various levels of S_2, will cross at the abscissa only if $K_1 = K_3$ and/or $K_2 = K_4$. Hence by this kinetic analysis the independence of the apparent affinity of one cation at its own site from the influence of the other cation at its site across the membrane not only rules out a ping-pong alternation of sites model, but the data also specify that at most one cation-binding affinity can change appreciably in the course of the reaction sequence (either K_1 or K_2, but not both). Recently Garrahan and Garay (*111*) argued also that the kinetic pattern limits "occluded conformations of the pump" (those in which the ions are not free for exchange) to small temporal fractions of the overall reaction sequence.

With this kinetic framework experiments to determine whether Na and K are released sequentially or simultaneously could be undertaken in terms of product inhibition (*116*). The enormous advantage of the cellular preparations in which the sidedness of the pump mechanism is preserved are obvious.

Finally, the interaction between cardioactive steroids and the Na/K pump provides opportunities for manipulation, identification, and, because of the high affinity and specificity of the binding, for quantitating the number of pump sites as well. Using ^{3}H-ouabain the number of Na/K pump sites has thus been determined in several tissues (118, 119, 120). In particular, this approach was illuminating in studies with HK and LK sheep and goat erythrocytes: certain genotypes lead to high sodium/low potassium (LK) erythrocytes, and other genotypes to more conventional low sodium/high potassium (HK) erythrocytes (14). These two classes of cells vary in pump and leak fluxes of Na and K, and (as shown by ^{3}H-ouabain binding) in density of pump units as well (118).

IDENTIFICATION OF THE NA/K–ATPASE WITH THE NA/K PUMP. In 1957 Skou (121) described an ATPase preparation from crab nerve that hydrolyzed ATP in the presence of Mg and that was markedly stimulated by the simultaneous presence of Na and K; on this basis he proposed that the Na- and K-activated ATPase were involved in the active transport of these ions. Post et al. (122) then demonstrated a convincing series of similarities between the properties of pump and ATPase, including (a) localization in the plasma membrane, (b) apparent affinities for Na and K, (c) (relative) specificity for ATP, and (d) sensitivity to ouabain. Subsequently, Glynn (123) and Whittam (124) showed that K activates the ATPase at sites on the external surface of the membrane, whereas Na activates on the internal surface. Bonting and Caravaggio (125) correlated Na/K–ATPase activity with transport rates in a variety of tissues, further strengthening the identification.

Such studies provide convincing evidence that the Na/K–ATPase is a part of—if not the entire machinery of—the Na/K pump. Although there may be slight activity in the absence of Na and K (an uncoupled ATPase) the bulk of the Mg-activated but Na- and K-independent ATPase activity of crude membrane preparations appears to be unrelated to the Na/K–ATPase and pump (93).

ISOLATION OF THE NA/K–ATPASE. Problems of purifying membrane-bound proteins, inherent in all approaches to isolating transport systems, have greatly slowed progress on the Na/K–ATPase. Most success has followed selection of starting materials with high endogenous specific activities of the Na/K–ATPase (e.g., electric organ, shark rectal gland, kidney medulla), use of chaotropic agents (e.g., NaI) to disrupt the membranes (in this case to elute impurities from the membrane matrix containing the enzyme), and fractionation of detergent solubilized material (126–134). Deoxycholate treatment often appears to increase the specific activity (128), but this procedure may alter the reaction process (135).

With these approaches highly purified preparations were recently obtained from kidney medulla (128, 129, 130, 131, 134) and shark rectal gland (132, 133). In the best preparations more than 90% of the final protein may represent the ATPase (132, 134), present in a lipid matrix necessary for reactivity (see below). Turnover numbers of 6500 (134) and 6300 (132) were calculated.

From highly purified preparations, gel electrophoresis after sodium dodecylsulfate (SDS) treatment reveals two protein bands (*129–134*). One band represents a protein with a molecular weight of about 100,000 daltons; values from 84,000 to 139,000 are found, depending on source and analytical technique. It is this protein that bears the ^{32}P after phosphorylation of the intact enzyme by ^{32}P-ATP, and thus presumably contains the active site. The other band represents a sialoglycoprotein with a molecular weight of about 50,000 daltons (35,000–57,0000); it is not certain whether this smaller protein is a functional component of the Na/K–ATPase although *in situ* it can be cross-linked to the larger protein (*131*). Since enzymatic activity cannot yet be restored to either or both gel fractions, this problem is unresolved. It is also uncertain whether the molar ratio of large to small proteins is 1:2 (*131*), 1:1 (*134*), or 2:1 (*133*). Attempts to reduce either protein to smaller subunits were unsuccessful.

Radiation-inactivation experiments indicate a molecular weight for the functional Na/K pump of at least 250,000 daltons (*136*). Consequently, it seems likely that the Na/K–ATPase functions as (at least) a dimer of the large proteins, consistent with formulations of allosteric effects and half-of-the-sites activity (*see* below). Calculations of size from the molecular weight estimates give dimensions sufficient to span the lipid core of the membrane (*92, 93*), an orientation that would seem necessary to satisfy the topology of the various binding sites and the function of the pump. Reports also vary on whether the ratio of ouabain-binding sites to phosphorylation sites on the enzyme is 1:2 (*129*), 1:1 (*137, 138*), or 2:1 (*137*).

Recently, antibodies against the Na/K–ATPase were prepared using both crude extracts and highly purified fractions (*139, 140*). Such antibodies should prove to be valuable tools for determining localization in membranes and for examining the reaction sequence since not all the steps appear to be affected equally.

Although some experiments on chemical modification (*79, 141, 142*) and studies of physical properties (*143, 144*) were undertaken with partially purified preparations, current progress toward isolating the Na/K–ATPase should make such aproaches both more meaningful and more productive. Shamoo (*145, 146*) recently described an ionophore—derived by proteolytic digestion of Na/K–ATPase preparations—that facilitates Na movement across lipid films.

INFLUENCE OF THE LIPID MATRIX. Removing a membrane-bound enzyme from its hydrophobic environment would seem likely to affect its properties, and early studies on Na/K–ATPase preparations subjected to lipid extractions or to digestion by phospholipases showed corresponding losses of enzymatic activity (*83, 147–151*). A more important issue is whether there is only a general requirement for a hydrophobic milieu or whether specific classes of lipids are needed, reflecting bonding requirements. Unfortunately, the issue is obscured by conflicting results. Several reports state that phosphatidyl serine alone among the phospholipids could restore activity to lipid-depleted enzyme preparations (*151, 152*); however, the unique ability of phosphatidyl serine may lie in its dual role as both a chelator of inhibitory metals and an amphiphilic lipid, while other common phospholipids serve only the latter function (*153*). A definitive experiment, treating a Na/K–ATPase preparation with phosphatidyl serine decarboxylase, produced the two opposite results in two studies (*84, 154*).

A more qualified issue than activity *vs.* no activity is defining how the properties of the ATPase reflect the characteristics of the lipid environment. Most attention has centered on the temperature-dependence of the enzyme, particularly the break in the Arrhenius plot at 20°C (*155, 156, 157*). Grisham and Barnett (*158*) showed that the motional freedom of the lipids in the enzyme preparation undergoes a transition at this temperature, and Tanaka and Teruya (*159*) found that partial replacement of the lipids in various Na/K–ATPase preparations altered the Arrhenius plots, independent of the enzyme source but dependent on the specific lipids.

Efforts to reconstiute enzyme preparations into lipoidal matrices in the form of black lipid films (*64*) or lipid vesicles (*133, 160*) have not yet achieved a primary goal—*viz.*, ATPase enzymatic activity of reasonable specific activity in a membrane of defined composition, oriented so that transport occurs and the sidedness of the cellular topology is preserved for study.

REACTION SEQUENCE OF NA/K–ATPASE. *Substrate Binding.* Nørby and Jensen (*161*) and Hegyvary and Post (*162*) measured the binding of radioactive ATP to enzyme preparations using the rate dialysis technique of Colowick and Womack (*163*), which permits more rapid measurements than equilibrium dialysis. Nevertheless, sufficient enzymatic hydrolysis can occur during the experiment, even in the absence of added Na and K, so that Mg had to be excluded in their studies. Under these conditions the dissociation constant, K_D, for ATP is about 0.1 μM. Two problems arise: (a) kinetic studies of Na/K–ATPase activity indicate that MgATP rather than free ATP is the substrate (*164, 165*), and (b) the K_m for MgATP is about 0.1–0.5 mM (*162, 165*), three orders of magnitude greater. Although K (absent in determining the K_D cited above) does increase the K_m (*166*) and K_D (*161, 162*) for ATP, the effect on K_D is largely prevented by the simultaneous presence of Na (*162*). Consequently, in determinations of K_m, which are performed in the presence of Na and K, no such drastic effect of K on K_D should occur (*165*).

Previous studies showed a biphasic substrate–velocity plot for ATP (*167, 168*), and it is possible that there are two classes of ATP sites: high-affinity sites with low K_D (perhaps with binding independent of Mg) and low-affinity sites where MgATP is the substrate (*165*). (The low-affinity sites correspond to the MgATP concentrations found *in vivo* and generally studied *in vitro*). This formulation bears on certain models for half-of-the-sites activity (*32, 165*) as well as some catalytic peculiarities seen at low substrate concentrations (*e.g.*, Na–ATPase activity) (*167*). In addition, discrimination between ATP and other nucleotides appears to be greater at the high-affinity sites (seen in rate-dialysis experiments) (*161, 162*) than at the low-affinity sites measured in terms of the K_m for hydrolysis *(115, 169)*. Free ATP, CaATP, and MgADP compete with MgATP at the low-affinity site *(164, 165, 170)*. Mn can substitute for Mg *(171, 172)*.

Recently, Brodsky and Shamoo *(173)* described an initial ATP–enzyme complex with linkage, suggested by pH stability, through a phosphoramido bond. They propose that the enzymatic sequence from this complex differs from that of the conventional acyl phosphate complex (*see* below).

Enzyme Phosphorylation. Albers *(126, 174)*, Post *(175, 176)*, and their associates demonstrated phosphorylation of the enzyme by ATP. This phosphorylation requires both Na and Mg and was subsequently shown to represent transfer of the terminal phosphate of ATP to the β carboxyl of an aspartyl residue forming an acyl phosphate *(123, 177)*. Although hydroxylamine releases phosphate from both native and denatured phosphoenzyme *(178, 179)*, in accord with the properties of an acyl phosphate, it does not block rephosphorylation of the native enzyme as would be expected from hydroxamate formation *(180, 181)*. This apparent discrepancy was explained as an action of hydroxylamine (or NH_4 released from it) on the native enzyme as an analog of K, thereby triggering dephosphorylation *(182)*. By this argument hydroxylamine is sterically barred from the acyl phosphate site, although with the denatured enzyme a conventional hydroxamate is formed. ^{18}O-exchange studies are consistent with formation of an acyl phosphate intermediate and its subsequent hydrolysis *(135, 183)*.

Enzyme Dephosphorylation. Adding K to the phosphoenzyme formed in the presence of MgATP and Na results in dephosphorylation *(126, 174, 176)*. Thus a simple reaction scheme encompassing Na-dependent formation of an acyl phosphate followed by K-dependent dephosphorylation was formulated *(79, 126, 176)*:

$$E + MgATP \xrightarrow{Na} E\text{—}P \xrightarrow{K} E + P_i$$

Recently, Fukushima and Tonomura *(184)* reported that adding K to such preparations leads not only to the liberation of P_i (as assumed) but that the dephosphorylation can result from phosphorylation of bound ADP to re-form ATP—these observations would require a reinterpreation of previous observations, and this issue currently awaits further study.

Another criticism of the reaction scheme as examined by sequential addition of activating ligands centers on whether in the simultaneous presence of the entire complement of reactants the same steps are followed *(173, 185, 186)*. The definitive answer to this possibility rests on rapid-kinetic studies in the presence of all activators, with sufficient time resolution to demonstrate successive intermediates.

ADP/ATP Exchange. Skou *(187)* noted that the ATPase preparation catalyzed an ADP/ATP exchange (ADP phosphotransferase), and this activity was subsequently explored in detail by Albers and associates *(79, 188)*. Na-dependent ADP/ATP exchange proceeds poorly in the presence of Mg concentrations (*ca.* 3 mM) optional for hydrolysis but increases as the Mg concentration is decreased (finite concentrations of Mg are required, however, on the order of 0.3 mM). Instead of lowering the Mg concentration the exchange reaction may also be demonstrated in the presence of certain inhibitors of the overall ATPase reaction (*e.g.*, NEM and oligomycin) (*79, 189*). With either the lowered Mg concentrations or the inhibitors enzyme phosphorylation can still be demonstrated in the presence of ATP and Na, and this phosphorylated intermediate was termed $E_1 \sim P$ (*189*), to indicate a high energy state suitable for phosphorylating ADP. In contrast, the phosphoenzyme complex formed at higher Mg concentrations (optimal for ATP hydrolysis) or the absence of these inhibitors was termed E_2–P (*189*). Transformation of $E_1 \sim P$ to E_2–P characteristics follows addition of Mg (*176, 189*); the criteria for distinguishing between these forms are ADP/ATP exchange *vs.* K-dependent dephosphorylation. When both $E_1 \sim P$ and E_2–P were shown by electrophoresis of peptide fragments to represent (probably) the same primary structure, the difference was attributed to alternative conformational states of the enzyme (*176*).

Na–ATPase Activity. At low ATP concentrations near the measured K_D for ATP, 0.1 μM, ATPase activity with Mg and Na can approach that with Mg, Na, and K or even be higher (*167, 190*), prompting the designation of Na-dependent K-inhibited ATPase activity. The interpretation of these experiments is unclear, although it seems likely that catalysis is by the Na/K–ATPase and that inhibition by K may reflect the ability of K to increase the K_D (*161, 162*). The low rate of Na–ATPase activity may be related to the slow dephosphorylation of E–P seen at higher ATP levels in the absence of K (*176*).

K–Phosphatase Activity. The Na/K–ATPase preparations also catalyze Mg- and K-dependent hydrolysis of several phosphatase substrates such as acetyl phosphate and *p*-nitrophenyl phosphate (*107, 114, 191–195*); Na is not required, and is generally inhibitory. K-phosphatase activity parallels Na/K–ATPase activity during purification and presumably represents the terminal hydrolytic steps of the overall ATPase reaction (*92, 93, 107, 195*). A notable discrepancy is the lower apparent affinity for K (*see* below).

With nitrophenyl phosphate as substrate product-inhibition studies indicate an ordered release of products with phosphate liberated last (*196*) so that an enzyme–phosphate intermediate should exist in this reaction sequence as well. At acid pH, where dephosphorylation is slowed, K-dependent labeling of the enzyme with ^{32}P-nitrophenyl phosphate could be shown (*197*), tentatively identified as being on a serine residue (*142*). Robinson (*142*) proposed that the terminal hydrolytic steps of the ATPase and the hydrolysis of phosphatase substrates include such an enzyme–phosphate intermediate, $E \cdot P'$

$$\text{ATP} + E \xrightarrow{\text{Mg, Na}} E-P \xrightarrow{K} E \cdot P' \longrightarrow E + P_i$$

$$XP + E \xrightarrow{\text{Mg, K}} E \cdot P' \longrightarrow E + P_i$$

with a reaction mechanism akin to that of the serine proteases.

In addition, Na-dependent phosphorylation sensitive to hydroxylamine was shown at neutral pH, suggesting that in the presence of Na acyl phosphate formation could occur *(197)*, permitting an overall reaction sequence comparable to that of the Na/K–ATPase *(197, 198)*:

$$XP + E \xrightarrow{Mg, Na} E - P \xrightarrow{K} E \cdot P' \longrightarrow E + P_i$$

Israel and Titus *(199)* showed Na-dependent phosphorylation of the enzyme with acetyl phosphate, apparently forming the same phosphoenzyme as with ATP.

Ouabain Inhibition. Binding of radioactive ouabain to the enzyme and the resultant inhibition were studied extensively by Albers, Schwartz, Brody, Yoda, and their colleagues *(92, 93, 137, 200–204)*. Since such binding has a relatively slow time course (minutes) with even slower dissociation, factors influencing these rates are readily measured.

Ouabain binding is favored by two groups of ligands: (a) Mg and P_i, and (b) Mg, Na, and ATP. (a) Binding favored by Mg and P_i is associated with phosphorylation of the enzyme by P_i to form the acyl phosphate intermediate *(137, 200, 205)*, the energetic requirements presumably met by the ouabain-binding equilibrium. On the other hand, phosphorylation does not seem to be required for binding *(92, 200)* although the issue is debated. The role of Mg is frequently related to its role in favoring certain conformational states of the enzyme, such as the $E_1 \sim P$ to $E_2–P$ transformation *(92, 93, 172, 176)*. In the presence of Mg and P_i binding is slowed by either Na or K *(92)*. (b) With the other ligand group (Mg, Na, and ATP) binding presumably follows formation of $E_2–P$ since these are the ligands required for acyl phosphate formation. K slows binding, perhaps by reducing steady-state levels of $E_2–P$ or by a direct effect on the enzyme *(92)*.

Dissociation of ouabain from the enzyme after binding in the presence of Mg and P_i is slower than after binding in the presence of Mg, Na, and ATP and is largely unaffected by monovalent cations *(92, 206, 207)*. In contrast, dissociation after binding in the presence Mg, Na, and ATP is slowed by monovalent cations, particularly K and its analogs, roughly in the rank order that they activate dephosphorylation *(208)*. Although in both cases ouabain apparently binds to the same site, different overall conformational states presumably result.

Examination of ouabain binding to intact cells (*e.g.*, erythrocytes) permits differentiation between actions on opposite sides of the membrane. For example, ouabain binding to erythrocytes may be favored by external Na *(209 vs. 210)*, which would not be predicted from studies on the enzyme preparations where Na slows ouabain binding with Mg and P_i and where the Na site for phosphorylation (with Mg and ATP) is on the inner membrane surface.

Formulations. Similar reaction schemes incorporating E_1 to $E_1 \sim P$ to $E_2–P$ to E_1 cycles were proposed by Albers, Post, and colleagues *(137, 176,*

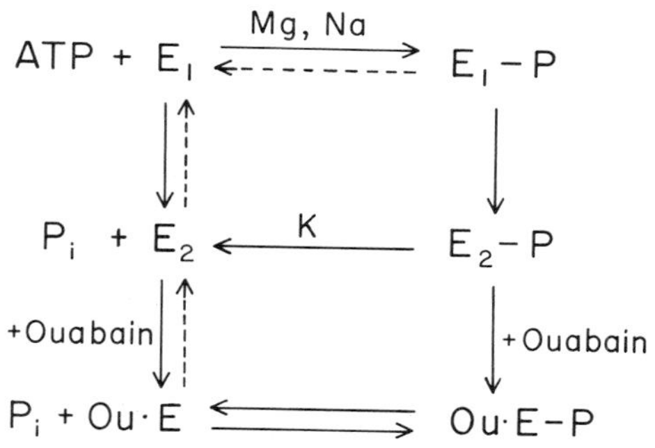

$$
\begin{array}{ccc}
ATP + E_1 & \xrightleftharpoons[]{Mg, Na} & E_1-P \\
\updownarrow & & \downarrow \\
P_i + E_2 & \xleftarrow{K} & E_2-P \\
+Ouabain \updownarrow & & \downarrow +Ouabain \\
P_i + Ou \cdot E & \rightleftharpoons & Ou \cdot E-P
\end{array}
$$

Journal of General Physiology

Figure 5. Reaction scheme for the Na/K–ATPase

The enzyme exists in two major forms, E_1 and E_2, and the reaction sequence involves a cyclical change between these forms. Ouabain binds to the E_2 and E_2-P forms, inhibiting the enzyme. (Redrawn from Post et al. (176)).

188); a representative scheme including ouabain binding is shown in Figure 5. Oligomycin is presumed to block conversion of $E_1 \sim P$ to E_2-P thus favoring ADP/ATP exchange (*79*) and Na–ATPase activity (*211*) while the E_2 forms favor ouabain binding (*176*).

From experiments on oligomycin inhibition of the Na/K–ATPase Robinson (*212*) proposed that oligomycin acts as an allosteric effector shifting the equilibrium between two states of the enzyme, E_I and E_{II}, having properties similar to the E_1 and E_2 forms above (but differing in not being necessarily sequential steps in the reaction process). Instead, catalytic activity could occur over two alternative (but interconvertible) routes (*172, 212*):

$$
\begin{array}{ccccc}
E_I + ATP & \xrightleftharpoons{Mg, Na} & E_I \sim P & \xrightarrow{K} & E_I + P_i \\
\updownarrow & & \updownarrow & & \updownarrow \\
E_{II} + ATP & \xrightarrow{Mg, Na} & E_{II} - P & \xrightarrow{K} & E_{II} + P_i
\end{array}
$$

In this scheme higher concentrations of Mg shift the equilibrium toward the E_{II} pathway (antagonistic to the effects of oligomycin), whereas in the conventional scheme there is a cyclical binding and release of Mg. Recent experiments indicate that for the conventional scheme a cyclical change in affinity for Mg of at least four orders of magnitude would be required (*165*). Skou (*213*) has also argued that ATPase forms having E_1 and E_2 properties represent alternative pathways, rather than sequential stages.

MONOVALENT CATION INTERACTIONS. Other cations cannot substitute for Na, which is required for enzyme phosphorylation, the ADP/ATP exchange reaction, and the overall hydrolysis of ATP (*92, 93, 189*). On the other hand, a number of monovalent cations can replace K (*92, 93, 189*) including the alkali metal series [probably even Na to a slight extent (*114*)], Tl (*107, 214*), and NH_4 (*107, 121*).

These substitutions aside, it is generally assumed that both Na and K are required. Low levels of activity when one ion is omitted may result in part from monovalent cation contamination (*215*) or may represent uncoupled activity, possibly induced by preparative stresses. By contrast, the Na–ATPase seen with low ATP concentrations (0.1 μM) may represent a distinct feature associated with those substrate concentrations (*see* above).

Sigmoidal curves relating ATPase activity to Na and K concentrations were noted by Skou and others (*121, 189*), and Squires (*216*) first proposed that the monovalent cations acted as homotropic allosteric activators. Subsequently, Robinson (*166*) also proposed heterotropic allosteric interactions between cation and ATP sites. Although the data are consistent with multiple interacting sites, models with independent sites cannot be rigorously excluded (*107, 166*). Several models propose distinct classes of coexisting K sites with differing affinities (*104, 217, 218*); however, formulations with one class of allosteric sites can accommodate the data as well (*107, 219*), and available information cannot discriminate between such alternatives.

Experiments on cation activation suggest that Na competes with K at the K sites and vice-versa (*107, 121, 122, 220*) although the kinetic responses are complex and may represent other interactions as well (*107, 219*). Such competition is also indicated in studies on the erythrocyte Na/K pump (*see* above) where the preserved sidedness of the transport system permits examination of each set of sites independently.

As noted before, the apparent affinity of the K-phosphatase activity for K, measured kinetically as the $K_{0.5}$ for activation—the concentration for half-maximal activation—is several fold lower than the apparent affinity of the Na/K–ATPase for K (Table I, *107, 114, 166*) although a similar sigmoidal response is present (*114*). At high K concentrations Na inhibits phosphatase activity with a K_i several fold higher than the $K_{0.5}$ for K; however, at low K concentrations Na stimulates (*114, 193, 217*). This peculiar response was interpreted (a) as the result of Na's promoting acyl phosphate formation by the substrate, and thus producing a pathway with high K affinity noticeable at low K concentrations (*107, 198, 219*), or (b) in terms of two classes of K sites (*217*).

In either case it is clear that the $K_{0.5}$ for K of the phosphatase activity is markedly reduced (an order of magnitude or more) by certain nucleotides in the presence of Na (*114, 194*), conditions that lead to acyl phosphate formation by the nucleotide (*221*). In these circumstances the $K_{0.5}$ for K approaches that of the overall Na/K–ATPase at comparable Na and nucleotide concentrations (*114, 115*). A plausible assumption is that acyl phosphate formation (which normally occurs in the ATPase reaction) markedly reduces the $K_{0.5}$ for K; with the phosphatase (in the absence of Na) acyl phosphate formation does not normally occur and the $K_{0.5}$ for K is higher. However, after forma-

Table I. Apparent Affinities of Various Forms of the Na/K–ATPase for Na and K[a]

from Enzyme Kinetics	$K_{0.5}$ for K (mM)	from Enzyme Inactivation	K_D for K (mM)
K–phosphatase	1.9	"free" enzyme	1.4
Na/K–ATPase with 90 mM NaCl, 3mM MgATP	0.8	—	
Na/K–ATPase with 10 mM NaCl, 0.4mM MgATP	0.1	phosphorylated enzyme with 10 mM NaCl, 0.2 mM MgCTP	0.06
from Enzyme Kinetics	K_i for Na (mM)	from Enzyme Inactivation	K_i for Na (mM)
K–phosphatase	6.0	"free" enzyme	6.0
from Enzyme Kinetics	$K_{0.5}$ for Na (mM)	from Enzyme Inactivation	K_D for Na (mM)
Na/K–ATPase with 10 mM KCl, 3 mM MgATP	5.0	—	
Na/K–ATPase with 2 mM KCl, 3 mM MgATP	2.8	phosphorylated enzyme with 1 mM MgATP	2.0
—		"free" enzyme	2.3

[a] Apparent affinities of the Na/K–ATPase are shown in terms of the kinetics of cation activation ($K_{0.5}$) of two enzymatic activities: K–phosphatase and Na/K–ATPase. In addition, the dissociation constants (K_D) are shown from experiments in which K-dependent inactivation of the enzyme by BeCl$_2$ or Na-modified inactivation by dicyclohexylcarbodiimide was examined. In these experiments the "free" enzyme refers to enzyme in the absence of substrate, and the phosphorylated enzyme refers to enzyme in the presence of Na and MgATP, conditions shown to result in acyl phosphate formation. The K_i for Na is in terms of competition for the K site activating the K-phosphatase or the BeCl$_2$-induced inactivation. (Refs. *34, 107, 169, 223*.)

tion of the acyl phosphate with nucleotide plus Na (or with phosphatase substrate plus Na) the $K_{0.5}$ for the phosphatase is reduced to similar values (Table I).

Clearly a more straightforward method of measuring affinities is desirable. Hokin and associates (*222*) used NMR relaxation techniques to measure Na binding to enzyme preparations and obtained values near the kinetically determined ones. A similar NMR study using Tl should provide information on K sites.

A different approach to measuring specific interactions with the enzyme is through cation-modified rates of inactivation of the Na/K–ATPase (*34, 115, 223*). In such experiments rate constants for inactivation by certain irreversible inhibitors of the enzyme are measured as a function of Na or K

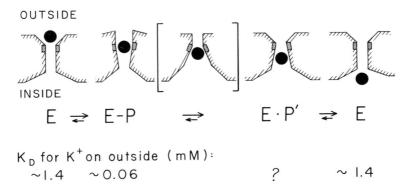

Figure 6. *Oscillating-pore mechanism for K transport*

A channel is shown for K transport, undergoing conformational changes correlated with the successive steps of the reaction sequence ($E \cdot P'$ is a terminal hydrolytic step corresponding to the enzyme–phosphate intermediate of the phosphatase reaction). The dissociation constants for K at the outer membrane face are cited from experiments with K-dependent enzyme inactivation. Presumably, a separate but analogous channel would transport Na (34).

concentration, permitting calculation of the K_D for Na or K binding. Consequently, affinity for Na or K can be determined for various ligand states corresponding to different stages of the enzymatic reaction sequence.

With this technique the K_D for K of the free enzyme is close to the $K_{0.5}$ for the K-phosphatase activity, but under conditions in which acyl phosphate formation occurs the K_D is reduced 20-fold to the range seen for the K–phosphatase or Na/K–ATPase under similar conditions (Table I, *34, 115*). In analogous experiments the K_D for Na is near the $K_{0.5}$ of the Na/K–ATPase and (in contrast to the K_D for K) is little changed by ligand states representing successive steps of the reaction sequence (*115, 223*). Since distinguishable Na and K sites are demonstrable for identical ligand states these data support models with separate coexisting sites for Na and K (*34, 115*).

The Shaw transport model with alternating Na and K sites, frequently associated with the conventional reaction scheme, cannot accommodate coexisting sites for Na and K. Earlier versions of this reaction scheme proposed that the E_1 form of the Na/K–ATPase had "inward facing" cation sites while the E_2 form had "outward facing" sites (*137, 176*); more recent proposals argue that K dissociates from an E_2 form (*225, 226*), and since K is released by the pump into the cytoplasm, an E_1 to E_2 to E_1 cycle cannot then represent the conformational changes effecting transport.

Several schemes incorporate separate coexisting sites for Na and K (*109, 110, 111, 113, 114, 115*), based on kinetic evidence and in accord with current formulations of Na/K pump kinetics (*see* above). In addition, recent studies suggest that these sites exist throughout most steps of the reaction cycle (*34, 115, 223*), unlike mobile carriers, and thus favor oscillating-pore mechanisms with high-affinity acceptance sites on one face and low-affinity discharge sites on the other (Figure 6). In such schemes the K–sites of the K–phosphatase reaction, which are also on the exterior surface (227), correspond to the K sites of the Na/K–ATPase before acyl phosphate formation (114, 115). Consequently, the phosphatase reaction in the absence of acyl phosphate formation would not be expected to effect ion transport (228).

Both the alternating-sites and coexisting-sites models fit schemes whereby Na binds first, initiating acyl phosphate formation with a resultant increase in affinity for K (sufficient to select K from the high Na extracellular fluid), this binding then lead to dephosphorylation:

$$\text{MgATP} + \underset{\underset{K}{\uparrow}}{\overset{\overset{Na}{\downarrow}}{E}} \longrightarrow E-P \longrightarrow E + P_i$$

The models differ in the significance of enzyme isomerization and the possibilities for simultaneous occupancy by both Na and K. The alternating-sites models require Na discharge before K binding, but the separate-sites models are vague about the sequence of ion discharge. Information about the properties of the discharge sites is lacking.

HALF-OF-THE-SITES ACTIVE MODELS. On the basis of kinetic patterns, Levitzki *et al.* (229) proposed that a significant class of enzymes operate with only half of their active sites functioning catalytically at a given stage in the

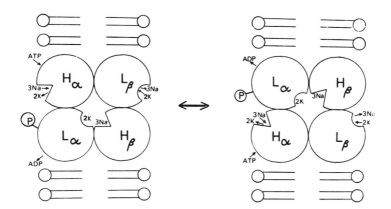

Proceedings of the National Academy of Science of the United States of America

Figure 7. Half-of-the-sites active model for Na and K transport

The Na/K–ATPase is conceived of as a tetrameric complex, $\alpha_2\beta_2$ with subunits L the opposite affinities. Active transport occurs (A to A') by rotation of the elements together with a change in their affinities (H to L and vice-versa). A and A' are proposed to be isoenergetic (32).

reaction sequence, a phenomenon termed half-of-the-sites activity. Stein et al. (32) suggest that the Na/K–ATPase is such an enzyme, incorporated into a tetrameric transport system (Figure 7). Their model features two dimeric components each with a catalytic site. Transport is by an alternation of sites process (the Shaw model, but with the systems in the two components out of phase with each other). Thus separate sites are present for Na and for K although on different halves of the ATPase. However, it is not clear how the moderate-affinity K sites of the free enzyme corresponding to the (external) K sites of the K-phosphatase activity might be incorporated. A similar model was presented in more detail by Repke et al. (230, 231).

In a somewhat different version by Robinson (115, 165, 224) a separate-sites transport scheme is proposed, incorporating the moderate-affinity K sites of the free enzyme. In this formulation low concentrations of MgATP fill the high-affinity substrate sites to produce E_1 (E_I) enzymatic characteristics, while high concentrations of MgATP (corresponding to those found *in vivo* and generally used *in vitro*) fill the low-affinity substrate sites as well to produce the E_2 (E_{II}) characteristics and the half-of-the-sites active system. Moreover, while one half-system catalyzes dephosphorylation the other half-system may be phosphorylated; high-affinity K sites thus may exist throughout (rather than oscillating between moderate- and high-affinity stages), such that Na and K could bind simultaneously for a concerted reaction process (185, 224, 232).

MAJOR UNRESOLVED ISSUES. To define the molecular processes associated with Na and K transport a number of topics still await investigation while most others require further clarification. Current progress in purifying the enzyme should permit investigation of certain previously unapproachable issues. For example, essentially no information has been obtained to define precisely what changes do take place although it has been obvious that conformational changes in the enzyme must occur. Careful physical studies on the isolated enzyme (and specific modifications of it) should now be feasible. In addition, a number of aspects of the reaction sequence remain vague, particularly the cation release. Enzyme preparations oriented in a membrane to preserve their "sidedness" would facilitate a number of approaches. Finally, applicability of half-of-the-sites active formulations awaits firm evidence.

The Ca Pump and Ca–ATPase of the Sarcoplasmic Reticulum.
CALCIUM TRANSPORT AND MUSCLE RELAXATION. Contraction and relaxation of skeletal muscle is regulated *in vivo* by the concentration of calcium ions in the vicinity of the contractile proteins (*see* Chapter 6). Muscle can contract when the free Ca concentration rises above a critical level (about 0.1 μM) and can relax when the concentration falls below this level. The series of steps leading to muscle contraction thus hinges on such an increase in cytoplasmic Ca through a process termed excitation-contraction coupling. The essential structures in this process are the plasma membrane (which conducts the action potential), the sarcoplasmic reticulum (an intracellular tubular network that releases and reaccumulates Ca), and the contractile proteins of the muscle. The muscle action potential initiated by depolarization at the neuromuscular junction spreads along the muscle membrane where invaginations of

the transverse tubules (T tubules) conduct the depolarization deep into the muscle cell. The T tubules pass near the terminal cysternae of the sarcoplasmic reticulum, and the electrical signal transmitted by the T tubules triggers the release of stored Ca in the vicinity of the contractile elements. Contraction then occurs in response to the elevated cytoplasmic Ca. Conversely, relaxation is accomplished by an active transport system in the sarcoplasmic reticulum membrane that reaccumulates the cytoplasmic Ca into this intracellular reservoir. The Ca pump thus permits muscle relaxation by reducing the Ca levels near the contractile system and also reaccumulates Ca for subsequent release. Although the mechanisms associated with Ca release remain obscure, considerable information about the Ca pump has been amassed, particularly through studies on transport phenomena and enzymatic activity in isolated preparations (233, 234).

Marsh (235) and Bendall (236) first described a "relaxing factor" in muscle homogenates which proved to be fragments of the sarcoplasmic reticulum in the crude microsomal fraction. The vesicles formed by shear forces from the tubules of the sarcoplasmic reticulum were shown in early experiments by Ebashi, Hasselbach, Weber, and their associates (237, 238, 239, 240) to accumulate Ca in the presence of Mg and an energy source (e.g., ATP) and to reduce Ca concentrations in the medium below 0.1 μM as required for relaxation *in vivo*. Thus, unlike microsomal preparations containing the Na/K–ATPase that cannot be used for transport studies because of their leakiness, the sarcoplasmic reticulum fragments may be exploited for both enzymatic and transport studies and may serve as a model system for general approaches to such processes.

THE CA PUMP AND CA–ATPASE. The Ca pump in sarcoplasmic reticulum fragments transports two Ca atoms into the vesicles for each ATP hydrolyzed to ADP and P_i (239, 240, 241). Conversely, ATP can be synthesized from ADP and P_i by allowing the pump to run backward dissipating a preexisting Ca gradient (242, 243), just as with the Na/K pump in erythrocytes (97). In the reverse direction the stoichiometry appears to be two Ca atoms per ATP. Mg is required for both the forward and reverse operation of the pump.

The ability to accumulate Ca is measured by two general procedures that result in quite different quantities of Ca transported (233, 234). In the presence of Mg, Ca, and ATP the pump transports Ca across the vesicular membrane where a major fraction may be bound to storage sites on the inner surface of the membrane or perhaps on intravesicular proteins (72, 244, 245). The concentration of free Ca is elevated, however, and a true gradient across the membrane is created. Leakage is relatively slow, and the vesicles retain the Ca in the absence of ADP and P_i. Beyond this, the capacity to accumulate Ca may be increased an order of magnitude by including certain permeable anions (e.g., oxalate) in the medium (233, 238). The oxalate equilibrates across the membrane, but the solubility product for calcium oxalate is exceeded within the vesicles where the Ca concentration is much higher as a result of active transport. Thus, calcium oxalate precipitates within the vesicles, reducing the concentration of free Ca, and permits the pump to function against a less steep gradient. Consequently, more Ca is transported as the calcium oxalate precipitates, and the capacity of the vesicle to amass total Ca is markedly increased. Unfortunately, different authors use different terminologies for these two procedures. (Note that there are also binding sites on the sarcoplasmic reticulum that accept Ca in the absence of an energy source, but the capacity there is much lower than that for energy-dependent accumulation.)

There is a distinguishable ATPase activity generally assumed to represent the Ca pump which requires both Ca and Mg (233, 234, 238) and corresponds to the Ca pump activity. This enzymatic activity is usually referred to as the Ca–ATPase or Ca/Mg–ATPase. In addition, sarcoplasmic reticulum preparations commonly contain some Mg–ATPase activity in the absence of Ca ("basic ATPase") which may represent a separate entity or an uncoupled version of the pump (234).

Although two Ca atoms are transported for each ATP hydrolyzed, the Ca–velocity plots are not sigmoidal, but they appear to follow Michaelis-Menten kinetics (246, 247). Similarly, plots of Ca–ATPase activity against Ca concentration are also hyperbolic (248, 249). The K_m's for both processes are essentially the same (on the order of 0.1–1.0 μM). The Ca in the medium is frequently buffered with the chelator ethylene glycol bis(β-aminoethyl ether)-N,N'-tetraacetic acid (EGTA) to improve accuracy in experiments which require low Ca concentrations. The Ca pump is sensitive to many general enzyme inhibitors (e.g., sulfhydryl reagents), but there are no specific inhibitors in the sense that the cardioactive steroids are specific inhibitors of the Na/K pump (233, 234). Ouabain does not affect the Ca pump.

COMPONENTS OF THE SARCOPLASMIC RETICULUM AND PURIFICATION OF THE CA–ATPASE. Centrifugation techniques permit the separation of fractions with high specific activities of transport and ATPase activity from muscle microsomal preparations (250). The Ca–ATPase may represent up to two-thirds of the protein of such vesicles, and the Ca pump seems to be the only major process. Such preparations appear as vesicles bounded by a conventional single trilaminar membrane in electron micrographs (251, 252, 253). However, negative staining (253, 254) reveals globular particles which extend from the outer surface (facing the medium in vitro or the cytoplasm in vivo) —these extensions are reminiscent of the mitochondrial ATPase (255) and Na/K–ATPase projections (132). Freeze–fracture techniques show large globular elements which span the thickness of the membrane (252, 254).

MacLennan et al. (254, 256) separated a functional Ca–ATPase with its lipid milieu from such preparations. It is represented by a protein with molecular weight of about 100,000 on gel electrophoresis. The protein is not composed of subunits (254, cf. 257), but it may be cleaved by proteolytic digestion into several fractions (254, 258). Correlations of morphological change with loss of activity following controlled proteolysis (254, 258) of both sarcoplasmic reticulum and enzyme preparations reconstituted into vesicles indicate that the globular particles seen on the surface by negative staining are part of the Ca–ATPase. Moreover, the fragments of the Ca–ATPase identified with these globular particles are also susceptible to iodination by lactoperoxidase (indicating a localization on the external surface) and include the site phosphorylated by ATP in the reaction process (258). Thus, the Ca–ATPase is pictured as a single polypeptide with a hydrophobic portion embedded in the membrane and a hydrophilic portion containing the ATPase site extending outward (which, by negative staining procedures, appears as a globular particle on the surface). Such a configuration would make rotational transport models unlikely.

In addition to the Ca–ATPase several other proteins may be identified in sarcoplasmic reticulum preparations, depending on the starting material and the approach. MacLennan et al. (72, 245) describe two intriguing proteins. Calsequestrin is an acidic protein of about 46,000 molecular weight (245, cf. 259), originally proposed as a storage element for Ca on the interior of the vesicular membrane (72). Although calsequestrin can bind about 900 nmoles Ca/mg protein with a K_D of 60 μM, in the presence of 100 μM KCl (reflecting intracellular conditions) the K_D is increased to 800 μM (245). In addition, iodination by lactoperoxidase suggests that calsequestrin is located on the outer surface of the membrane (258). The other notable protein (molecular weight slightly higher than 55,000) binds only 20 nmoles Ca/mg protein (about 1 mole of Ca/mole protein), but it has a much greater affinity for Ca: a K_D in 100 mM KCl of 2–4 μM (245).

Radiation inactivation studies on the vesicles indicate a molecular weight of the functional Ca–ATPase *in situ* of about 200,000 (260), twice that of the identified Ca–ATPase protein. Thus, this ATPase (like the Na/K–ATPase) may function as a dimer (*see* below).

INFLUENCE OF THE LIPID MATRIX. Lipids are essential for Ca transport and for the Ca–ATPase (82, 233, 261, 262, 263, 264). Extraction of the membrane lipids or digestion with phospholipase A or C can block both transport and hydrolysis, but activity can be restored with a variety of lipids (233, 262). Thus, there is less evidence for a specific bonding requirement between lipid and protein than there is with the Na/K–ATPase. Nevertheless, the sarcoplasmic reticulum membrane is notable for the high degree of unsaturation of the fatty acids in the phospholipids and the almost total absence of cholesterol (264, 265), both of which contribute to the high degree of fluidity of the membrane interior. Like the Na/K–ATPase the break in the Arrhenius plots of Ca–ATPase activity can be correlated with a change in motional freedom of the membrane lipids (266).

Because of the high specific activity of transport elements in the native sarcoplasmic reticulum, it is an ideal candidate for physical studies on a functional membrane. Particularly interesting information has been obtained by using ESR spin label probes (267, 268) and by NMR relaxation techniques (67, 269). Both approaches demonstrate that the motional freedom of about three-fourths of the membrane lipids is consistent with their being in a bilayer formation (67, 267) and that there seems to be little hindrance to lateral diffusion of the lipids (268, 269). However, the methyl protons of the choline headgroup in phosphatidyl choline are not visible in ^{1}H NMR spectra (67, 270), and the relaxation times of the methyl carbons in ^{13}C NMR studies also indicate a restrictive environment, suggesting an association with membrane proteins (67).

RECONSTITUTION OF THE CA PUMP. Racker (271) and Meissner and Fleisher (272, 273) recently assembled functional transport systems from the purified Ca–ATPase protein plus phospholipid dispersions, and reconstituted detergent solubilized sarcoplasmic reticulum into vesicles. The technique involves removing the suspending detergent by dialysis, which permits self-assembly into membrane form. The resultant vesicles have Ca–ATPase and Ca-accumulating capacities near those of the starting material. The striking importance of these experiments lies in their promise for future investigations: the study of isolated components that may be specifically selected, labeled, and modified.

REACTION SEQUENCE. *Substrate Binding and Specificity.* Direct studies of ATP binding to the Ca–ATPase have not been reported, but the K_m for MgATP has been examined. Yamamoto and Tonomura (248) found a biphasic substrate-velocity plot, like that of the Na/K–ATPase (167, 168); the high-affinity site has a K_m of 0.7 μM while the low-affinity site (if that is the correct interpretation of the experiment) has a K_m of 50 μM. Both values are quite below the concentration of MgATP found in muscle *in vivo*. Although free Mg can slow phosphorylation of the enzyme by ATP (presumably by competition with Ca) (274, 275), MgATP may be the actual substrate. On the basis of kinetic studies Yamamoto and Tonomura (248) considered MgATP to be the true substrate, and Inesi and Almendares (276) found more ^{14}C-ATP bound to the sarcoplasmic reticulum in the presence of Mg than in its absence.

A number of nucleoside triphosphates may substitute for ATP both for transport and enzymatic activity (233, 234, 277). In addition, acetyl phosphate, carbamyl phosphate, and p-nitrophenyl phosphate fulfill both functions (233, 234), in contrast to their reported inability to energize Na and K transport. ATP may act both as substrate and as heterotropic allosteric effector—a dual role not as readily played by alternative substrates. The biphasic substrate–velocity plot may represent an increased reactivity arising from ATP binding at a second site (248), and monovalent cations inhibit the ATPase at low substrate concentrations but not at higher concentrations corresponding to MgATP levels found *in vivo* (247, 277). Spin labels bonded to the enzyme show a conformational change in the presence of high concentrations of ATP that is not directly correlated with the translocation process (90) although interpretations of such experiments must be hesitant.

Phosphorylation of the Enzyme. In the presence of Ca the enzyme is phosphorylated by ATP (248, 277, 278, 279, 280), to form an acyl phosphate on the β-carboxyl of an aspartyl residue (281, 282). Recently, Post et al. (281) reported that the tripeptide containing the acyl phosphate, obtained from a proteolytic digest of the phosphorylated enzyme, is identical to that obtained under similar conditions from the Na/K–ATPase: ser (or thr)-asp-lys. Other peptide fragments differed, of course. There is one phosphorylation site/100,000 daltons (283).

The Ca–ATPase also catalyzes an ADP/ATP exchange reaction (274, 280, 284), presumably with the participation of this acyl phosphate and in accord with an apparent equilibrium constant for phosphorylation near 1 (274). Increasing concentrations of Ca accelerate this exchange (284). Under comparable conditions the rate of ADP/ATP exchange is considerably faster than that of ATP hydrolysis (280).

The enzyme also can be phosphorylated by inorganic phosphate in the presence of Mg and a Ca gradient from vesicle to medium (285, 286, 287). This corresponds to experiments showing a net synthesis of ATP from ADP and P_i energized by a Ca flux driving the pump backward (242). However, Masuda and de Meis (288) recently reported that P_i can phosphorylate the enzyme in the absence of a Ca gradient (the vesicles made leaky by treatment with diethyl ether); the energetics of such a phosphorylation are obscure. Although phospholipase A and C inactivate the Ca–ATPase, enzyme phosphorylation is affected little (289, 290). Dephosphorylation in such preparations, however, is inhibited.

Dephosphorylation. Dephosphorylation of the acyl phosphate is accelerated by Mg (274, 275), and Mg is often stated to be necessary for this process. Liberation of P_i appears to be slower than phosphorylation under conditions of substrate saturation, but the actual rate-limiting step may be an isomerization that must occur before hydrolysis of the acyl phosphate (290). As cited above, lipid depletion slows dephosphorylation much more than phosphorylation, which may reflect participation of the lipid environment in such an isomerization (290).

There is no direct chemical evidence for P_i accumulation within the vesicles (291), but histochemical studies showed phosphate-heavy metal deposits on the interior (253). Nevertheless, these deposits may represent subsequent migration of the phosphate, and the release of P_i is commonly assumed to be extravesicular.

Although de Meis and de Mello (277) suggested a heterotropic allosteric effect of nucleotides on dephosphorylation, Martonosi *et al.* (290) found in rapid-quench experiments no change in the rate of acyl phosphate hydrolysis with ATP concentrations up to 0.5 mM (still below levels *in vivo* but higher than the K_m for the low-affinity site).

Ion Binding to the Enzyme. Dissociation constants for Ca have been measured directly as well as approximated from kinetic effects. Ikemoto (292) found by equilibrium dialysis three classes of sites for Ca with approximately one of each class per 100,000 daltons: α sites with $K_D = 0.25 \mu M$, β sites with $K_D = 25 \mu M$, and γ sites with $K_D = 1$ mM. Adding 1.5 mM ATP increased the affinity of all sites slightly, but, suprisingly, reduced the capacity of the α- and β sites. In correlations with Ca–ATPase studies, binding at the α site activated the enzyme while binding at the γ site inhibited it; the β site appeared uninvolved. The K_D for the α-site is in reasonable agreement with the K_m for Ca–ATPase activation and for the Ca pump (233, 234), but the

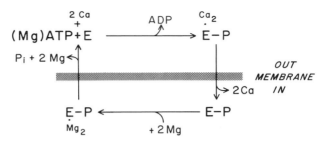

Figure 8. A reaction sequence for the Ca-ATPase

This scheme incorporates an exchange of Ca for Mg across the sarcoplasmic reticulum membrane (adapted from Refs. 249, 278)

stoichiometry of the pump would require two Ca sites, assuming that the Ca to be transported binds to the Ca–ATPase protein itself. Thus the finding of only one α-site per 100,000 daltons is unexpected; perhaps, this bears on formulations of the functional Ca pump as a dimer of two ATPase proteins (292), in accord with radiation-inactivation estimates of molecular weight (see above). It would also seem plausible that the low-affinity γ sites might be discharge sites for transported Ca.

Yamada and Tonomura (275) measured competition between Ca and Mg for enzyme phosphorylation and obtained a K_m for Ca of 0.35 μM and a K_i for Mg of 11 mM. For dephosphorylation they found that Ca competes with Mg to hinder hydrolysis, and the ratio of K_m for Mg to K_i for Ca is then 2.5.

In addition to changes in affinity for Ca cited above (248, 292), substrates may influence cation interactions with enzyme further; de Meis and Hasselbach (247) showed that with acetyl phosphate as substrate the Ca–velocity plot is sigmoidal although they (as others) found that with ATP the plot is hyperbolic.

Reaction Schemes. The essential features of the Ca–ATPase are incorporated into a plausible reaction scheme in Figure 8. Ca activates phosphorylation to form an acyl phosphate intermediate, and Mg activates dephosphorylation. Kinetic studies suggest that Ca binds after ATP (248), in accord with ATP-induced increases in affinity for Ca (292); Mg antagonizes Ca binding (275) but also may serve with ATP as the substrate, MgATP (248).

Recently, Makinose (287) found in a system where phosphorylation by P_i is being driven by Ca efflux that adding ADP increased Ca flux across the membrane in both directions but decreased net phosphorylation of the enzyme. On this basis he argues that in the normal functioning of the pump Ca must be released before the acyl phosphate is hydrolyzed:

$$2\,Ca + E \cdot ATP \xrightarrow{} \overset{Ca_2}{E-P} \xrightarrow{\text{transport}} \overset{Ca_2}{E-P} \xrightarrow{} E-P + 2\,Ca$$
$$E + ATP \qquad\qquad\qquad\qquad\qquad\qquad\qquad E + P_i$$

If P_i were released before Ca (in the forward direction of the pump) then an increase in Ca efflux must increase phosphorylation of the enzyme, which it did not. It should be noted that adding ADP to this system initiates a Ca/Ca exchange: increased Ca flux in both directions, with the phosphorylated enzyme shuttling Ca across the membrane. An analogy with the Na/Na exchange in erythrocytes, also dependent on ADP (*96*), is obvious.

The available data do not specify that Ca must be released before Mg binds, although this is a frequent assumption on the plausible model that Ca sites become in the course of the reaction Mg sites. This, however, need not be the case (*cf.* the Na/K–ATPase).

The formulation in Figure 8 is strikingly similar to many proposals for the Na/K-ATPase. Ca plays the role of Na in promoting acyl phosphate formation, but Mg corresponds to K in activating hydrolysis. Beyond this, both enzymes have catalytic proteins with similar molecular weights, and the primary structure at the phosphorylation site is the same. ^{18}O-exchange experiments suggest a similar mechanism for both (*135, 183, 249*). The high-affinity sites bind ATP with essentially the same K_D, and in both cases nucleotide binding may influence catalytic events through heterotropic allosteric interactions, perhaps in a dimeric enzyme operating on a half-of-the-sites active principle.

TRANSPORT MECHANISM. Unfortunately, firm information that might distinguish between alternative formulations is largely lacking. Although it is clear that Ca enters the vesicle, it is uncertain whether *in vivo* it is accompanied by anions to preserve electrical neutrality or is exchanged for other cations. There is no indication that the Ca pump is electrogenic. If the physiological release of Ca to initiate muscle contraction occurs through a transient leakiness, then the interior of the sarcoplasmic reticulum could equilibrate with free Mg in the cytoplasm, which is in the millimolar range; this Mg would then be available for subsequent exchange. Moreover, the exchange of Mg for Ca seems plausible in light of proposed reaction schemes (Figure 8); hence, variations on the Shaw model for a Na/K exchange pump were advanced (*249, 278*). Experiments like those of Hoffman and Tosteson (*109*) and Garay and Garrahan (*110*) on the Na/K pump have yet to be performed on the Ca pump. Such information seems essential, since separate-sites models for Ca and Mg exchange can be formulated as well as the alternating-sites models.

No information relevant to a carrier is available although Blondin (*293*) recently described a divalent cation ionophore prepared by proteolytic digestion of mitochondria. On the other hand, the presence of low-affinity Ca sites (γ sites) together with high affinity α sites suggests that discharge and acceptance sites for Ca may coexist, in accord with a pore model for transport. It would be helpful to have a tabulation of Ca and Mg affinities for successive stages of the reaction scheme. It should be noted that the fluidity of the membrane and the ease of lateral diffusion for lipids may be largely irrelevant to arguments for rotational transport models (*234, 290*): major "frictional" forces would be expected in driving polar superficial groups through the hydrophobic membrane interior. If—as some suggest—the stoichiometry of Ca to ATP can range far higher than two (*233, 294, 295*), then obvious limitations on possible models would be imposed.

A half-of-the-sites active formulation (*see* above) could incorporate plausibly the proposed allosteric effects, dimeric enzyme models, and biphasic substrate–velocity plots. Transport by separate units out of phase with each other—so that while one cation moves in the other moves out—has particular esthetic appeal for exchange reactions, but currently there is no other compelling justification here.

To date physical studies have been unrewarding. Circular dichroism studies showed no evidence of conformational changes with transport (*296*), fluorescent probe experiments seemingly relate to the accumulated Ca rather than the transport process (*297*), and x-ray diffraction studies—while demonstrating an asymmetric membrane structure (*298*)—give no indication of functional changes in conformation. Since it is obvious that structural changes affect transport, the crucial need is not merely to document that such changes occur but to specify precisely what it is that changes, and by how much.

Literature Cited

1. Wilbrandt, W., Rosenberg, T., *Pharmacol. Rev.* (1961) **13**, 109.
2. Stein, W. D., "The Movement of Molecules across Cell Membranes," Academic Press, New York, 1967.
3. Dowben, R. M., Ed., "Biological Membranes," Little Brown, Boston, 1969.
4. Christensen, H. N., *Adv. Enzymol.* (1969) **32**, 1.
5. Lieb, W. R., Stein, W. D., *Biochim. Biophys. Acta* (1972) **265**, 187.
6. Oxender, D. L., *Ann. Rev. Biochem.* (1972) **41**, 777.
7. Hokin, L. E., Ed., "Metabolic Transport," Academic Press, New York, 1972.
8. Nystrom, R. A., "Membrane Physiology," Prentice-Hall, Englewood Cliffs, 1973.
9. Dick, D. A. T., "Cell Water," Butterworths, Washington, 1966.
10. Forster, R. E., *Curr. Top. Membr. Transport* (1971) **2**, 41.
11. Mauro, A., *Circulation* (1960) **21**, 845.
12. Staverman, A. J., *Trans. Faraday Soc.* (1952) **48**, 176.
13. Leaf, A., *Ann. N.Y. Acad. Sci.* (1959) **72**, 396.
14. Tosteson, D. C., Hoffman, J. F., *J. Gen. Physiol.* (1960) **44**, 169.
15. Robinson, J. D., *J. Theoret. Biol.* (1968) **19**, 90.
16. Van Slyke, D. D., Wu, H., McLean, F. C., *J. Biol. Chem.* (1923) **56**, 765.
17. Goldman, D. E., *J. Gen. Physiol.* (1943) **27**, 37.
18. Hodgkin, A. L., Katz, B., *J. Physiol. Lond.* (1949) **108**, 37.
19. Dainty, J., *Adv. Bot. Res.* (1963) **1**, 279.
20. Everitt, C. T., Redwood, W. R., Haydon, D. A., *J. Theoret. Biol.* (1969) **22**, 20.
21. Andreoli, T. E., Dennis, V. W., Weigl, A. M., *J. Gen. Physiol.* (1969) **53**, 133.
22. Wilson, F. A., Dietschy, J. M., *J. Clin. Invest.* (1972) **51**, 3015.
23. Gutknecht, J., Tosteson, D. C., *Science* (1973) **182**, 1258.
24. Lieb, W. R., Stein, W. D., *Biophys. J.* (1972) **10**, 585.
25. Burgen, A. S. V., *Canad. J. Biochem. Physiol.* (1957) **35**, 569.
26. Crane, R. K., *Fed. Proc.* (1965) **24**, 1000.
27. Kimmich, G. A., *Biochim. Biophys. Acta* (1973) **300**, 31.
28. Thomas, R. C., *Physiol. Rev.* (1972) **52**, 563.

29. Post, R. L., Jolly, P. C., *Biochim. Biophys. Acta* (1957) **25**, 118.
30. Mitchell, P., *Nature* (1957) **180**, 134.
31. Opit, L. J., Charnock, J. S., *Nature* (1965) **209**, 471.
32. Stein, W. D., Lieb, W. R., Karlish, S. J. D., Eilam, Y., *Proc. Nat. Acad. Sci. U.S.* (1973) **70**, 275.
33. Jardetzky, O., *Nature* (1966) **211**, 969.
34. Robinson, J. D., *Arch. Biochem. Biophys.* (1973) **156**, 232.
35. Garrahan, P. J., Glynn, I. M., *J. Physiol. Lond.* (1967) **192**, 159.
36. Halvorson, H. O., Okada, H., Gorman, J., "The Cellular Functions of Membrane Transport," Hoffman, J. F. (ed.), 171, Prentice-Hall, Englewood Cliffs, 1964.
37. Wiley, J. S., *Brit. J. Haemat.* (1972) **22**, 529.
38. Kedem, O., "Membrane Transport and Metabolism," Kleinzeller, A., Kotyk, A. (eds.), 87, Academic Press, New York, 1961.
39. Katchalsky, A., Curran, P. F., "Nonequilibrium Thermodynamics in Biophysics," Harvard, Cambridge, Mass., 1965.
40. Solomon, A. K., *J. Gen. Physiol.* (1952) **36**, 57.
41. Bangham, A. D., Standish, M. M., Watkins, J. C., *J. Mol. Biol.* (1965) **13**, 238.
42. Ussing, H. H., Zerahn, K., *Acta Physiol. Scand.* (1951) **23**, 110.
43. Leaf, A., Anderson, J., Page, L. B., *J. Gen. Physiol.* (1958) **41**, 657.
44. Kleinzeller, A., "Metabolic Transport," L. E. Hokin, Ed., 91, Academic Press, New York, 1972.
45. Whittam, R., "Transport and Diffusion in Red Blood Cells," Williams and Wilkins, Baltimore, 1964.
46. Lucké, B., McCutcheon, M., *Physiol. Rev.* (1932) **12**, 68.
47. Osterhout, W. J. V., *Proc. Natl. Acad. Sci. U.S.* (1935) **21**, 125.
48. Baker, P. F., Hodgkin, A. L., Shaw, T. I., *J. Physiol. Lond.* (1962) **164**, 330.
49. Brinley, F. J., Mullins, L. J., *J. Gen. Physiol.* (1968) **52**, 181.
50. Hoffman, J. F., *J. Gen. Physiol.* (1962) **45**, 837.
51. Palade, G. E., Siekevitz, P., *J. Biophys. Biochem. Cytol.* (1956) **2**, 171.
52. de Belleroche, J. S., Bradford, H. F., *Prog. Neurobiol.* (1973) **1**, 275.
53. Kaback, H. R., *Biochim. Biophys. Acta* (1972) **265**, 367.
54. Peachey, L. D., Rasmussen, H., *J. Biochem. Biophys. Cytol.* (1961) **10**, 529.
55. Cereijido, M., Curran, P. F., *J. Gen. Physiol.* (1965) **48**, 543.
56. Sharp, G. W. G., Leaf, A., *Physiol. Rev.* (1966) **46**, 593.
57. Diamond, J. M., Bossert, W. H., *J. Gen. Physiol.* (1967) **50**, 2061.
58. Curran, P. F., MacIntosh, J. R., *Nature* (1962) **193**, 347.
59. Mueller, P., Rudin, D. O., *Curr. Topics Bioenerg.* (1969) **3**, 157.
60. Haydon, D. A., Hladky, S. B., *Quart. Rev. Biophys.* (1972) **5**, 187.
61. Tosteson, D. C., *Fed. Proc.* (1968) **27**, 1269.
62. Storelli, C., Vogeli, H., Semenze, G., *FEBS Lett.* (1972) **24**, 287.
63. Jain, M. K., White, F. P., Strickholm, A., Williams, E., Cordes, E. H., *J. Membr. Biol.* (1972) **8**, 363.
64. Bangham, A. D., *Prog. Biophys. Mol. Biol.* (1968) **18**, 29.
65. Reeves, J. P., Dowben, R. M., *J. Membr. Biol.* (1970) **3**, 123.
66. Kornberg, R. D., McConnell, H. M., *Biochemistry* (1971) **10**, 1111.
67. Robinson, J. D., Birdsall, N. J. M., Lee, A. G., Metcalfe, J. C., *Biochemistry* (1972) **11**, 2903.
68. Henderson, R., Wang, J. H., *Biochemistry* (1972) **11**, 4565.
69. Benzer, T. I., Raferty, M. A., *Proc. Nat. Acad. Sci. U.S.* (1972) **69**, 3634.
70. Pardee, A. B., Prestidge, L. S., Whipple, M .B., Dreyfuss, J., *J. Biol. Chem.* (1966) **241**, 3962.
71. Pardee, A. B., *Science* (1968) **162**, 632.
72. MacLennan, D. H., Wong, P. T. S., *Proc. Nat. Acad. Sci. U.S.* (1971) **68**, 1231.

73. Fox, C. F., Kennedy, E. P., *Proc. Nat. Acad. Sci. U.S.* (1965) **54**, 891.
74. Kennedy, E. P., "The Lac Operon," J. Beckwith, D. Zipser, Eds., 49, Cold Spring Harbor, New York, 1970.
75. Kundig, W., Ghosh, S., Roseman, S., *Proc. Nat. Acad. Sci. U.S.* (1964) **52**, 1067.
76. Roseman, S., "Metabolic Transport," L. E. Hokin, Ed., 41, Academic Press, New York, 1972.
77. Kennedy, E. P., Rumley, M. K., Armstrong, J. B., *J. Biol. Chem.* (1974) **249**, 33.
78. Razin, S., *Biochim. Biophys. Acta* (1972) **265**, 241.
79. Fahn, S., Hurley, M. R., Koval, G. J., Albers, R. W., *J. Biol. Chem.* (1966) **241**, 1890.
80. Schoner, W., Schmidt, H., Erdmann, E., *Biochem. Pharmacol.* (1972) **21**, 2413.
81. Anderson, B. E., Weigel, N., Kundig, W., Roseman, S., *J. Biol. Chem.* (1971) **246**, 7023.
82. Balzer, H., Makinose, M., Fiehn, W., Hasselbach, W., *Naunyn-Schmiedebergs Archiv. Pharmakol.* (1968) **260**, 456.
83. Goldman, S. S., Albers, R. W., *J. Biol. Chem.* (1973) **248**, 867.
84. Roelofsen, B., van Deenen, L. L. M., *Eur. J. Biochem.* (1973) **40**, 245.
85. Worthington, C. R., *Curr. Top. Bioenerg.* (1973) **5**, 1.
86. Levine, Y. K., *Prog. Surface Sci.* (1973) **3**, 279.
87. Anet, F. A. L., Levy, G. C., *Science* (1973) **180**, 141.
88. Kornberg, R. D., McConnell, H. M., *Proc. Nat. Acad. Sci. U.S.* (1971) **68**, 2564.
89. Linden, C. D., Wright, K. L., McConnell, H. M., Fox, C. F., *Proc. Nat. Acad. Sci. U.S.* (1973) **70**, 2271.
90. Landgraf, W. C., Inesi, G., *Arch. Biochem. Biophys.* (1969) **130**, 111.
91. Wu, C. W., Stryer, L., *Proc. Nat. Acad. Sci. U.S.* (1972) **69**, 1104.
92. Schwartz, A., Lindenmayer, G. E., Allen, J. C., *Curr. Top. Membr. Transport* (1972) **3**, 1.
93. Hokin, L. E., "Metabolic Transport," L. E. Hokin, Ed., 269, Academic Press, New York, 1972.
94. Baker, P. F., "Metabolic Transport," L. E. Hokin, Ed., 243, Academic Press, New York, 1972.
95. Hoffman, J. F., *Circulation* (1962) **26**, 1201.
96. Glynn, I. M., Hoffman, J. F., Lew, V. L., *Phil. Trans. Roy. Soc. Lond. B.* (1971) **262**, 91.
97. Garrahan, P. J., Glynn, I. M., *J. Physiol. Lond.* (1967) **192**, 237.
98. Lant, A. F., Whittam, R., *J. Physiol. Lond.* (1968) **199**, 457.
99. Lew, V. L., Glynn, I. M., Ellory, J. C., *Nature* (1970) **225**, 865.
100. Sen, A. K., Post, R. L., *J. Biol. Chem.* (1964) **239**, 345.
101. Garrahan, P. J., Glynn, I. M., *J. Physiol. Lond.* (1967) **192**, 217.
102. Thomas, R. C., *J. Physiol. Lond.* (1969) **201**, 495.
103. Armando-Hardy, M., Jr., Ellory, J. C., Lew, V. L., *J. Physiol. Lond.* (1973) **231**, 110P.
104. Garrahan, P. J., Glynn, I. M., *J. Physiol. Lond.* (1967) **192**, 175.
105. Sachs, J. R., Welt, L. G., *J. Clin. Invest.* (1967) **46**, 65.
106. Garrahan, P. J., *Nature* (1969) **222**, 1000.
107. Robinson, J. D., *Arch. Biochem. Biophys.* (1970) **139**, 17.
108. Schneider, R. P., *Arch. Biochem. Biophys.* (1974) **160**, 552.
109. Hoffman, P. G., Tosteson, D. C., *J. Gen. Physiol.* (1971) **58**, 438.
110. Garay, R. P., Garrahan, P. J., *J. Physiol. Lond.* (1973) **231**, 297.
111. Garrahan, P. J., Garay, R. P., *Ann. N.Y. Acad. Sci.* (1974) **242**, 445.
112. Glynn, I. M., *J. Physiol. Lond.* (1956) **134**, 278.
113. Skou, J. C., *Curr. Top. Bioenerg.* (1971) **4**, 357.

114. Robinson, J. D., *Biochemistry* (1969) **8**, 3348.
115. Robinson, J. D., *Ann. N.Y. Acad. Sci.* (1974) **242**, 185.
116. Cleland, W. W., *Enzymes* (1970) **2**, 1.
117. Dixon, M., Webb, E. C., "Enzymes," 70, Academic Press, New York, 1964.
118. Dunham, P. B., Hoffman, J. F., *Biochim. Biophys. Acta* (1971) **241**, 399.
119. Baker, P. F., Willis, J. S., *J. Physiol. Lond.* (1972) **224**, 441.
120. Landowne, D., Ritchie, J. M., *J. Physiol. Lond.* (1970) **207**, 529.
121. Skou, J. C., *Biochim. Biophys. Acta* (1957) **23**, 394.
122. Post, R. L., Merritt, C. R., Kinsolving, C. R., Albright, C. D., *J. Biol. Chem.* (1960) **235**, 1796.
123. Glynn, I. M., *J. Physiol. Lond.* (1962) **160**, 18 P.
124. Whittam, R., *Biochem. J.* (1962) **84**, 110.
125. Bonting, S. L., Caravaggio, L. L., *Arch. Biochem. Biophys.* (1963) **101**, 37.
126. Albers, R. W., Fahn, S., Koval, G. J., *Proc. Nat. Acad. Sci. U.S.* (1963) **50**, 474.
127. Nakao, T., Tashima, Y., Nagano, K., Nakao, M., *Biochem. Biophys. Res. Commun.* (1965) **19**, 755.
128. Jorgensen, P. L., Skou, J. C., *Biochem. Biophys. Res. Commun.* (1969) **37**, 39.
129. Jorgensen, P. L., *Ann. N.Y. Acad. Sci.* (1974) **242**, 36.
130. Kyte, J., *J. Biol. Chem.* (1971) **246**, 4157.
131. Kyte, J., *J. Biol. Chem.* (1972) **247**, 7642.
132. Hokin, L. E., Dahl, J. L., Deupree, J. D., Dixon, J. F., Hackney, J. F., Perdue, J. F., *J. Biol. Chem.* (1973) **248**, 2593.
133. Hokin, L. E., *Ann. N.Y. Acad. Sci.* (1974) **242**, 12.
134. Lane, L. K., Copenhaver, J. H., Jr., Lindenmayer, G. E., Schwartz, A., *J. Biol. Chem.* (1973) **248**, 7197.
135. Dahms, A. S., Boyer, P. D., *J. Biol. Chem.* (1973) **248**, 3155.
136. Kepner, G. R., Macey, R. I., *Biochim. Biophys. Acta* (1968) **163**, 188.
137. Albers, R. W., Koval, G. J., Siegel, G. J., *Mol. Pharmacol.* (1968) **4**, 324.
138. Kyte, J., *J. Biol. Chem.* (1972) **247**, 7634.
139. Jorgensen, P. L., Hansen, O., Glynn, I. M., Cavieres, J. D., *Biochim. Biophys. Acta* (1973) **291**, 795.
140. Askari, A., *Ann. N.Y. Acad. Sci.* (1974) **242**, 372.
141. Banerjee, S. P., Wong, S. M. E., Khanna, V. K., Sen, A. K., *Mol. Pharmacol.* (1972) **8**, 8.
142. Robinson, J. D., *Nature* (1971) **233**, 419.
143. Nagai, K., Lindenmayer, G. E., Schwartz, A., *Arch. Biochem. Biophys.* (1970) **139**, 252.
144. Yoda, A., Hokin, L. E., *Mol. Pharmacol.* (1972) **8**, 30.
145. Shamoo, A. E., Albers, R. W., *Proc. Nat. Acad. Sci. U.S.* (1973) **70**, 1191.
146. Shamoo, A. E., *Ann. N.Y. Acad. Sci.* (1974) **242**, 389.
147. Schatzman, H. J., *Nature* (1962) **196**, 677.
148. Tanaka, R., Abood, L. G., *Arch. Biochem. Biophys.* (1964) **108**, 47.
149. Noguchi, T., Freed, S., *Nature* (1971) **230**, 148.
150. Hokin, L. E., Hexum, T. D., *Arch. Biochem. Biophys.* (1972) **151**, 453.
151. Wheeler, K. P., Whittam, R., *J. Physiol. Lond.* (1970) **207**, 303.
152. Fenster, L. J., Copenhaver, J. H., Jr., *Biochim. Biophys. Acta* (1967) **137**, 406.
153. Specht, S. C., Robinson, J. D., *Arch. Biochem. Biophys.* (1973) **154**, 314.
154. de Pont, J. H. M., Van Prooijen-Van Eeden, A., Bonting, S. L., *Biochim. Biophys. Acta* (1973) **323**, 487.
155. Gruener, N., Avi-Dor, Y., *Biochem. J.* (1966) **100**, 762.
156. Robinson, J. D., *J. Neurochem.* (1967) **14**, 1143.
157. Charnock, J. S., Cook, D. A., Casey, R., *Arch. Biochem. Biophys.* (1971) **147**, 323.
158. Grisham, C. M., Barnett, R. E., *Biochemistry* (1973) **12**, 2635.
159. Tanaka, R., Teruya, A., *Biochim. Biophys. Acta* (1973) **323**, 584.

160. Slack, J. R., Anderton, B. H., Day, W. A., *Biochim. Biophys. Acta* (1973) **323**, 547.
161. Nørby, J. G., Jensen, J., *Biochim. Biophys. Acta* (1971) **233**, 104.
162. Hegyvary, C., Post, R. L., *J. Biol. Chem.* (1971) **246**, 5234.
163. Colowick, S. P., Womack, F. C., *J. Biol. Chem.* (1969) **244**, 774.
164. Hexum, T. D., Samson, F. E., Jr., Himes, R. H., *Biochim. Biophys. Acta* (1970) **212**, 322.
165. Robinson, J. D., *Biochim. Biophys. Acta* (1974) in press.
166. Robinson, J. D., *Biochemistry* (1967) **6**, 3250.
167. Neufeld, A. H., Levy, H. M., *J. Biol. Chem.* (1969) **244**, 6493.
168. Kanazawa, T., Saito, M., Tonomura, Y., *J. Biochem. Tokyo* (1970) **67**, 693.
169. Robinson, J. D., unpublished observations.
170. Epstein, F. H., Whittam, R., *Biochem. J.* (1966) **99**, 232.
171. Atkinson, A., Hunt, S., Lowe, A. G., *Biochim. Biophys. Acta* (1968) **167**, 469.
172. Robinson, J. D., *Biochim. Biophys. Acta* (1972) **266**, 97.
173. Brodsky, W. A., Shamoo, A. E., *Biochim. Biophys. Acta* (1973) **291**, 208.
174. Fahn, S., Koval, G. J., Albers, R. W., *J. Biol. Chem.* (1968) **243**, 1993.
175. Charnock, J. S., Post, R. L., *Nature* (1963) **199**, 910.
176. Post, R. L., Kume, S., Tobin, T., Orcutt, B., Sen, A. K., *J. Gen. Physiol.* (1969) **54**, 306 S.
177. Post, R. L., Kume, S., *J. Biol. Chem.* (1973) **248**, 6993.
178. Nagano, K., Kanazawa, T., Mizuno, N., Tashima, Y., Nakao, T., Nakao, M., *Biochem. Biophys. Res. Commun.* (1965) **19**, 759.
179. Hokin, L. E., Sastry, P. S., Galsworthy, P. R., Yoda, A., *Proc. Nat. Acad. Sci. U.S.* (1965) **54**, 177.
180. Chignell, C. F., Titus, E., *Proc. Nat. Acad. Sci. U.S.* (1966) **56**, 1620.
181. Schoner, W., Kramer, R., Seubert, W., *Biochem. Biophys. Res. Commun.* (1966) **23**, 403.
182. Charnock, J. S., Opit, L. J., Potter, H. A., *Biochem. J.* (1967) **104**, 17 C.
183. Dahms, A. S., Kanazawa, T., Boyer, P. D., *J. Biol. Chem.* (1973) **248**, 6592.
184. Fukushima, Y., Tonomura, Y., *J. Biochem. Tokyo* (1973) **74**, 135.
185. Skou, J. C., *Physiol. Rev.* (1965) **45**, 596.
186. Skou, J. C., Hilberg, C., *Biochim. Biophys. Acta* (1969) **185**, 198.
187. Skou, J. C., *Biochim. Biophys. Acta* (1960) **42**, 6.
188. Fahn, S., Koval, G. J., Albers, R. W., *J. Biol. Chem.* (1966) **241**, 1882.
189. Albers, R. W., *Ann. Rev. Biochem.* (1967) **36**, 727.
190. Blostein, R., *J. Biol. Chem.* (1970) **245**, 270.
191. Judah, J. D., Ahmed, K., McLean, A. E. M., *Biochim. Biophys. Acta* (1962) **65**, 472.
192. Bader, H., Sen, A. K., *Biochim. Biophys. Acta* (1966) **118**, 116.
193. Nagai, K., Izumi, F., Yoshida, H., *J. Biochem. Tokyo* (1966) **59**, 295.
194. Yoshida, H., Nagai, K., Ohashi, T., Nakagawa, Y., *Biochim. Biophys. Acta* (1969) **171**, 178.
195. Askari, A., Koyal, D., *Biochem. Biophys. Res. Commun.* (1968) **32**, 227.
196. Robinson, J. D., *Biochim. Biophys. Acta* (1970) **212**, 509.
197. Robinson, J. D., *Biochem. Biophys. Res. Commun.* (1971) **42**, 880.
198. Robinson, J. D., *Arch. Biochem. Biophys.* (1970) **139**, 164.
199. Israel, Y., Titus, E., *Biochim. Biophys. Acta* (1967) **139**, 450.
200. Schwartz, A., Matsui, H., Laughter, A. H., *Science* (1968) **160**, 323.
201. Lindenmayer, G. E., Schwartz, A., *J. Biol. Chem.* (1973) **248**, 1291.
202. Akera, T., Brody, T. M., *J. Pharmacol. Exp. Ther.* (1971) **176**, 545.
203. Yoda, A., *Mol. Pharmacol.* (1973) **9**, 51.
204. Yoda, A., Yoda, S., Sarrif, A. M., *Mol. Pharmacol.* (1973) **9**, 766.
205. Siegel, G. J., Koval, G. J., Albers, R. W., *J. Biol. Chem.* (1969) **244**, 3264.
206. Allen, J. C., Harris, R. A., Schwartz, A., *Biochem. Biophys. Res. Commun.* (1971) **42**, 366.

207. Van Winkle, W. B., Allen, J. C., Schwartz, A., *Arch. Biochem. Biophys.* (1972) **151**, 85.
208. Tobin, T., Brody, T. M., *Biochem. Pharmacol.* (1972) **21**, 1553.
209. Gardner, J. D., Conlon, T. P., *J. Gen. Physiol.* (1972) **60**, 609.
210. Sachs, J. R., *J. Gen. Physiol.* (1974) **63**, 123.
211. Blostein, R., Whittington, E. S., *J. Biol. Chem.* (1973) **248**, 1772.
212. Robinson, J. D., *Mol. Pharmacol.* (1971) **7**, 238.
213. Skou, J. C., Butler, K. W., Hansen, O., *Biochim. Biophys. Acta* (1971) **241**, 443.
214. Britten, J. S., Blank, M., *Biochim. Biophys. Acta* (1968) **159**, 160.
215. Goldfarb, P. S. G., Rodnight, R., *Biochem. J.* (1970) **120**, 15.
216. Squires, R. F., *Biochem. Biophys. Res. Commun.* (1965) **19**, 27.
217. Albers, R. W., Koval, G. J., *J. Biol. Chem.* (1973) **248**, 777.
218. Lindenmayer, G. E., Schwartz, A., Thompson, H. K., Jr., *J. Physiol. Lond.* (1974) **236**, 1.
219. Robinson, J. D., *Biochim. Biophys. Acta* (1973) **321**, 662.
220. Ahmed, K., Judah, J. D., Scholefield, P. G., *Biochim. Biophys. Acta* (1966) **120**, 351.
221. Tobin, T., Baskin, S. I., Akera, T., Brody, T. M., *Mol. Pharmacol.* (1972) **8**, 256.
222. Ostroy, F., James, T., Noggle, J., Hokin, L. E., *Fed. Proc.* (1972) **31**, 432 abs.
223. Robinson, J. D., *FEBS Lett.* (1974) **38**, 325.
224. Robinson, J. D., *Ann. N.Y. Acad. Sci.* (1975) **243**, 60.
225. Siegel, G. J., Goodwin, B., *J. Biol. Chem.* (1972) **247**, 3630.
226. Post, R. L., Hegyvary, C., Kume, S., *J. Biol. Chem.* (1972) **247**, 6530.
227. Rega, A. F., Garrahan, P. J., Pouchan, M. I., *J. Membr. Biol.* (1970) **3**, 14.
228. Garrahan, P. J., Rega, F., *J. Physiol. Lond.* (1972) **223**, 595.
229. Levitzki, A., Stallcup, W. B., Koshland, D. E., Jr., *Biochemistry* (1971) **10**, 3371.
230. Repke, K. R. H., Schön, R., *Acta Biol. Med. Germ.* (1973) **31**, K19.
231. Repke, K. R. H., Schön, R., Henke, W., Schönfeld, W., Streckenbach, B., Dittrich, F., *Ann. N.Y. Acad. Sci.* (1974) **242**, 203.
232. Whittam, R., Chipperfield, A. R., *Biochim. Biophys. Acta* (1973) **307**, 563.
233. Martonosi, A., *Curr. Top. Membranes Transport* (1972) **3**, 83.
234. Inesi, G., *Ann. Rev. Biophys. Bioeng.* (1972) **1**, 191.
235. Marsh, B. B., *Biochim. Biophys. Acta* (1952) **9**, 247.
236. Bendall, J. R., *Nature* (1952) **170**, 1058.
237. Ebashi, S., Lipmann, F., *J. Cell Biol.* (1962) **14**, 389.
238. Hasselbach, W., Makinose, M., *Biochem. Z.* (1961) **333**, 518.
239. Hasselbach, W., *Progr. Biophys. Mol. Biol.* (1964) **14**, 167.
240. Weber, A., *Curr. Top. Bioenerg.* (1966) **1**, 203.
241. Hasselbach, W., Makinose, M., *Biochem. Z.* (1963) **339**, 94.
242. Makinose, M., Hasselbach, W., *FEBS Lett.* (1971) **12**, 271.
243. Kanazawa, T., Yamada, S., Tonomura, Y., *J. Biochem. Tokyo* (1970) **68**, 593.
244. Carvalho, A. P., Leo, B., *J. Gen. Physiol.* (1967) **50**, 1327.
245. Ostwald, T. J., MacLennan, D. H., *J. Biol. Chem.* (1974) **249**, 974.
246. Worsfold, M., Peter, J. B., *J. Biol. Chem.* (1970) **245**, 5545.
247. de Meis, L., Hasselbach, W., *J. Biol. Chem.* (1971) **246**, 4759.
248. Yamamoto, T., Tonomura, Y., *J. Biochem. Tokyo* (1967) **62**, 558.
249. Kanazawa, T., Boyer, P. D., *J. Biol. Chem.* (1973) **248**, 3163.
250. Meissner, G., Conner, G. E., Fleischer, S., *Biochim. Biophys. Acta* (1973) **298**, 246.
251. Hasselbach, W., Elfvin, L.-G., *J. Ultrastruct. Res.* (1967) **17**, 598.
252. Deamer, D. W., Baskin, R. J., *J. Cell Biol.* (1969) **42**, 296.
253. Ikemoto, N., Sreter, F. A., Nakamura, A., Gergely, J., *J. Ultrastruct. Res.* (1968) **23**, 216.
254. Stewart, P. S., MacLennan, D. H., *J. Biol. Chem.* (1974) **249**, 985.

255. Kagawa, Y., Racker, E., *J. Biol. Chem.* (1966) **241**, 2475.
256. MacLennan, D. H., *J. Biol. Chem.* (1970) **245**, 4508.
257. Pucell, A. G., Martonosi, A., *Arch. Biochem. Biophys.* (1972) **151**, 558.
258. Thorley-Lawson, D. A., Green, N. M., *Eur. J. Biochem.* (1973) **40**, 403.
259. Ikemoto, N., Bhatnagar, G. M., Nagy, B., Gergely, J., *J. Biol. Chem.* (1972) **247**, 7835.
260. Vegh, K., Spiegler, P., Chamberlain, C., Mommaerts, W. F. H. M., *Biochim. Biophys. Acta* (1968) **163**, 266.
261. Martonosi, A., *Biochem. Biophys. Res. Commun.* (1963) **13**, 273.
262. Martonosi, A., Donley, J., Halpin, R. A., *J. Biol. Chem.* (1968) **243**, 61.
263. McFarland, B. H., Inesi, G., *Arch. Biochem. Biophys.* (1971) **145**, 456.
264. Fiehn, W., Hasselbach, W., *Eur. J. Biochem.* (1970) **13**, 510.
265. Meissner, G., Fleischer, S., *Biochim. Biophys. Acta* (1971) **241**, 356.
266. Inesi, G., Millman, M., Eletr, S., *J. Mol. Biol.* (1973) **81**, 483.
267. McConnell, H. M., Wright, K. L., McFarland, B. G., *Biochem. Biophys. Res. Commun.* (1972) **47**, 273.
268. Scandella, C. J., Devaux, P., McConnell, H. M., *Proc. Nat. Acad. Sci. U.S.* (1972) **69**, 2056.
269. Lee, A. G., Birdsall, N. J. M., Metcalfe, J. C., *Biochemistry* (1973) **12**, 1650.
270. Davis, D. G., Inesi, G., *Biochim. Biophys. Acta* (1971) **241**, 1.
271. Racker, E., *J. Biol. Chem.* (1972) **247**, 8198.
272. Meissner, G., Fleischer, S., *Biochem. Biophys. Res. Commun.* (1973) **52**, 913.
273. Meissner, G., Fleischer, S., *J. Biol. Chem.* (1974) **249**, 302.
274. Panet, R., Pick, U., Selinger, Z., *J. Biol. Chem.* (1971) **246**, 7349.
275. Yamada, S., Tonomura, Y., *J. Biochem. Tokyo* (1972) **72**, 417.
276. Inesi, G., Almendares, J., *Arch. Biochem. Biophys.* (1968) **126**, 733.
277. de Meis, L., de Mello, M. C. F., *J. Biol. Chem.* (1973) **248**, 3691.
278. Kanazawa, T., Yamada, S., Yamamoto, T., Tonomura, Y., *J. Biochem. Tokyo* (1971) **70**, 95.
279. Martonosi, A., *J. Biol. Chem.* (1969) **244**, 613.
280. Makinose, M., *Eur. J. Biochem.* (1969) **10**, 74.
281. Bastide, F., Meissner, G., Fleischer, S., Post, R. L., *J. Biol. Chem.* (1973) **248**, 8385.
282. Degani, C., Boyer, P. D., *J. Biol. Chem.* (1973) **248**, 8222.
283. MacLennan, D. H., Seeman, P., Iles, G. H., Yip, C. C., *J. Biol. Chem.* (1971) **246**, 2702.
284. Makinose, M., *Biochem. Z.* (1966) **345**, 80.
285. Yamada, S., Tonomura, Y., *J. Biochem. Tokyo* (1972) **71**, 1101.
286. Makinose, M., *FEBS Lett.* (1972) **25**, 113.
287. Makinose, M., *FEBS Lett.* (1973) **37**, 140.
288. Masuda, H., de Meis, L., *Biochemistry* (1973) **12**, 4581.
289. Martonosi, A., *J. Biol. Chem.* (1969) **244**, 74.
290. Martonosi, A., Lagwinska, E., Oliver, M., *Ann. N.Y. Acad. Sci.* (1974) **227**, 549.
291. Yamada, T., Yamamoto, T., Tonomura, Y., *J. Biochem. Tokyo* (1970) **67**, 789.
292. Ikemoto, N., *J. Biol. Chem.* (1974) **249**, 649.
293. Blondin, G. A., *Biochem. Biophys. Res. Commun.* (1974) **56**, 97.
294. Martonosi, A., Feretos, R., *J. Biol. Chem.* (1964) **239**, 659.
295. Sreter, F. A., *Arch. Biochem. Biophys.* (1969) **134**, 25.
296. Mommaerts, W. F. H. M., *Proc. Nat. Acad. Sci. U.S.* (1967) **58**, 2476.
297. Vanderkooi, J. M., Martonosi, A., *Arch. Biochem. Biophys.* (1971) **144**, 99.
298. Dupont, Y., Harrison, S. C., Hasselbach, W., *Nature* (1973) **244**, 555.
299. Alvarado, F., Crane, R. K., *Biochim. Biophys. Acta* (1962) **56**, 170.

Work supported by U. S. Public Health Service grant NS-05430.

Chapter

4

Chemical Reactions in Oxidative Phosphorylation

D. Rao Sanadi and Hartmut Wohlrab, Department of Cell Physiology, Boston Biomedical Research Institute, 20 Staniford Street, Boston, Mass. 02114

THE PRINCIPAL SOURCE of energy for biological functions in cells is the oxidation of low molecular weight carbon compounds derived from the breakdown of carbohydrates, lipids, and proteins by way of the tricarboxylic acid cycle. The terminal oxidative event in the process is the transport of reducing equivalents (electrons and H^+) *via* a series of carriers that results in the reduction of oxygen to water. Some of the oxidative steps which involve a large change in free energy are coupled to the synthesis of ATP from ADP and inorganic phosphate (P_i) (oxidative phosphorylation) so that the net loss of free energy is minimal. The reservoir of ATP thus generated becomes the main currency for energy exchange in the cell. The oxidative energy can also be used directly in mitochondria without involvment of ATP for energy linked reactions, such as ion transport across membranes, nicotinamide nucleotide transhydrogenation, and reversed electron flow.

ATP production coupled to oxidation/reduction reactions of the respiratory chain and ATP utilization by various energy dependent metabolic reactions are not 100% efficient, and the attendent loss of free energy is responsible for the maintenance of body temperature of the organisms. These processes are part of a network of chemical reactions which are exquisitely regulated to maintain homeostasis.

The forerunner to the discovery of phosphorylation coupled to oxygen reduction was the observation of Englehardt [1], who described esterification of P_i during respiration of lysed erythrocytes. An excellent historical treatment has been presented by Kalckar [2]. The significance of the observations was realized when Kalckar [3] showed that phosphorylation of glucose, glycerol, or AMP occurred during oxidation of citrate and other substrates in kidney homogenates. The phosphorylation depended on the presence of oxygen, which excluded participation of glycolysis. Soon after, Belitzer and Tsibakowa [4] and Ochoa [5] reported that two to three molecules of P_i were esterified per oxygen atom consumed (or P:O values of 2 to 3) and postulated that phosphorylation occurred not only during dehydrogenation of the substrate (substrate-level phosphorylation) but also during transport of the hydrogen atoms (or electrons) to oxygen. The isolation of mitochondria by Hogeboom, Schneider and Palade (6) permitted the next important development—*viz.*, the unequivocal demonstration that phosphorylation was coupled to the oxidation of NADH by the respiratory chain (7). These observations have led to the current phase of research centered on the mechanism of the reactions. Recent reviews (*8, 9, 10, 11, 12, 13, 14*) should be consulted for more complete discussion.

Mitochondria are intracellular organelles with double membranes (Figure 1). The inner membrane has several in-foldings (cristae) which

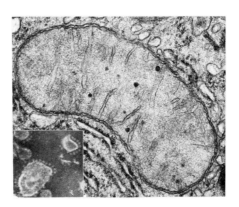

Figure 1. A mitochondrion in a rat liver parenchymal cell as seen in thin section, ×37,500. Inset shows submitochondrial particles prepared by sonic disruption of bovine heart mitochondria. The preparation was negatively stained with phosphotungstate ×152,000. The vesicle is surrounded by 90 A spheres which appear to be attached to the membrane by slender rod-like structures (courtesy of J. Hall).

are densely packed in mitochondria from tissues with high respiratory activity (*e.g.*, myocardium, insect flight muscle, and avian salt gland) and more loosely packed in other tissues (*e.g.*, liver). Methods for separating the two membranes and their properties have been reviewed recently (*15*). Outer membrane markers are monoamine oxidase and cytochrome b_5 reductase, which is insensitive to antimycin A, and the inner membrane contains the respiratory chain and related activities. The matrix space confined by the inner membrane has several of the soluble dehydrogenases which are released by rupture of the mitochondria. The outer membrane is permeable to low molecular weight solutes (*e.g.*, sucrose) and ions (*e.g.*, adenine nucleotides) (*16*), but the inner membrane has specific transport systems. The matrix space varies in volume with its ion concentration, and as a result, the inner membrane occurs in various configurations described as condensed, orthodox, etc. (*17, 18, 19*).

When mitochondria are disrupted by exposure to sonic oscillations, vesicular submitochondrial particles of approximately 1000-A diameter are formed. Ernster and Lee (*20*) have demonstrated that the polarity of the membrane in these particles is reversed with respect to intact mitochondria—*i.e.*, the matrix side faces outwards. As a result, the 90-A spheres attached to the inner membrane and normally projecting into the matrix space (or on the M side) are on the outside of submitochondrial particles in contact with the medium.

Respiratory Chain

The respiratory chain is a group of membrane-associated redox centers that are directly responsible for catalyzing the oxidation of various metabolites by molecular oxygen (Figure 2). Its primary purpose is not just the oxidation of the metabolites and the reduction of oxygen but also the transduction of the released free energy into another form such as ATP, ion gradients, nicotinamide nucleotide transhydrogenation, or configurational changes of the inner mitochondrial membrane. These energy linked reactions can be uncoupled from the electron transport reactions either by mechanical damage or uncoupling agents. Many studies on the redox carriers have been carried out on such uncoupled systems. In the last few years, much effort has been directed at identifying the energy transducing components of the respiratory chain.

Identification of Membrane Associated Redox Centers. OPTICAL ABSORPTION SPECTRA. It is practically impossible to determine the absolute optical absorption spectrum of a mitochondrial suspension since the apparent optical density of the suspension from light scattering (which increases dramatically at shorter wavelengths) is much larger than that of the redox centers. However, when the absolute spectrum of a reduced suspension is subtracted from the absolute spectrum of an oxidized sample, the high apparent optical density from light scattering is eliminated, and one obtains a difference spectrum. This subtraction process is done automatically in a double beam (one beam for each cuvette) spectrophotometer. Only those double beam spectrophotometers that have their photomultipliers placed very close and equidistant to both sample cuvettes are optimally suitable for difference spectra.

These principles and others have been utilized by Britton Chance and his co-workers in designing and constructing the first double beam spectrophotometer (*21, 22, 23*). Their results provide the groundwork for all optical analyses of mitochondria.

In the visible light region, mitochondria show typical absorption bands (Figure 3). The absorption from the heme proteins (cytochromes) occurs in three regions of the spectrum: α, β, and γ. The α and β bands disappear almost completely, while the γ bands shift to shorter wavelengths upon oxi-

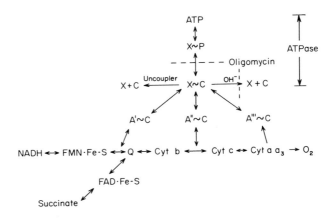

Figure 2. A scheme for electron transport and oxidative phosphorylation according to the chemical hypothesis

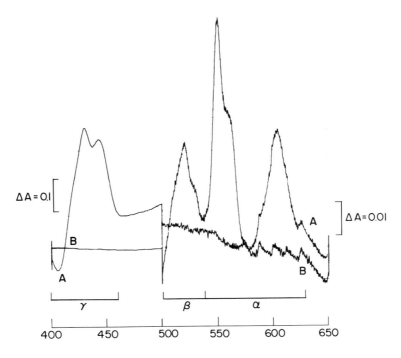

Figure 3. Reduced minus oxidized optical absorption spectrum of a suspension of partly solubilized mitochondria from the flight muscle of the blowfly Sarcophaga bullata. *Sodium dithionite was used as a reductant. At wavelengths below 500 nm the absorbance scale is 0.1, while above 500 nm it is 0.01. The spectrum (A) is superposed on the baseline (13).*

dation of the cytochromes. The relative optical extinctions are smallest in the β region ($E_{mM} \sim 10$, 500–540 nm), larger in the α-region ($E_{mM} \sim 20$, 540–630 nm), and are largest in the γ region ($E_{mM} \sim 80$, 400–460 nm). Each reduced cytochrome has an absorption band at room temperature in each region. When the temperature of the mitochondrial sample is lowered to that of liquid nitrogen (77°K), the absorption bands increase in intensity, become narrower in half-band width, and the α and β absorption bands of some cytochromes split into closely spaced bands. The main absorption peaks shift a few nanometers to the blue (shorter wavelength). At liquid helium temperature (4°K), little additional resolution is obtained. Liquid nitrogen temperatures are frequently utilized to differentiate better between absorption spectra of the cytochromes and to trap mitochondria—and thus, their redox state—in defined metabolic states. (For a summary of specific optical absorption data, see Ref. 24).

Optical absorptions are also shown by the dehydrogenases (flavoproteins) of the respiratory chain. The flavoproteins show maximal absorptions around 450 nm only in the oxidized state. Their absorption bands are very broad compared with the relatively narrow ones of the cytochromes and have much smaller extinction coefficients ($E_{mM} \sim 11$). Non-heme iron proteins also absorb in this region of the optical spectrum. Their optical density changes upon oxidation and reduction are, however, still smaller than those of the flavoproteins ($E_{mM} \sim 5$, near 440 nm). The changes are in the same relative direction as the cytochromes. Copper ions of the mitochondrial membrane (cytochrome c oxidase) absorb light at 830 nm in the oxidized state ($E_{mM} \sim 1.4$) (25), while ubiquinone (coenzyme Q) has a millimolar extinction coefficient (289–280 nm) of 8.8 (26).

These are the primary optical absorption bands that are readily observed. More careful analysis permits one to detect interactions between cytochromes, cytochromes and copper ions, cytochromes and inhibitors of electron transport or energy transduction (27), and the high and low energy forms of cytochromes (28).

ELECTRON PARAMAGNETIC RESONANCES. Electron paramagnetic resonance (EPR) has been introduced to the study of mitochondrial redox centers (Figure 4) (29) in several ways: the identification and characterization of iron–sulfur centers (previously referred to as non-heme iron), cytochromes, flavoproteins, copper ions, and membrane components modified by attachment of spin labels. The pioneering studies on redox centers have all been conducted at liquid nitrogen temperatures; however, recently the introduction of studies near liquid helium temperatures has contributed enormously to the discovery and the characterization of many new iron–sulfur centers of the respiratory chain (30).

The kinetic compatibility of EPR changes with electron transport has yet to be rigorously established. At liquid nitrogen temperatures, essentially only signals at $g = 1.94$ and 2.02 of iron–sulfur centers can be detected. At or near liquid helium temperatures, signals with g values from 2.103 to 1.863 were observed (30). Cupric ions show g values of about 2.00 (31). Cytochromes show g values of 3.33 for c_1 (29), 3.44 for b_k (29), 3.78 for

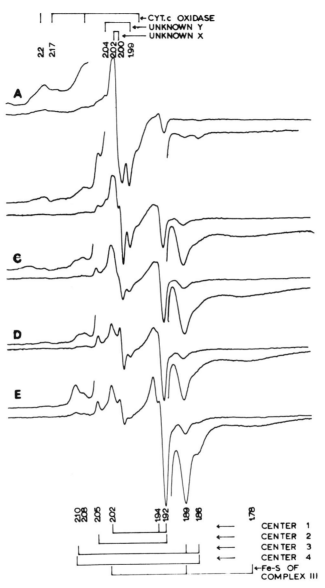

Figure 4. EPR spectra (2900 to 3900 G at 9.2 GH_z) of samples at 13°K from a reductive titration of submitochondrial particles (ETP_H) with NADH; g-values of the absorptions of the redox species are indicated. Centers 1 through 4 refer to non-heme iron species in the oxidative phosphorylation site 1 region of the respiratory chain (29).

b_T (*29*), 3.00 for *c*, 3 and 6 for *a* and 3 and 6 for a_3 (*32*). Changes from high to low spin ferric iron of the cytochrome *a* heme can be seen by changes in *g* value from 6 to 3 (*33, 34*).

CHEMICAL DETERMINATIONS. In most cases, chemical determinations have preceded the EPR studies. Iron–sulfur centers were discovered as part of various respiratory chain complexes and have been proposed as participants in electron transport. Much larger amounts of non-heme iron are found by chemical analysis than by EPR. However, the EPR signals have yet to be fully analyzed quantitatively. A similar problem exists with the copper ions of the terminal portion of the respiratory chain (cytochrome oxidase), where more copper ions are seen by chemical methods than can be detected by EPR (*33*).

Hemes of cytochromes have been determined chemically. The hemes of the *b*-type cytochromes are non-covalently bound to the lipoprotein, while the *c*-type cytochromes have their hemes bound covalently to the (lipo-) proteins (*24*).

Chemical methods have been used to determine rates of oxidation of iron–sulfur centers and copper ions (*31*). Rapid freezing is inherently superior for EPR measurements, but it is difficult at the high concentration of membranes required. The resulting high viscosity reduces the maximum mixing rate.

OXYGEN TITRATIONS. The terminal portion of the respiratory chain has been studied to determine whether those redox centers that have been discovered account for all the redox centers in a given region of the respiratory chain. The method involves titration of a redox center with O_2, together with a determination of the amount of O_2 required to oxidize a cytochrome of the terminal region of the respiratory chain at maximal rate. This titration gives values consistent with the number of redox centers already known and has established their kinetic competence within the respiratory chain. The results demonstrate that seven redox centers on the oxygen side of the antimycin A inhibition site are kinetically compatible with electron transport (*35*).

Sequence of Redox Centers. OXIDATION KINETICS. The redox centers that have been identified have been characterized extensively on a kinetic basis. A necessary but not sufficient, requirement for a redox carrier to be a member of the respiratory chain is that it can be oxidized at a sufficiently rapid rate to be compatible with the maximum respiration rate. Since the carrier could be on a side path in rapid equilibrium with a member of the respiratory chain, this is not a sufficient condition.

Oxidation kinetics have permitted the arrangement of the respiratory chain components as shown in Figure 5 (*36*). Oxidation rates of Cu ions show their kinetic competence in isolated membrane fragments in comparison to overall respiration rates (*31*). The kinetic studies also permitted the placement of b_{566} (b_T) on the oxygen side of b_{562} (b_k) (*37*). These kinetic studies also associated the α, β, and γ absorption bands with their respective redox centers (cytochrome) (*37*), *i.e.*, those absorption bands with the

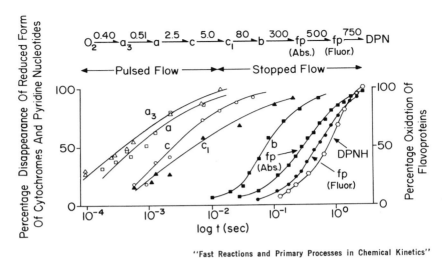

"Fast Reactions and Primary Processes in Chemical Kinetics"
Figure 5. Kinetics of oxidation of redox centers of the mitochondrial respiratory chain during the anaerobic–aerobic transition (36)

same rates of appearance upon oxidation or reduction belong to the same redox center.

REDUCTION KINETICS. Reduction kinetics have recently been utilized to establish a sequence of the iron–sulfur centers located on the substrate side of ubiquinone. Orme-Johnson and co-workers (*29, 38*) titrated the oxidized mitochondrial membrane fragments with dithionite (a very strong reductant) and established a likely sequence. The most recent formulation of the iron–sulfur centers and their sequence in the respiratory chain is that reported by Ohnishi et al. (*39*).

REDOX-POTENTIAL MEASUREMENTS. The redox midpoint potentials of the electron carriers were determined several decades ago on purified complexes of the respiratory chain. Experiments with intact mitochondria under conditions that overcame the problem of spectrophotometric interference by the redox dyes were reported only recently and are shown in Figure 6. Redox dyes absorb light in the visible region of the spectrum. Thus, to identify the redox state of a cytochrome during a redox titration, special care must be exercised to eliminate any possible interference from the dyes used as redox buffers. Other precautions that have to be strictly observed are (1) that the suspension is anaerobic—*i.e.*, no electron transport to O_2 is occurring, and (2) that the system is as close to equilibrium as possible (*40*). One way of determining the first condition is by using the fluorescent bacterium *photobacterium phosphoreum*, *i.e.*, the bacteria fluoresce in the presence of very low concentrations of oxygen, (*41*). The second condition can largely be satisfied by varying the redox dye concentrations and by carrying out a reductive as well as an oxidative titration. Even under these conditions the interpretation of the redox results have been challenged since the redox centers are associated with a semipermeable membrane which actively transports ions (and redox dyes) under energized conditions.

A major contribution of the redox titrations lies in the first clear demonstration of two cytochrome b species in the mitochondrial membrane and the apparent localization of the phosphorylation sites (42, 43). This specific location of the energy transduction sites provided the primary

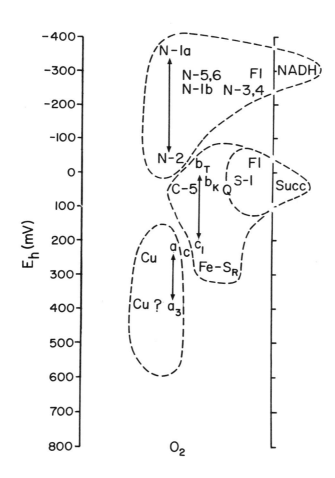

Annual Reviews of Biophysics and Bioengineering

Figure 6. A diagramatic presentation of mitochondrial redox centers on a half reduction potential scale. The dashed lines enclose redox centers that have been associated with the complexes of the dissociated respiratory chain. N-1a to N-6 refer to the non-heme iron centers of complex I (different nomenclature from center 1 to 4 of Figure 4); S-1 to the non-heme irons of complex II, C-5 and $Fe-S_R$ are non-heme iron centers of the complex III region (57).

impetus for the isolation and reconstitution of the first and third phosphorylation sites (*44, 45*).

Controversies have arisen about the actual number of b-cytochromes in the respiratory chain. Slater and co-workers suggest three or more species (*46*), and Wikstrom and co-workers suggest three (*47*). Dutton and co-workers (*48*) observed three different b-type cytochromes in submitochondrial particles. Davis and co-workers (*49*) suggest three from purification studies. One is thought to be associated with complex II and two with complex III. The different b's have been characterized on the basis of oxidation rates (*37*), reduction rates (*50*), low temperature absorption spectra (*51*), association studies with antimycin A and other inhibitors, and the above mentioned purification studies (*49*).

INHIBITORS OF ELECTRON TRANSPORT. It is often difficult to identify a compound that uniquely inhibits electron transport without affecting energy transduction. Essentially all respiratory chain inhibitors act in the cytochrome a–a_3 region (oxidative phosphorylation site III), the cytochrome b–c_1 region (site II), and the iron–sulfur center 2-center 5 region (site I) (*see* also Figure 4). Thus cyanide, azide, carbon monoxide, and hydroxylamine act at or near site III; antimycin A, HOQNO, 2,3-propanediol, hydrolapachol act at or near site II; and rotenone and piericidin A act at or near site I. Many of these inhibitors can be shown to have secondary effects; however, their inhibition at certain concentrations is uncoupler insensitive—*i.e.*, the inhibition is not relieved by uncouplers.

If the inhibitor traps a redox center in a high (or low) energy form and prevents the uncoupler from discharging the high energy form, electron transport is blocked. At this point, the difference, if any, between an energy-transduction inhibitor at the level of the redox centers and an inhibitor specific to transport only is unclear. In fact, there exists no molecular model for any electron transport inhibitor (except for CO which acts competitively with O_2 as a ligand of the iron of cytochrome a_3).

Inhibitors are very useful in supplementing the information obtained by other methods on the sequence of redox centers in the respiratory chain. Thus, when antimycin A blocks the respiratory chain, the b cytochromes are reduced, while the c and a cytochromes are oxidized during steady state respiration.

EXTRACTION AND REPLACEMENT STUDIES. The ideal way of identifying and characterizing the sequential steps of the respiratory chain is to remove or extract one component, without disturbing the others, to determine the resulting crossover, and to replace the component to show complete reconstitution of the activity. This process, however, is extremely difficult in practice since nearly all redox centers are an integral part of the inner mitochondrial membrane, *i.e.*, they are lipoproteins. Cytochrome c is the only exception. It can be obtained as a positively charged protein from mitochondria by exposing the intact mitochondria first to a hypotonic medium and then washing the mitochondria with a salt solution (0.15M KCl). Cytochrome c has been extracted and replaced, the crossover identified, and reconstitution studies carried out (52). Actually the reconstitution has only been partially successful, since the efficiency of oxidative phosphorylation, as indicated by the respiratory control ratio, has not been quite completely restored.

Ubiquinone (Q) is a lipid soluble redox center that can be readily extracted from the mitochondrial membrane and, under appropriate conditions, reactivates electron transport reactions between the dehydrogenases and the b-type cytochromes (53). Energy transduction has not been reconstituted in this system. In fact, controversies still exist whether ubiquinone is a necessary redox intermediate between b-type cytochromes or between dehydrogenases and b-type cyctochromes (49, 54, 55). Also, Q may function as several pools of active redox centers. This latter hypothesis is not supported by Kröger and Klingenberg (55), who have presented elegant results demonstrating that the membrane-associated ubiquinone is one pool of reducing equivalents that is kinetically competent in the respiratory chain.

The respiratory chain contains several complexes: I, II, III, and IV (56, 57, see also Figure 6), and also complex V (58). Complexes I to IV are groups of oxidation–reduction components from various sections of the respiratory chain. Complex V is composed of coupling factors and hydrophobic proteins and is probably the same as the oligomycin sensitive ATPase (see below). These studies again support the sequence of redox carriers established by kinetic, redox, and inhibitor studies. The complexes have been studied in a detailed way, yet it is not clear to what extent they represent the true picture of the situation in the membrane. The original observations were that complex I can oxidize NADH and reduce ubiquinone, complex II can oxidize succinate and reduce Q, complex III can oxidize ubiquinone and reduce cytochrome c, and complex IV can oxidize cytochrome c and reduce molecular oxygen.

Complex IV, cytochrome c oxidase, has been extensively characterized, but the molecular basis for the action of inhibitors (except of CO) has not been established. Whether cytochromes a and a_3 are identical is not clear, and the role of the two Cu ions has not been established, except that they are kinetically competent and have a redox midpoint potential in the range of those of the a hemes.

Complex III may consist of two b-type cytochromes (49), non-heme iron (59), and cytochrome c_1. Peculiar kinetic properties have been observed and related to the energy transduction phenomena (60, 61). These are discussed below with oxidative phosphorylation.

Complex II has been the subject of much controversy. It is made up of succinate dehydrogenase and a b-type cytochrome (49). The number of non-heme irons that are required for enzymatic activity has not been rigorously established. Another preparation of succinate dehydrogenase that is reconstitutively active has eight non-heme iron centers per flavin (62, 63); other preparations have fewer. Oxalacetate may exert a regulatory function on the succinic dehydrogenase (64). The redox state of ubiquinone may also regulate the activity of succinate dehydrogenase (65).

A low molecular weight NADH-dehydrogenase that can oxidize NADH and reduce an artificial redox dye or quinones has been isolated. A more intact system (complex I) is one whose oxidation of NADH can be inhibited by rotenone or piericidin A. The most functionally intact system yet resolved from the respiratory chain is one that is able to oxidize NADH and is capable of coupling this oxidation to the phosphorylation of ADP (44). All systems that are coupled to phosphorylation are also sensitive to the inhibitors. In fact, site I phosphorylation has never been demonstrated in a system that lacks rotenone or piericidin A sensitivity (30).

Organization of the Respiratory Chain. IN-MEMBRANE-PLANE ORGANIZATION. Very little is known about the physical spacing of redox centers. Density figures, *i.e.*, how many cytochromes of a certain type exist per unit membrane, have been estimated by Kröger and Klingenberg (66). Their studies determined the stoichiometries between dehydrogenases and cytochromes: about one NADH-dehydrogenase per 10 cytochrome c oxidases. A similar figure holds for the other dehydrogenases. Ubiquinone, on the other hand, exists in 10-fold excess over the cytochrome c oxidase. This suggested that ubiquinone permits the rapid equilibration of reducing equivalents between a dehydrogenase and the many respiratory chains (55). A similar role has been ascribed to cytochrome c (67).

Thus, the respiratory chains are not separate entities or complexes but are able to communicate rapidly one with another, as represented in Figure 7, (66, 67). It has not been established over how many respiratory chains the ubiquinone and cytochrome c pools can equilibrate reducing equivalents nor how the equilibration between cytochrome c molecules occurs, *i.e.*, whether the cytochrome c molecules move on the membrane or whether they are stationary and equilibrate among themselves.

Evidence has been presented on the proximity of the a-hemes of cytochrome oxidase. The a-hemes of neighboring cytochrome oxidases do not appear to be close enough to interact (68). Within a single cytochrome c oxidase heme–heme interaction (69) and/or heme–copper interactions (70) may occur. All other redox centers in the respiratory chain lie far enough apart that they do not significantly perturb each other. A possible exception may be the cytochrome c_1–cytochrome b_T interaction. Cytochrome b_T can be fully reduced only when cytochrome c_1 is oxidized (60, 61). The question of the structure and function relationships in a redox center arises: what role does the lipoprotein play in the redox reactions of the prosthetic group?

Molecular models on the role of the protein in cytochrome redox reactions have been postulated only for cytochrome c (71). Aromatic amino acids appear to play a primary role in the protein–heme interactions.

TRANS-MEMBRANE-PLANE ORGANIZATION. Mitchell and his co-workers (54) attribute much significance to the asymmetric, trans-membrane-plane organization of the respiratory chain. Extensive experimental evidence supports the basic concept of asymmetry in the organization of the redox centers in the inner mitochondrial membrane (66, 72, 73). These results are summarized in Figure 8. Whether it is a necessary configuration for

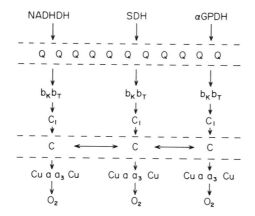

Figure 7. Scheme for in-membrane-electron flow based on (66, 67)

mitochondrial oxidative phosphorylation has not been established, but reconstitution experiments indicate that this may be the case.

Two experimental systems have been utilized to establish the sidedness of the mitochondrial membrane: (1) the intact mitochondria, with or without the outer membrane, and (2) sonic submitochondrial particles (SMP), electron transport particles (ETP), or electron transport particles from heavy layer beef heart mitochondria (ETPH) that are able to transduce energy.

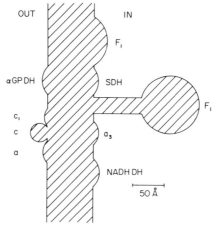

Figure 8. Trans-membrane-plane positions of some of the enzymes and electron carriers of the mitochondrial inner membrane. (The position of F_1 (whether in the membrane or extended away from the membrane) has not yet been definitely established.)

Dominant among experiments supporting the idea that SMP have the matrix side of the inner membrane facing the medium are: (1) SMP take up protons during energy linked electron transport, instead of transporting them out like the mitochondria (74), (2) antibodies against F_1 inhibit ATPase of SMP but not of mitochondria (75), (3) ATP driven reactions of SMP are insensitive to atracyloside (inhibitor of adenine nucleotide translocase) while those driven by extramitochondrial ATP are sensitive, and (4) cytochrome c can be extracted from mitochondria but not from SMP.

Antibodies have been used to identify the location of cytochrome oxidase (75): cytochrome a_3 is on the M side (matrix side or inner side of the inner mitochondrial membrane of intact mitochondria) and cytochrome a on the C side (cytochrome c side or outside of inner mitochondrial membrane of intact mitochondria) of inner mitochondrial membrane. The interpretations of these results have been challenged since the antibody–antigen complex may alter the distribution of enzymes within the membrane (76). Studies on the inhibition of cytochrome oxidase by azide suggest that a_3 is on the M side (77), but this interpretation has been challenged (78). Extraction of cytochrome c with high salt and its replacement can be carried out without loss of matrix proteins or mitochondrial NAD–NADH, which is impermeable to the intact mitochondrial membrane (52). This places cytochrome c on the C side. SMP are able to oxidize exogenous c, but in an uncoupled reaction. When c is incorporated into SMP (by sonicating mitochondria in the presence of excess cytochrome c and washing the resulting SMP with high salt to remove cytochrome c bound to the outside), it can be oxidized in an energy conserving manner (73).

Succinate dehydrogenase (SDH) has been purified and recombined with SDH-depleted sonic mitochondrial particles (63). The resulting succinate oxidation is able to transduce energy. This experiment suggests that SDH is rebound, as in intact mitochondria, to the M side of the inner mitochondrial membrane.

Various dyes have been utilized to establish a difference between the two sides of the inner mitochondrial membrane. Thus, ANS fluorescence is quenched upon energization of the inner mitochondrial membrane (C side out) by respiration or ATP, while it is enhanced when SMP (M side out) are energized (79). A difference in auramine O binding has been used to indicate a difference in polarity between the two sides of the inner mitochondrial membrane (80). Electrophoresis is able to distinguish mitochondria from SMP by their differing mobilities (81).

Other methods that have been used successfully to establish sidedness of the membrane, utilize redox dyes (or artificial electron acceptors) and membrane–impermeable protein labels. Chance and co-workers suggested, from ferricyanide reduction kinetics, that ferricyanide is only accessible to a fraction of the total redox sites of sonic mitochondrial particles—*i.e.*, that normal SMP are a mixture of M side in and M side out membrane fragments (82). Ferricyanide reduction by succinate in intact mitochondria is sensitive to antimycin A, while the same reduction by SMP is not sensitive to antimycin A. This implies that the antimycin A-insensitive ferricyanide reduction site in SMP is not accessible to ferricyanide in intact mitochondria because of the inner membrane barrier.

Schneider and co-workers (83) used ^{35}S-diazobenzenesulfonate to characterize further the sidedness of the inner mitochondrial membrane. This probe does not permeate membranes, yet it binds covalently to proteins exposed to the medium. The method had originally been used to label surface proteins of the erythrocyte membrane (84, 85, 86). Schneider and co-workers (83) showed that the mitochondrial ATPase (F_1) was not labeled by intact mitochondria, while cytochrome c was labeled. The reverse labeling pattern occurred in submitochondrial particles. Carroll and Eytan (87) extended the ^{35}S-diazobenzenesulfonate labeling method to the subunits of cytochrome oxidase and concluded that of the six subunits attributed to cytochrome oxidase (molecular weights 39,000(I), 21,500(II), 15,500(III), 11,500(IV), 9,500(V), 7,500(VI)), II, V, and VI were primarily exposed on the C side, while III was primarily exposed on the M side.

Oxidative Phosphorylation

Energy Transduction Sites. ATP:O RATIOS. When metabolites are oxidized by molecular oxygen *via* the respiratory chain, ADP is phosphorylated. The number of ATP molecules synthesized per pair of electrons which flow down the chain from substrate to oxygen is defined as the P/O or ATP/O or ATP/$2e^-$ ratio. ATP:O ratios approaching whole integers have been found for the NADH oxidation (ATP:O of 3), succinate oxidation (ATP:O of 2) and α-glycerolphosphate oxidation (ATP:O of 2). Metabolites such as glutamate–malate must be oxidized *via* NADH, and their oxidation yields an ATP:O ratio of 3. Extramitochondrial NADH can be oxidized by yeast and plant mitochondria in a rotenone-insensitive manner (*i.e.*, by-passing site I) and thus gives an ATP:O ratio of only 2. ATP:O ratios can most easily be determined from the amount of oxygen consumed during the time required to phosphorylate a given amount of ADP. During the time of ADP phosphorylation, the respiration is faster, perhaps by as much as ten times (state 3), than in the absence of ADP (state 4) (88). The relative efficiency of energy transduction of a mitochondrial preparation is indicated by the ratio of state 3:state 4, which is also called respiratory control ratio (RCR) (*see* Figure 9). The higher the RCR, the higher is the energy transduction efficiency of the mitochondrial preparation.

The first basic question was: are specific sites of the respiratory chain involved in these phosphorylation reactions; *i.e.*, two sites for succinate oxidation and three sites for NADH oxidation. Crossover studies suggested specific energy transduction sites (88).

To understand the crossover principle, site II will be analyzed. During state 4, the *b* cytochromes are highly reduced while the *c* and *a* cytochromes are highly oxidized. When ADP is added (in the presence of phosphate), the *b* cytochromes become much more oxidized, while the *c* and *a* cytochromes become more reduced. The *b–c* region is thus implicated in a phosphorylation site since the absence of ADP prevents e^- flow through this

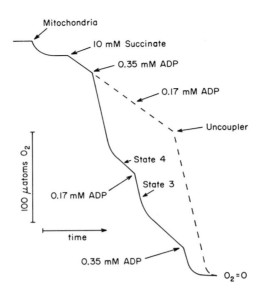

Figure 9. Rates of succinate oxidation by coupled mitochondria. Oligomycin is present in the experiment with the dashed line, i.e., ADP has no effect on the respiration rate. At $O_2 = 0$, the medium has reached anaerobiosis.

section of the chain. Similar crossovers were found in the cytochrome c-oxidase region (site III) and the flavoprotein–ubiquinone region (site I).

Competition between NADH oxidation and succinate oxidation suggests that sites II and III are shared by NADH and succinate oxidation, *i.e.*, there are no separate chains for succinate oxidation and NADH oxidation (*89*). Each cytochrome-type is reduced completely and as a homogeneous pool by NADH or succinate, again suggesting the same cytochrome chains function for both NADH and succinate oxidation. Jacobs discovered that TMPD with ascorbate as reductant, in the presence of antimycin A yields an ATP:O ratio approaching one (*90*). This result identifies concretely the region of site III energy transduction. Other artificial redox mediators have been utilized since those pioneering studies to demonstrate an ATP:O ratio approaching one for the fumarate (*91*) or UQ_1 (*92*) reduction by NADH and for the ferricyanide reduction by succinate (*93*).

All these studies, however, only indicated regions of energy transduction. Even in those cases that established a crossover between two of the cytochromes, it was still to be resolved which of the two cytochromes is the primary energy transducing cytochrome.

MIDPOINT REDOX POTENTIALS AND PHOSPHATE POTENTIALS. All the above studies have utilized respiration for characterizing energy-transduction phenomena. However, the analysis can also be carried out by utilizing

ATP-driven reactions. The reversibility of electron transport and other energy utilizing phenomena make this type of analysis possible.

The reversal of electron transport was first established by the observation that intramitochondrial NAD^+ can be reduced in the presence of succinate and ATP or TMPD, ascorbate, and ATP in cytochrome oxidase inhibited systems (94). These experiments on reversed electron transport are thoroughly discussed below.

Wilson and co-workers observed that the midpoint redox potential of cytochrome a_3 shifts by about 230 mV when ATP is added to a well-coupled mitochondrial suspension (42), as shown in Figure 10 (95). An attempt was made to find a redox center similarly affected by energy in-put at site II. Cytochrome b appeared a likely candidate, but it had two midpoint redox potentials (b_K and b_T), one of which (b_T) underwent an ATP dependent midpoint redox potential shift (43). (The subscript T indicates an energy transducing component of the chain.) More recently, implication of iron–sulfur centers $1a$ and 2 has been indicated for energy transduction site I by correlating an EPR spectral change with redox potential changes determined before the sample was frozen (see Ref. 30).

Much controversy still exists as to the significance of the ATP-dependent midpoint redox potentials. Heme protein studies as model systems for cytochromes would suggest that large changes in the optical absorption spectra might be expected. An energy state-dependent spectral absorption change has in fact been reported for cytochrome a_3 that would suggest a different cytochrome state than the classical oxidized and reduced states (28). The observed changes are small and no such changes have been observed with cytochrome b_T.

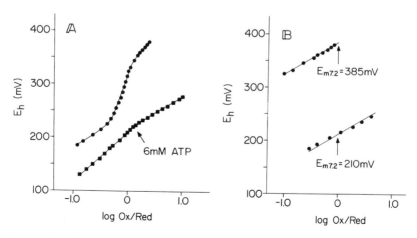

Figure 10. Typical plot of the redox potential of the suspending medium vs. the log of the ratio of oxidized to reduced cytochrome oxidase as determined spectrophotometrically. Graph A shows the plots plus and minus ATP. Graph B shows a mathematical resolution of the minus ATP trace into cytochrome a and a_3 (95).

Attempts have been made to detect energy dependent conformational changes of the cytochromes to correlate definite changes in the cytochromes with changes in midpoint redox potential. Wohlrab and Ogunmola (68) found that the affinity of cytochrome a_3 for CO was decreased in the presence of ATP. This is significant since CO is an electrically neutral molecule whose distribution across the inner mitochondrial membrane should not be influenced by potential gradients across the inner mitochondrial membrane. The observed change in affinity, however, was small. Storey and Lee have measured conformational changes by CD spectra and were unable to detect a high energy conformational form of cytochrome b_T (96).

Dutton and Lindsay (97) have found that the energy linked cytochrome a_3 and b_T changes caused by ATP are actually measured under equilibrium conditions, since the change was independent of redox buffer concentrations and direction of redox titration (i.e., reductive and oxidative titrations give the same results). However, the ATP sets up reversed electron transport, and this phenomenon may, in fact, be responsible for the apparent change in midpoint redox potential.

Lambowitz and co-workers (98) concluded from redox titrations of plant mitochondria that the midpoint redox potential shift of b_T (b-566) produced by ATP results from insufficient equilibration of the redox dyes with the b_T. The reduction of b_T by reducing equivalents from cytochrome c or c_1, which requires ATP, is short-circuited by PMS. PMS oxidizes b_T and reduces c_1 or c. A cyclical electron transport system is set up which hydrolyzes ATP. This ATP hydrolysis is antimycin A-sensitive since antimycin A blocks the electron transport from c_1 to b_T. In animal mitochondria, the rate of b_T reduction by ATP is much faster than the rate of PMS reduction by b_T. Lambowitz and co-workers (98) suggest therefore that the lack of rapid equilibration of all redox centers with each other (b_T, c_1, c, PMS, etc.) which is required for redox potential determinations, is responsible for the apparent ATP dependent midpoint potential shift of cytochrome b_T. Wikstrom and Lambowitz (99) have extended these studies to rat liver mitochondria which had been used in the original observations by Wilson and co-workers (42, 43).

The ATP-dependent midpoint potential shift has also been challenged since the redox titrations may have been incorrectly related to the absorption bands (100). Wikstrom reinterpreted the original potentiometric results of Wilson and Dutton (42) and suggested that, in the correlation of spectral changes in the redox potential, cytochrome a could change by -60 mV and a_3 by -110 mV, instead of the original interpretation of a_3 by -220 mV and a by $+50$ mV (57). Larger shifts in midpoint redox potentials are required to be significant for oxidative phosphorylation. Possibly the third phosphorylation site does not occur between c and a_3 but between a_3 and oxygen (101). Phosphorylation does occur in the presence of oxygen and one does not know what the ligand on a_3 is during phosphorylation. All the midpoint redox potential measurements are carried out under anaerobic conditions with a water ligand (or no ligand) on a_3 instead of an oxygen or some intermediate in the oxygen reduction.

Much effort has been spent correlating the phosphate potential with the midpoint redox potential of b_T and a_3 (*102*). The potentials of the cytochromes of the respiratory chain in state 3 and state 4 correlate well with the predicted thermodynamic behavior (*57*). Various theoretical schemes involving the postulated high energy forms of the cytochromes are well discussed in several publications (*103, 104, 105*).

Inhibitors of Energy Transduction. There are several types of inhibitors of the energy transduction process (1) uncouplers, (2) energy transfer inhibitors, and (3) metabolite transport inhibitors. Their action and their significance in elucidating mitochondrial energy transduction will be discussed in reference to Figure 2.

UNCOUPLERS. Uncouplers are substances that permit maximum rates of electron transport to occur irrespective of the energy state of the mitochondria. In fact, those substances that inhibit respiration in the presence of uncouplers can properly—in the classical sense—be called electron transport inhibitors.

Much experimental evidence suggests that uncouplers act as proton conductors across the inner mitochondrial membrane. The more thorough experiments on black lipid membranes attempting to correlate the efficiency of uncouplers in inducing proton-permeability in black lipid membranes with uncoupling of mitochondrial respiration were unsuccessful (*106*). They were, however, successful when liposomes (lipid vesicles) were utilized (*107*).

Experiments by Wilson and co-workers suggest that the effectiveness of uncouplers depends on the pK_a of their dissociable groups and that their effffectiveness follows a Henderson–Hasselbach dissociation curve when certain assumptions are made (*108*). Recent experiments by Hanstein and Hatefi (*109*) also suggest that the role of an uncoupler is not necessarily related to proton transport. They discovered that trinitrophenol (picrate) is not able to uncouple mitochondria but is able to uncouple all the energy linked reactions of submitochondrial particles. The proton conduction induced by trinitrophenol is much less than that of a classical uncoupler, and the molecule appears not to permeate membranes as expected for uncouplers from the Mitchell hypothesis.

Very recently Lee (*110*) reported that trinitrophenol is quite ineffective in relieving the oligomycin inhibition of NADH oxidation in submitochondrial particles.

The very high efficiency of the salicylanilides (*e.g.*, S-13) in uncoupling respiration-linked energy transduction suggested a stoichiometry between uncoupler and respiratory chain redox centers (*111, 112*). The stoichiometry, however, depends on the rate of formation of high energy intermediates and the amount of uncoupler may thus vary from 1/300 at low oxidation rates (*113*) to 1.0 of the cytochrome a at maximum oxidation rates (*111, 114*). The number of binding sites for uncouplers has recently been determined, and competitive bindings of the uncouplers have been carried out (*115*). The sites have not been identified on a molecular basis.

Hanstein and Hatefi (*115*) have utilized the uncoupler, 2-azido-4-nitrophenol (NPA), as a photoaffinity label. The label is associated with proteins in the 20,000 to 30,000 dalton range which do not appear to be part of the ATPase (F_1). Copeland and co-workers (*116*) utilized the sulfhydryl reacting uncoupler 2,4-dinitro-5-(bromoacetoxyethoxy)phenol (DNBP). They suggest that the labeled proteins may be involved in the primary energy transduction process.

ENERGY TRANSFER INHIBITORS. These substances block the utilization of high energy intermediates when they are bound to the mitochondrial inner membrane. They do not, like the uncouplers, discharge the energized state since in the presence of the energy transfer inhibitors, respiration-linked reactions such as ion transport can still be carried out.

Oligomycin is the classical inhibitor of this type. Its inhibition of state 3 respiration can be released by uncoupler. It inhibits the uncoupler-activated ATPase, ATP-driven ion-transport, ATP-driven transhydrogenase, and ATP-driven reversed electron-transport. It has no effect on respiration-driven ion-transport or respiration-driven transhydrogenase activity (Figure 2). Rutamycin acts in essentially the same manner. DCCD acts also in a similar way, but it binds covalently to the inner mitochondrial membrane, and this property has been utilized to isolate the DCCD binding protein (*117*).

Aurovertin behaves phenomenologically like oligomycin; it however, appears to alter the affinity of the adenine nucleotides to the mitochondrial ATPase (*118*). Alkyltins have been shown to act as energy transfer inhibitors. Because trialkyltins are also able to catalyze a $OH^-:Cl^-$ exchange across the mitochondrial membrane, their inhibitory action is more difficult to analyze (*119*). Tetradifon, a much smaller molecule than oligomycin, has been reported to act like oligomycin (*120*). The azide inhibition of coupled respiration at low concentrations can be eliminated by uncouplers. From these latter experiments, it has been suggested that the uncoupler acts specifically at site III. Azide causes energy dependent spectral changes in cytochrome oxidase. Hydrolapachol appears to act similarly, but its site of action is at or near the second phosphorylation site (*121*).

These latter type of inhibitors (azide and hydrolapachol) may actually act at a site closer to the respiratory chain than the uncoupler sites and differently from oligomycin-type inhibitors. Energy linked ion transport does not occur in the presence of these inhibitors.

METABOLITE TRANSPORT INHIBITORS. Energy linked respiration can be inhibited in several additional ways: (1) inhibition of metabolite transport into the matrix where interaction with the dehydrogenase occurs (*e.g.*, succinate or NAD-linked substrates), (2) absence of oxygen, and (3) inhibition of transport of phosphate or adenine nucleotides. The first case is only indirectly related to the energy transduction process. Case (3) behaves phenomenologically like oligomycin when respiration rates are measured, since the inhibition can be overcome by uncouplers. The principal difference, however, is that in the presence of the transport inhibitors intramitochondrial ADP is phosphorylated.

Hypothetical Mechanisms—Concepts and Basic Supporting Observations

Chemical Hypothesis. The first formulation of an hypothesis for explaining the intermediary steps in oxidative phosphorylation was based on Warburg's mechanism of gylcolytic phosphorylation, *i.e.*, the substrate (AH_2) reacted with P_i and was then oxidized to a high energy form ($A \sim P$). The current version (*8*) resulted from a better understanding of the mechanism of ATP synthesis coupled to the oxidation of glyceraldehyde-3-phosphate and other substrate level phosphorylations which cause NAD reduction. The original proposal of Slater, further modified in view of additional data, is shown in Reactions 2–5 and Figure 2.

$$AH_2 + B + C \rightarrow A \sim C + BH_2 \tag{2}$$

$$A \sim C + X \rightarrow X \sim C + A \tag{3}$$

$$X \sim C + P_i \rightarrow X \sim P + C \tag{4}$$

$$\underline{X \sim P + ADP \rightarrow X + ATP} \tag{5}$$

$$AH_2 + B + ADP + P_i \rightarrow A + BH_2 + ATP \tag{6}$$

Other letters (*e.g.*, I and Y in place of C and X respectively) have been used in the above reactions. The symbol I was introduced to suggest an inhibited state ($A \sim I$) which arose during state 4 respiration and required ADP and phosphate for release of A for cycling through Reactions 2 and 3.

There are several experimental observations which formed the basis for these formulations. The most important ones follow:

(1) Disrupted mitochondria carry out substrate oxidations at maximal rates in the virtual absence of phosphate. The respiration stimulated by 2,4-dinitrophenol (DNP) did not require P_i (*122*). These observations did not fully exclude rapid turnover of a small amount of endogenous phosphate. The best evidence that mitochondrial oxidations can occur in the absence of phosphate came from the experiments of Chappell and Greville (*123*), who showed that even when phosphate esterification was excluded by oligomycin, DNP still stimulated respiration. In further confirmation of the findings, arsenate, unlike DNP, does not release respiratory inhibition by oligomycin (*124, 125*). In analogy to the better understood reactions in substrate level phosphorylations, arsenate was believed to substitute for P_i, and, in fact, competition between these two ions has been demonstrated in respiration-linked phosphorylations also. These latter observations indicate that DNP acts before P_i enters the reactions.

(2) Reversed electron flow and nicotinamide nucleotide transhydrogenation driven by oxidation of substrate does not require phosphate, as shown by the use of oligomycin, which in fact stimulated the reactions contrary to the inhibition expected if phosphate turnover was necessary (*10*). These reactions are discussed later.

Many observations have been made that are consistent with the chemical hypothesis. However, the search for chemical intermediates in the process, which would provide unequivocal proof of the validity of the hypothesis, has not been successful. Conceivably, there could be at least two intermediates—a non-phosphorylated and a phosphorylated one. Several possible reasons for the consistent failure to demonstrate convincingly the involvement of intermediates include: (1) the intermediates could be protein-bound and the search for these has been inadequate since the purified proteins participating in the process have not all been isolated, and (2) the coupling reactions may occur in the hydrophobic phase of the membrane, an environment which might be essential for the stability of the intermediate. The evidence advanced to discount the hypothesis has been of a negative type which is not entirely satisfactory.

Chemiosmotic Hypothesis. Since its original formulation by Mitchell (*126*), the chemiosmotic hypothesis has undergone several revisions. An

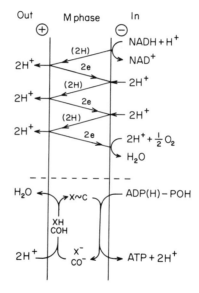

Figure 11. Mitchell's chemiosmotic hypothesis

excellent review has been published by Greville (*14*). The basic principles of the hypothesis, shown in Figure 11, are that the mitochondrial membrane has selective permeability and that reactions across it are vectorial. The primary energy conserving reaction coupled to oxidations in the respiratory chain is regarded as a translocation of protons from inside to the outside of the mitochondria. Since the coupling membrane (inner membrane or M-phase in Figure 11) is impermeable to protons and most other ions, an electrochemical gradient and associated membrane potential results, so that energy is stored as a combination of a pH gradient and membrane potential in varying ratios. Because of this, the hypothesis can be more appropriately called an electrochemical hypothesis. The coupling membrane has another property, *viz.*, the capacity to transport protons, cations, and anions by exchange diffusion carriers. The introduction of the concept of energy stored in the form of a membrane potential overcame the objection that the pH gradient of 3.5 units necessary for the synthesis of ATP is too high for biological systems (in fact, unknown in biological systems) and could damage the mitochondrial proteins.

The second part of the overall process is mediated by a reversible, proton-translocating ATPase which utilizes the electrochemical energy derived from respiration for driving the dehydration of ADP and P_i. In essence, the loss of OH^- from P_i during ATP formation is promoted by the high H^+ concentration on the opposite side of the membrane. The oxidation-driven proton translocation and subsequent formation of ATP are distinct, independent processes. According to this hypothesis, the only link between them is the electrochemical gradient. Several versions of the mechanisms of the reversible ATPase have been suggested (*14*), one of which is shown in Figure 11. F_1-ATPase was considered as the reversible ATPase, although the reaction catalyzed by the isolated F_1 is essentially irreversible; the reversible membrane-bound ATPase system is more complex, as shown by its oligomycin sensitivity, and probably involves other proteins also. A detailed chemical mechanism for the ATP synthesis has been proposed recently (*127*).

To explain the vectorial nature of the proton translocation, and to account for the extensive evidence for three coupling sites, Mitchell (*14*) has proposed that the respiratory chain is folded in three oxido-reduction loops (Figure 11). In each loop, a reduced hydrogen carrier (*e.g.*, flavin H_2 or quinol) transports the two H^+ and two electrons from the inner surface (or matrix side) of the coupling membrane to its outer surface, and an electron carrier (*e.g.*, Fe^{2+} or an uncharged species together with a negatively charged carrier) transports electrons from the outer to the inner side. The three hydrogen carriers were postulated to be F_{pD1} (flavoprotein 1), F_{pD2} (flavoprotein 2), and ubiquinone. The electron carriers were considered to be Fe-S protein (non-heme iron in older terminology), cytochrome *b*, and cytochrome oxidase. Recent data no longer support the proposal that F_{pD1} and F_{pD2} act sequentially.

Proton translocation associated with substrate oxidation and ATP breakdown in mitochondria has been demonstrated, but mechanistic interpretations of the data have differed. Mitchell and Moyle (*14, 128*) measured pH changes in the suspending medium after the addition of small pulses of oxygen to anaerobic mitochondria in the presence of β-(OH)butyrate and succinate and obtained H^+/O ratios of close to 6 and 4 respectively. Thus, consistent with the hypothesis, two protons were translocated at each coupling site. Furthermore, these authors have shown that, as expected, ATP hydrolysis in mitochondria is accompanied by the ejection of protons. Although the stoichiometry was more difficult to establish because of interfering reactions, the final estimate of H^+/ATP was 1.9, which was remarkably close to that predicted by the hypothesis. Similar results were obtained with submitochondrial particles but the direction of proton translocation was opposite to that seen with whole mitochondria. That a pH gradient can drive ATP synthesis was unequivocally demonstrated in the acid-bath experiments of Jagendorf and Uribe (*129*). They found that if broken chloroplasts are placed in a medium at pH 5.0 and then brought rapidly to pH 8.0, ATP synthesis occurred even in the dark and in the presence of inhibitors of electron flow.

Chance and Mela (*130*) and Chappell (*131*) have provided an alternative explanation for the proton translocation, viz., that the X $\sim$ C intermediate drives a secondary Ca^{2+} or H^+ pump. Experimental evidence supporting these explanations is also available.

Mitchell attributes the action of uncouplers to their ability to dissolve in the hydrophobic membrane, to render it permeable to protons, and to discharge the electrochemical gradient. In fact, trifluoromethoxycarbonylcyanide phenylhydrazone (FCCP) discharged the pH gradient established by oxidation or by ATP hydrolysis (*128*). It is also discharged when the membrane structure is disrupted by detergents.

The Mitchell hypothesis requires that the coupling membrane should have low proton permeability. The effective proton conductance (sum of H^+ and OH^- conductances) measured by the slow drift of the pH gradient established by substrate oxidation or ATP breakdown was 0.11 μg ion of H^+/second $\times$ pH $\times$ gm mitochondrial protein, remarkably low for a biological membrane (*132*). On the other hand, the value obtained by Chance et al. (*133*) using the NAD/NADH equilibrium as a pH indicator was 80. This rate is 10 times that expected if proton permeability limited respiration in state 4.

Chappell (*131*) has produced evidence for the existence of exchange diffusion systems (e.g., $H_2PO_4^-$ for OH^-) in mitochondria, meeting yet another requirement of the Mitchell hypothesis.

A number of observations conflict with the chemiosmotic hypothesis. One of these relates to measurements of the membrane potential and pH gradient. Tedeschi (*134*) determined the membrane potential across the inner mitochondrial membrane of Drosophila mitochondria with the fluorescent indicator 3,3'-dihexyl-2-2'-oxacarbocyanine. He found a potential of 10 to 30mV with the inside of the mitochondria positive. These measurements confirm his earlier direct determinations of the membrane potential with a microelectrode. They found only small changes in membrane potential between different metabolic states. These results basically support the earlier data of Harris and Pressman (*135*) on rat liver mitochondria. Most of these experiments, thus, show a passive potential of 30 mV (positive inside rather than negative required by the chemiosmotic hypothesis), irrespective of the type of permeant anion used (monovalent, divalent, or trivalent).

Mitchell and Moyle (*136*) estimated the proton motive force from the transmembrane pH gradient and the transmembrane electrical potential. The latter was determined from the potassium distribution in the presence of valinomycin. Their calculations yield a transmembrane proton motive force of about 250 mV (negative inside). They estimate a difference of 30 mV between the respiratory state in the absence of a phosphate acceptor (state

4) and that in the presence of a phosphate acceptor (state 3). The pH gradient across the mitochondrial membrane was estimated at less than one pH unit (state 4), while state 6 (calcium uptake without an anion) yielded a pH gradient larger than one pH unit. Schuldine and co-workers (137) observed that a proton concentration gradient of 2.5 to 3.0 pH units across the chloroplast membrane could not alone support ADP phosphorylation. When they superimposed a diffusion potential with potassium chloride in the presence of valinomycin or with sodium chloride in the presence of nonactin, ADP phosphorylation was obtained.

Van Dam and Engel (138) demonstrated that mitochondria can be energized by a diffusion potential. This energization is only temporary since both types of ions are permeable, but the anion is much more permeable than the cation. They used cytochrome b_{566} and ANS fluorescence changes as indicators of energization.

Wilson and Erecinska (139) interpret their midpoint redox potential measurements of the mitochondrial cytochromes as an indication of alkalinization of 0.75 pH unit with respect to the suspending medium in the presence of phosphate potentials greater than $10^2\ M^{-1}$.

A second basic mechanistic question is what is the role of the inner mitochondrial membrane in oxidative phosphorylation? Is it required for exclusion of water in the hydrophobic region of the high energy intermediate, or for the compact arrangement of the respiratory chain and the coupling factors, or for the metabolic compartmentation of the cell. Although much effort has gone into attempts to carry out oxidative phosphorylation in the absence of a membrane structure, these experiments have not yet been successful.

A dramatic result has recently been published by Hunter and co-workers (140), who treated beef heart submitochondrial particles with lysolecithin. Electron microscopy suggests that most membraneous structures have been eliminated. Even more important, this preparation is no longer sensitive to uncoupling by the ionophores, nigericin plus valinomycin in the presence of potassium chloride. However, the energy linked reactions are still sensitive to uncouplers and oligomycin. These experiments do not necessarily demonstrate that membrane structures are not required for uncoupling or energy-linked reactions, but they could suggest that sites of action for ionophores may be lost in the lysolecithin treatment.

Conformational Coupling. A number of investigators had considered a mechanism for oxidative phosphorylation based on its analogy to conversion of chemical into mechanical energy during muscle contraction. This germinal conformational coupling hypothesis was formalized by Boyer (11) when he speculated that the energized state would be brought about by coupling oxidation to the production of a conformational change in a protein. The energized protein could be a respiratory carrier or an associated structure, and the conformational change could result from the formation or cleavage of a covalent bond. This version is only a slightly modified form of the chemical hypothesis. In a recent revision of the hypothesis (141), involvement of a covalent bond has been abandoned and conformational changes are considered to occur through multiple non-covalent bonding in a protein. Boyer postulates that ATP is synthesized from ADP and P_i bound to neighboring sites on the ATP synthetase protein in a reaction (Reaction 2, Figure 12) that is independent of the oxidative steps. The energy for this

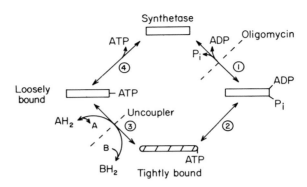

Figure 12. Boyer's conformational coupling hypothesis

synthesis is derived from a conformational change in the protein yielding a form that is capable of binding ATP tightly. In fact, there is experimental evidence for similar reactions in actomysin. The input of energy from the oxidative steps in the respiratory chain is considered to restore the conformation of the synthetase protein to the original form with weak affinity for ATP (Reaction 3). Uncouplers are believed to act on this process.

Several important observations have prompted the above hypothesis. Eisenhardt and Rosenthal (*142*) had observed esterification of a small amount of P_i to form ATP in a reaction that was uncoupler insensitive. Their findings have been confirmed recently (*141*), although the kinetics have yet to be documented. Investigators in Boyer's laboratory have also found that P_i–$H^{18}OH$ exchange catalyzed by mitochondria was less sensitive to uncouplers than the overall oxidative phosphorylation (*143, 141*). This can be readily explained by the following set of reactions, although alternative explanations are also available (*144*):

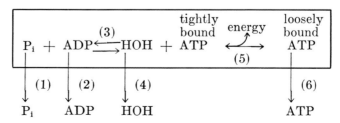

Reactions 3 and 5 occur at the catalytic sites within the membrane, and the exchanges between the membrane bound and medium components are Reactions 1, 2, 4, and 6. Uncouplers, acting on Reaction 5, should have less effect on the P_i–HOH exchange which is mediated by Steps 1, 3, and 4. The exchanges involving ATP would be more sensitive, since they encompass the uncoupler sensitive Step 5 (*see* later).

Finally, tightly-bound nucleotides have been observed in phosphorylating particles and in ATPase (*145, 146, 147*). Their properties will be discussed in later sections of this article.

The above experiments do not shed any light on the mechanism of the proposed conformational change in a respiratory enzyme coincident with oxidation reduction. Its feasibility has been pointed out by Chance (*148*) who called it a membrane-Bohr effect analogous to the changes in hemoglobin.

If the primary energized state associated with oxidation is indeed a conformational change, it must be communicated directly to ATP synthetase by an interlocking protein network, according to the conformational coupling hypothesis. Such direct communication is not required for the chemiosmotic hypothesis where the communication is through a proton gradient, or for the chemical hypothesis where diffusible intermediates could be operating. The generation of the energized state by a mechanism according to one hypothesis and its utilization according to another also needs consideration.

A general statement that can be safely made at this time is as follows: all of the hypotheses have experimental support and offer scope for experimentation. The chemiosmotic hypothesis has strong esthetic appeal, particularly in unifying cell transport phenomena, but has yet to meet satisfactorily some of the above objections.

Energy Linked Reactions

Exchange Reactions. Isotope exchange reactions involving only some of the components of an overall enzyme reaction have often proved useful in determining the sequence of the steps, identifying the rate limiting reactions, and gaining insight into the catalytic mechanisms. The exchange of ^{18}O from water with the oxygen of P_i in mitochondria, observed by Cohn (*149*) in 1953, led to the development of these reactions as tools for the study of the mechanisms of oxidative phosphorylation. To explain the above exchange, Boyer and co-workers (*150*) and Swanson (*151*) looked for and found ATP–$^{32}P_i$ exchange. The ATP–$H^{18}OH$ exchange in mitochondria was demonstrated by Cohn and Drysdale (*152*) and by Boyer *et al.* (*153*). The ADP–ATP exchange reaction was discovered subsequently, and strong evidence was presented by Cooper and Lehninger (*154*) that at least part of the exchange observed with mitochondria is related to oxidative phosphorylation.

One of the pitfalls in these studies has been the presence of enzymes unrelated to oxidative phosphorylation, which catalyze similar exchanges (*155*). Inhibition of the reactions by uncouplers and oligomycin, and the presence of activity in submitochondrial particles have been used as distinguishing characteristics. An important finding from the exchange experiments is that the β–γ bridge oxygen (*i.e.*, the O between the last two P atoms) is derived from ADP (*156*). This is inconsistent with the formation of an adenylated precursor of ATP (*e.g.*, X ~ ADP) which then reacts with P_i. The results are consistent with the formation of a phosphorylated precursor or with the formation of ATP by a concerted reaction.

The ATP–P_i exchange reaction has been used extensively because of its experimental simplicity. It is clearly a part of the oxidative phosphorylation system, since it is inhibited by uncouplers (*153*) and by low levels of oligomycin (*157*). Arsenate is also an inhibitor, particularly at low levels of P_i (*143*). With intact mitochondria, the rates of ATP–P_i exchange exceed net phosphorylation rates (*156*), but they are variable in different types of particles. A combination of cyanide, rotenone and antimycin A with or without substrate present, inhibit the exchange only slightly (*11*) indicating that the ATP–P_i exchange is independent of the redox state of the respiratory chain. The important useful implication of the exchange reaction is the maintenance of the energy of the ATP in a form free of phosphate (E^*) or one in which the phosphate can be exchanged with P_i in the medium. The reaction is explained by the equation:

$$E + ATP + HOH \leftrightarrows E^* + ADP + {}^{32}P_i \qquad (7)$$

when (E^*) is X ~ I or a proton gradient or an energized conformation of a protein.

The exchange of ^{18}O between ATP and $H^{18}OH$ in mitochondria and submitochondrial particles is inhibited by uncouplers and oligomycin (*144*). Equation 7 alone does not explain this exchange since, under certain conditions, the ATP-HOH exchange exceeds by several times the ATP–P_i exchange measured simultaneously (*158*). The ATP–HOH exchange is unique to mitochondrial oxidative phosphorylation and photophosphorylation in chloroplasts (*11*).

The relationship of the exchange of ^{18}O between P_i and HOH to oxidative phosphorylation also rests on its sensitivity to uncouplers (*149*) and oligomycin (*158*), but certain features have been difficult to explain. Lardy and co-workers (*159*) found that oligomycin and aurovertin inhibited the ATP–P_i and P_i–HOH exchanges equally and that their effects were additive, establishing that similar sites were involved in these reactions. An interesting difference, on the other hand, is the lower sensitivity of the P_i–HOH exchange to uncouplers in mitochondria (*149*) and submitochondrial particles (*158*), although the resistant exchange depends upon ATP cleavage (*156*). Similarly, the P_i–HOH exchange is more resistant to *p*-chloromercuribenzoate (*158*). These observations support the explanation (*158*) that P_i–HOH exchange is independent of the dynamic reversal of oxidative phosphorylation and are also considered supportive of the conformational coupling hypothesis.

Reversed Electron Flow. In 1955, Chance (*160*) found that addition of succinate to aerobic mitochondria caused reduction of endogenous NAD, as shown by the increase in $A_{340\text{-}370nm}$ in a dual wavelength spectrophotometer. Strong support for this indirect reduction came when Klingenberg and Schollmeyer (*161*) showed by chemical analysis that coincident with glycerol-l-phosphate (instead of succinate) oxidation, a greater reduction of NAD occurred than with β-hydroxybutyrate in state 3 or 4. Since anaerobiosis or addition of electron transport inhibitors (such as antimycin, CN⁻, or rotenone) or of uncouplers inhibited the reduction of NAD by succinate, the energy conserved at sites 2 and 3 during the oxidation of succinate could be used to drive the thermodynamically unfavorable reduction of NAD by succinate (Figure 2). Energy linked reduction of cytochrome b by cytochrome c involving reversal of only site 2, and that of NAD by cytochrome c involving reversal of sites 2 and 1 have been demonstrated in both mitochondria and submitochondrial particles (8, 9). Submitochondrial particles are particularly suitable for these demonstrations since they have no permeability barrier for NAD, and added NAD can be reduced in substantial amounts rendering the measurements quite simple (162).

The reversed electron flow driven by oxidative energy is not inhibited by oligomycin. In fact, in submitochondrial particles, an increase in the rate of reversed electron flow occurs at concentrations of oligomycin which inhibit oxidative phosphorylation, presumably because the antibiotic inhibits the energy loss due to ATPase activity. On the other hand, when ATP provided the energy for the reversed electron flow, which can be observed readily in submitochondrial particles inhibited by CN⁻, oligomycin was a strong inhibitor at concentrations that blocked oxidative phosphorylation. These effects of oligomycin further establish that an energized state generated before the formation of ATP provided the energy for the reversed electron flow (Figure 2). The effects of uncouplers and of oligomycin also demonstrate that the pathway for energy utilization in reversed electron flow is the same as the pathway in oxidative phosphorylation.

Ernster and Lee (163) observed a low-level effect of oligomycin in submitochondrial particles prepared in the presence of EDTA. At concentrations roughly one-fifth those necessary for inhibition of oxidative phosphorylation, oligomycin stimulated ATP production or ATP-driven reactions. This has been attributed to blocking a secondary energy leak (or hydrolysis of $X \sim C$) in the energy transfer reactions (Figure 2).

Since the energized state can be used either for ATP synthesis from ADP and P_i or for NAD reduction by reversed electron flow, there should be competition between these two processes. In fact, this does occur, and, by comparing the decrease in ATP formation with the amount of NAD reduced, i.e., by trapping it in a reaction with α-ketoglutarate and NH_3 as glutamate, one ATP equivalent is utilized per NAD reduced as in Equation 8 (164). Consistent with this stoichiometry, one mole of ATP is formed by the anaerobic oxidation of one mole of NADH by fumarate, i.e., a P/2e ratio of 1 (165).

$$\text{Succinate} + \text{NAD} + \text{ATP} \rightleftarrows \text{Fumarate} + \text{NADH} + \text{H}^+ + \text{ADP} + \text{P}_i \quad (8)$$

Reversed electron flow driven by ATP breakdown in submitochondrial particles has demonstrated the concept of phosphate potential, ATP/(ADP ·

P_i), as a controlling factor in regulating the rate of oxidative phosphorylation. From the extent of cytochrome c reduction by cytochrome a in the presence of increasing amounts of added ATP, the equilibrium constant (or phosphate potential) was calculated to be $8\ \mu M^{-1}$ (*161*). The G'_0 derived from this value was 12.8 kcal for ATP breakdown in mitochondria. Other similar measurements have given values in the range of 15 to 16 kcal (*166, 167*).

Nicotinamide Nucleotide Transhydrogenation. Enzymes that catalyze the transhydrogenation from NADPH to NAD (Equation 9) have been purified from several bacteria (*168, 169, 170*) and higher organisms.

$$NADPH + NAD^+ \rightleftarrows NADP^+ + NADH \tag{9}$$

Mitochondria also contain a similar transhydrogenase (*171*) which catalyzes the reaction in both directions, but with a vastly greater rate in the NADPH → NAD direction (*172*). Krebs and Kornberg (*173*) postulated that input of energy to the transhydrogenase system could stimulate NADP reduction in mitochondria, and this suggestion was experimentally substantiated by the rapid and complete reduction of intramitochondrial NADP during the oxidation of a substrate (*174, 175*). Danielson and Ernster (*176*) finally demonstrated reduction of $NADP^+$ by NADH in submitochondrial particles driven by ATP-energy or energy from succinate oxidation. The apparent equilibrium constant for the energy independent reaction is 0.8 (*171*) and that for the energy dependent reaction (Equation 10) is of the order of 500 (*177*).

$$NADH + NADP + \stackrel{\sim}{\rightleftarrows} NAD + NADPH \tag{10}$$

The same transhydrogenase catalyzed the reaction in both directions, coupled in one direction and uncoupled in the other. This proposal was supported by the observation that the addition of antibody to purified energy independent transhydrogenase also inhibited the energy dependent reaction (*170*). The reversibility of the energy dependent transhydrogenase—*i.e.*, the reduction of NAD by NADPH, coupled to energy transduction, has been observed using several methods, including net ATP synthesis from ADP and P_i (*178*) and proton translocation (*179*). The mechanism for utilization of the energy has not yet been elucidated.

The stereochemistry of H atom transfer in the energy linked and the energy independent transhydrogenase has been studied (*180*). The reaction in both directions involves the 4A H atom of NADH and 4B H atom of NADPH. The reactions involve no exchange of H atoms between the reduced nucleotide and water.

Energy-dependent reduction of $NADP^+$ by NADH has also been observed in photosynthetic bacteria, with ATP or light as the energy source. Orlando (*181*) has isolated a soluble factor which stimulates the reaction and has proposed that reduction of a protein disulfide to a dithiol was involved in the transhydrogenation.

Kinetic studies on the reaction in both directions have indicated formation of a ternary complex *via* the Theorell–Chance mechanism (*172*).

Inhibition of the energy-linked reaction is competitive with respect to NAD^+ and NADPH, suggesting that the conformation of the enzyme in the energized and non-energized states may be different (*182*).

The metabolic role of the mitochondrial energy dependent transhydrogenation is not clear, and speculations that it supplies NADPH for extramitochondrial biosynthetic reactions (*e.g.*, fatty acid synthesis) have yet to be substantiated.

Ion Transport. The inner mitochondrial membrane has high selectivity towards ions. Many ions are able to penetrate the membrane under certain conditions. One can detect uptake or efflux of mono- and multivalent cations and mono- and multivalent anions. It may be improper to subdivide this section into separate anion and cation transport processes, since significant interactions between the various transport systems have been observed. Investigations have, however, been directed towards the transport of specific ions and the influence of other ions on their transport. This section is thus divided into: (1) multivalent cation transport, (2) monovalent cation transport, and (3) anion transport.

The more rigorous criteria for carrier mediated transport across the mitochondrial membrane are considered in the anion transport section. There never was much doubt in the minds of biochemists about the carrier function in Ca^{2+} transport since such a dramatic stimulation of ADP-limited respiration occurs with Ca^{2+}. This type of stimulation does not generally occur with the metabolic anion, and thus much more thorough investigations had to be carried out.

MULTIVALENT CATION TRANSPORT. Dominant among multivalent cation transport systems that have been studied is Ca^{2+} transport. The Ca^{2+} uptake by mitochondria requires energy, either from respiration, ATP, or the energy of other ion gradients.

When limited amounts of Ca^{2+} are added to mitochondria that are respiring in state 4 in the absence of phosphate, a sudden increase in respiration occurs which lasts until the mitochondria are fully loaded or until the medium concentration of Ca^{2+} becomes limiting. After this has occurred, the respiration returns to an apparent state 4 and further addition of Ca^{2+} causes no increase in respiration. This state is called state 6 (*183*). Rat liver mitochondria are fully loaded in state 6 with Ca^{2+} at about 100 nmoles/mg protein (*184*).

A similar stimulation of respiration with Ca^{2+} will occur in the presence of up to 2 mM phosphate. The respiration will not return to a controlled state when higher concentrations of phosphate are present, unless ATP and Mg^{2+} are also present in the medium. Phosphate is accumulated in mitochondria together with Ca^{2+}, at a ratio of Ca^{2+}:P of about 1.7. Maximum loading of mitochondria with up to about 2500 nmoles Ca^{2+} per mg protein can be accomplished in the presence of phosphate, ATP, Mg^{2+}, and a respiratory substrate.

Other anions that show effects similar to phosphate in energy-linked Ca^{2+} transport are able to carry a proton across the mitochondrial membrane —*e.g.*, acetate or arsenate. Permeable anions, like nitrate or thiocyanate that are anions of strong acids, have no effect on Ca^{2+} transport. Lehninger calls the weak acid anions B-type and the strong acid anions A-type (185).

The efficiency of Ca^{2+} transport has been estimated. One can calculate a ratio of Ca^{2+} taken up for two reducing equivalents transported per phosphorylation site. This ratio attains a value approaching 2.0 in the absence of phosphate or in the presence of phosphate plus ATP and Mg^{2+}. In the presence of phosphate, but absence of ATP and Mg^{2+}, this ratio is reduced.

During the uptake of Ca^{2+}, the interior of mitochondria becomes alkaline and the medium becomes acidic. The $H^+:Ca^{2+}$ ratio in the absence of phosphate approaches 1.0. This ratio implies incomplete charge compensation. In the presence of phosphate, this ratio drops to 0.6 and, in the presence of acetate, to 0.3 (184).

What can be said about the molecular basis of the transport process? One of the requirements for transport by a carrier-type mechanism rather than by a physical diffusion process is that the former shows saturation kinetics, while the diffusion process does not. This criterion has been satisfied in several studies, most recently by Chen and co-workers (186), with mitochondria from the crab. Since there is a correlation between stimulation of respiration by Ca^{2+} and Ca^{2+} uptake, a K_m can be determined from the stimulated respiration. This K_m reaches a value of 90 μM in crab mitochondria (186) and 45 μM in pigeon heart mitochondria (183). The V_{max} ranges from 600 nmoles Ca^{2+}/min/mg protein (186) to 1100 nmoles/min/mg protein (25°C) in the presence of phosphate (183). A recent reevaluation of Ca^{2+} uptake by rat liver mitochondria demonstrated a Ca^{2+} concentration for half maximal uptake rate of 50 to 70 μM, with a V_{max} of about 490 nmoles Ca^{2+} per min per mg protein (187).

Ruthenium red (188) and lanthanides (189) inhibit the uptake of Ca^{2+} at very low concentrations and provide further evidence for a carrier-type mechanism. The inhibitors are effective at a concentration which is equivalent to that of the mitochondrial cytochromes (0.1 nmole/mg protein).

Efforts have been made to identify the first reaction step in the Ca^{2+} transport process. Binding studies suggest that two types of binding can be observed, a high affinity binding which can be eliminated by uncouplers, and a low affinity binding site which appears to have no relation to energy-linked Ca^{2+} transport. The dissociation constant of the high affinity binding site is less than 1 μM, while that of the low affinity binding site is about 50 μM (184). Very recently, the existence of the high affinity binding sites has been questioned since respiratory inhibitors that inhibit respiration completely eliminate the high affinity binding sites (190, 191). The question as to the significance of the high affinity binding sites thus remains.

Two types of glycoproteins have been isolated from mitochondria. One is water insoluble (192) while the other is soluble (193). Both show high affinity Ca^{2+} binding which is sensitive to ruthenium red and lanthanides, but is insensitive to respiratory inhibitors. The role of these gylcoproteins in the respiration-linked Ca^{2+} transport is not known.

The study of Ca^{2+} transport has recently been aided by the discovery of a divalent cation ionophore which makes the mitochondrial membrane permeable to Ca^{2+} ions (194). Safranin is a dye that appears to act at the Ca^{2+} binding site, and spectral changes in safranin absorption have been utilized to characterize the Ca^{2+} binding site (195). Mn^{2+} is transported like Ca^{2+} in an energy-linked manner. It is a paramagnetic ion which permits EPR spectroscopy to be utilized in the characterization of the Ca^{2+} transport process (196).

Many experiments suggest that the Ca^{2+} pump of the sarcoplasmic reticulum rather than the mitochondrial Ca^{2+} pump is primarily responsible for controlling muscle contractions. Intramitochondrial Ca^{2+} levels may, however, control various enzymatic activities, and the Ca^{2+} pump may exert its physiological function in this manner.

MONOVALENT CATION TRANSPORT. The inner mitochondrial membrane in its resting or low energy state is impermeable to monovalent cations including protons. There are, however, conditions under which very high rates of monovalent cation transport can be observed. These conditions can most conveniently be discussed in terms of permeation in the presence and absence of energy. Permeation in the absence of energy will be discussed first since that is the system with inherently fewer variables.

Table I. Monovalent Cation Transport in the Absence of Energy

		Spontaneous Swelling
NH_4NO_3		no
	+ uncoupler	yes
KNO_3		no
	+ valinomycin	yes
NH_4Ac		yes
KAc		no
	+ valinomycin	no
	+ valinomycin + uncoupler	yes
	+ nigericin	yes

Table I demonstrates the various types of experiments that form the basis of some of the theories of mitochondrial ion transport.

Evidence has been presented that anions of strong acids, if they have sufficient lipid solubility, penetrate the mitochondrial membrane in the anion form (SCN^-, NO_3^-). Weak acids permeate the membrane in their undissociated form, such as acetic acid. K^+ is impermeable to the mitochondrial membrane under non-energized conditions. Valinomycin, which has a very high specificity for K^+ over Na^+, makes the membrane permeable to K^+ (197). The membrane is impermeable to NH_4^+, but NH_3 can readily permeate the mitochondrial membrane (198). The extent of NH_3 permeation is, however, limited by the resulting charge or pH differential across the membrane. When the permeant anion NO_3^- is added, the mitochondria will swell only when a proton conducting uncoupler is also added to the mitochondria. NH_4^+ will cross the mitochondrial membrane in the presence of an uncoupler and an anion such as NO_3^- or SCN^-. Nigericin catalyzes the electroneutral exchange of potassium against protons. Schemes A, B, and C (Figure 13) demonstrate the postulated mechanisms for passive mitochondrial swelling in ammonium acetate, potassium acetate, and ammonium nitrate.

A high concentration of sodium acetate, in the presence of EDTA will also cause passive swelling. Under these conditions, an endogenous ionophore may be activated. When an impermeable anion such as chloride (impermeable at neutral pH, but not at alkaline pH) is present, spontaneous swelling will not occur. The mitochondria will, however, swell at concentrations of the salt which are less than isoosmolar.

The selective permeability of the inner mitochondrial membrane for monovalent cations can be dramatically changed by: (1) the presence of an ionophore (valinomycin for K^+, nigericin or nonensin for K^+ vs. H^+, gramicidin for Na^+ or K^+); (2) the presence of heavy metals, (e.g., Zn^{2+}); (3) the presence of thiol group reagents such as chloromercuribenzoate; (4) lowering the Mg^{2+} content of the membrane; (5) addition of an uncoupler; (6) addition of parathyroid hormone; (7) addition of certain histones; or (8) addition of a nonionic detergent such as Triton X-100 (for K^+) (199).

The question has been raised as to why the thiol group reagents should activate the non-discriminate passage of K^+ or Na^+ into mitochondria. Blondin and co-workers (200) have isolated from mitochondria an ionophore that catalyzes the passive flux of K^+ and Na^+ through the inner mitochondrial membrane. The physiological role of this ionophore has not been established.

MONOVALENT CATION TRANSPORT IN THE PRESENCE OF ENERGY. Coupled and respiring mitochondria swell in the presence of potassium acetate. Two

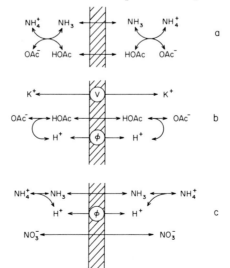

Figure 13. Schemes of passive ion fluxes through the mitochondrial inner membrane. V = valinomycin, ϕ = uncoupler.

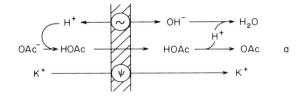
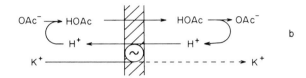

Figure 14. Schemes for energy linked ion accumulation by the mitochondria. $\sim$ = proton pump, ψ = membrane potential induced transport.

basically different mechanisms have been postulated to explain this observation. Scheme a (Figure 14) represents the anion pump model (*201*), while Scheme b represents the proton pump model (*202*).

The experimental results at this time are not able to differentiate between these two schemes. In scheme a, the electron transport chain generates OH^- inside the mitochondria. This permits the acetic acid to dissociate, and the resulting regative charge pulls K^+ electrophoretically across the mitochondrial membrane. It has been questioned whether the negative potential is sufficient to draw the K^+ across the membrane. The proton pump model (Scheme b) couples K^+ transport directly to H^+ transport.

Attempts have been made to determine the efficiency of K^+ transport. K^+ transported per two reducing equivalents per phosphorylation site has been determined. This ratio reaches a maximum of 3 in the presence of valinomycin and other K^+ flux facilitating ionophores (*199*). The presence of nigericin lowers this ratio dramatically. Earlier results suggested K^+ uptake ratios of 3.2 per 2 e^- per phosphorylation site (*144*). This ratio would be more in line with the ratio for Ca^{2+} of 2. The latter ratio permits transfer of 4 cationic charges per energy conservation site.

Other monovalent cations that can also be transported but in a less efficient manner include Li^+ and $Tris^+$.

ANION TRANSPORT. The inner mitochondrial membrane is highly selective towards anions and possesses a number of anion specific carrier or transport systems. Almost all substrates that are metabolized within the mitochondrion are anions at neutral pH. This can bring their transport under the control of the potential gradient generated by the respiratory chain across the inner mitochondrial membrane.

The inner membrane is also permeable to several anions for which there exists no obvious metabolic role. These have already been discussed in association with the energy linked uptake of mono- and divalent cation.

These anions are generally thought to penetrate the membrane without a specific transporter.

Evidence has been presented in the past ten years that the inner membrane of rat liver mitochondria contains at least eight anion transporters. Most of these anion transporters are exchange diffusion carriers or antiporters with a 1:1 molecular stoichiometry. In other words, the passage of an anion into the mitochondria occurs simultaneously with the exit of another anion, e.g., like a rapid equilibrium random bi–bi mechanism of Cleland (203). Evidence has been presented for some carriers that the charge of the two exchanging anions is the same during the actual transport process. The anions may, of course, associate with or dissociate protons once they have left the membrane and thus can act as proton carriers.

CRITERIA FOR CARRIER CATALYZED TRANSPORT. Little is known about the molecular mechanism of a transporter. Several criteria had to be used to determine whether anions pass through the membrane *via* transporters. The four generally used criteria are: (1) substrate specificity, (2) saturation kinetics, (3) specific inhibition, and (4) activation energy. Substrate specificity is the most generally accepted criterion. The specificity of almost all carriers however is not nearly as high as that of a typical enzyme. The specificities will be discussed with the individual carriers. The requirement for saturation kinetics originates also from enzyme kinetics with a finite turnover number as opposed to diffusion with a transport rate proportional to the concentration difference between the two sides of the membrane. A specific inhibition is supportive evidence on the same basis as substrate specificity. The inhibition has however a much more important role in transporter studies, since it may be common to a number of related transport processes. The activation energy again becomes important as it permits this carrier process to be differentiated from a non-facilitated diffusion process.

METHODS FOR IDENTIFYING TRANSPORTER SYSTEMS. Several methods have been utilized to identify and characterize the inner membrane transporters. Pyridine nucleotides cannot permeate the mitochondrial membrane. Metabolites that can permeate the mitochondrial membrane and that can be metabolized by the mitochondrial enzymes are generally able to reduce the pyridine nucleotides. The method involves monitoring of absorption or fluorescent changes caused by the reduction of the pyridine nucleotides (204). The results of this method are difficult to interpret because the rate limiting step in the metabolic sequence often cannot be readily identified.

The swelling of mitochondria has been used very successfully in most of the pioneering studies of metabolite transport (205). Mitochondria that are suspended in isoosmolar media (ideally, the concentration of an impermeant substrate equivalent to the osmolar concentration of substances within the mitochondria) scatter significant amounts of light. The light scattering is determined as changes in transmission at 700 nm (or 546 nm), where interferences from light absorbing redox carriers is minimal. As the mitochondria take up metabolite, they swell and cause the transmission of light to increase. This method can be used in experiments of net uptake of metabolites but not for exchange reactions. It can be used in the coupling of exchange and transport, as in malate uptake linked to the phosphate–OH^- antiporter. In this system, phosphate enters *via* the phosphate carrier by exchange against OH^-. The intramitochondrial phosphate can then exchange against the extramitochondrial malate. Thus, catalytic amounts of phosphate are required to catalyze a net uptake of malate. A definitive disadvantage of this method lies in the need for higher than natural or physiological amounts of metabolite.

The swelling method can sometimes easily differentiate between electrogenic transport or electroneutral exchange. Non-respiring mitochondria can be placed into an isoosmotic medium of the ammonium salt. NH_3 has been demonstrated to penetrate readily the mitochondrial membrane (*198*). Three possible situations may occur (1) Swelling in isoosmotic ammonium salt implies electroneutral exchange, *i.e.*, the anion exchanges for OH^- which neutralizes the H^+ left by the ammonia; (2) swelling in isoosmotic potassium salt in the presence of valinomycin, which makes the membrane selectively permeable to K^+, suggests electrogenic transport of the anion; (3) swelling of mitochondria in isoosmotic ammonium salt in the presence of proton conducting uncoupler suggests electrogenic transport of the anion. Variations in these basic methods yield useful information on anion transport.

The determination of metabolite transport into mitochondria, especially by the neutral exchange processes, can be carried out by centrifugation through silicone oil or high sucrose solutions (*206*). This method basically involves mixing mitochondria with the metabolite in a solution which is layered on top of silicone oil. The silicone oil may be layered on top of a perchloric acid solution. After a definite incubation time, the mixture is centrifuged. The mitochondria pass through the silicone and are quenched in the perchloric acid. The mitochondria are thus separated from the suspending medium and the mitochondrial enzymes are inactivated (quenched) by the perchloric acid. The mitochondrial metabolites can thus be determined. One drawback of this method is the rather long times in handling.

An attempt has been made to shorten the incubation time by separating the mitochondrial suspension from the metabolite suspension by an air bubble (*207*), the whole layering being carried out in a glass capillary which can be centrifuged. The incubation time is thus determined by the time it takes for the mitochondria to be centrifuged through the metabolite layer.

A very rapid sampling method is the filtration of a mitochondrial suspension under pressure through a filter with pore sizes of about 0.45μ (*208*). Mitochondria can be mechanically and very rapidly mixed with metabolites and immediately filtered. Mixing effectiveness however is limited by shearing forces on the filter and the leakage of metabolites from mitochondria on the filter from breakage or anaerobiosis.

A most successful method for rapid kinetic analyses is the inhibitor stop method (*209*). This method requires an inhibitor which reacts very rapidly and is specific. Mitochondria are mixed with the metabolite, allowed to react for a definite time and then rapidly quenched. This method does not stop the metabolites from being metabolized. However, the mitochondria, once the transport has been stopped, can then be centrifuged and metabolites of the supernatant and pellet determined.

MITOCHONDRIAL ANION TRANSPORTERS. At least eight anion transporters have been identified, but their relationship to each other is not clear at this time. Phosphate can be exchanged against OH^- (or transported with H^+) on the phosphate carrier (210). It can be exchanged for dicarboxylates on the dicarboxylate carrier (211). NEM is an effective inhibitor of the phosphate carrier, while n-butylmalonate inhibits the dicarboxylate carrier. The dicarboxylate carrier exchanges malate, fumarate, succinate, malonate, or oxaloacetate for each other or phosphate.

The tricarboxylate carried exchanges the physiological substrates citrate, cis-aconitate, and threo-D_8-isocitrate (212). Phosphoenolpyruvate is also efficiently transported. This carrier has been investigated with substrate analogs and some postulates have been presented as to possible groups that are important at the tricarboxylate binding site. The above tricarboxylic metabolites can be exchanged for mitochondrial L-malate. Benzene-1,2,3-tricarboxylate is a very effective inhibitor of the carrier. The specificity of the carrier has been questioned, since it is able to exchange dicarboxylates such as succinate and fumarate. They are, however, exchanged at much lower rates comparable with those of the nonphysiological dicarboxylate D-malate.

The citrate–malate exchange is electrogenic. Robinson and coworkers (213) observed that citrate (tricarboxylate) was unable to completely exchange for malate (dicarboxylate). The explanation was that an unbalanced charge or pH was responsible. Utilizing K^+ carrying valinomycin, H^+ carrying uncouplers, phosphate, or exogenous pH variations, malate^{2-} is exchanged against citrate^{2-}. Once the ions leave the carrier, they dissociate to the proper charge at the pH of the medium and thus generate the potential. This experiment demonstrates the role of charges in controlling metabolite fluxes across the inner mitochondrial membrane.

The 2-ketoglutarate carrier has been extensively studied and detailed kinetic analyses (203) demonstrate that the carrier behaves according to a rapid equilibrium random bi–bi mechanism. The carrier is able to exchange all other dicarboxylates. The electrogenic transport of pyruvate has been identified on the basis of the specific inhibitor α-cyano-4-hydroxycinnamate (214). Glutamate can be exchanged for OH^- (215) or, electrogenically, for asparate (216).

ADENINE NUCLEOTIDE TRANSPORT. The adenine nucloetide carrier was the first transporter of the inner mitochondrial membrane to be characterized extensively (217). Phenomenologically, the carrier catalyzes a 1:1 exchange of exogenous ADP or ATP for endogenous ATP or ADP. Evidence exists that the Mg^{2+}–ATP or Mg^{2+}–ADP complexes are not substrates for the adenine nucleotide translocase (208). At neutral pH, ADP and ATP are differently charged. This exchange has thus an electrogenic component which during oxidative phosphorylation is compensated by the phosphate carrier (218). The potential difference can, however, also be nullified by a proton carrying uncoupler.

ADP will exchange in coupled mitochondria readily for endogenous ATP, while exogenous ATP will not readily exchange for endogenous ADP. The presence of a proton conducting uncoupler will eliminate this difference.

Highly specific inhibitors have been identified. Atractyloside is a competitive inhibitor for ADP binding to the carrier. Carboxyatractyloside is a noncompetitive inhibitor for ADP transport. Bongkrekic acid increases the affinity of ADP to the carrier binding site. These inhibitor studies lead to postulates for various rotational positions of the carrier in the membrane as functions of inhibitory state (219).

The transporters may be located in clusters in the inner membrane. Evidence for this conclusion results from the sigmoidal binding curve of ^{35}S-carboxyatractyloside to the inner mitochondrial membrane (220).

Attempts to identify the ADP binding site with a protein consists of a decreased binding upon trypsin treatment, photooxidation, and heat treatment (220). Very recently a protein fraction has been isolated which binds ADP, and this ADP binding is sensitive to atractyloside (221). Racker and co-workers (222) incorporated a possibly similar protein fraction into soybean phospholipid vesicles and were able to demonstrate atractyloside- and bongkrekic acid-sensitive adenine nucleotide transport.

Resolution and Reconstitution

The resolution of the mitochondrial oxidative phosphorylation system into its component parts and reconstitution of the reaction with purified proteins is probably the most direct way of studying the details of its mechanism. The success with this approach has been limited because of the intimate association of structural and functional roles of the proteins and lipids and the probable occurrence of reactions within the non-polar membrane phase. In spite of these, it has been possible to deplete submitochondrial particle preparations of phosphorylation activity by a variety of procedures, although unfortunately the depletion is hardly ever complete or selective with respect to the factors. The depleted activity in energy-linked reactions could be restored partially by the addition of soluble protein preparations, which are designated coupling or energy transfer factors (Table II). The factors have been purified to varying degrees from mitochondrial extracts (see Beechey and Cattell (223) for an excellent review) and their properties have been studied. With the exception of one preparation (F_1 or Factor A), the coupling factors have no intrinsic enzymatic activity, which leaves recombination with the depleted particle as the only available assay. Since several unidentified components are interacting and interlocked in the overall process, it is difficult to determine where the isolated factor fits in and exactly what it does in restoring the activity. Some of the factors may function indirectly by improving the structure of the membrane, which may stabilize the integrated enzyme system as well as permit more efficient

Table II. Coupling Factors and their Properties

Factor	Assay Particle[a]	Intrinsic Activity	Properties
F_1	N-particle A-particle	High ATPase	Stimulates phosphorylation and ATP–P_i exchange activity of particle. Dissociated into subunits and inactivated in the cold. M.W. 360,000 (approx.) Tightly bound ADP and ATP.
Factor A	Urea particle	Latent ATPase that can be activated by heat	Increases P/O ratios, ATP–P_i exchange, ATP-dependent NAD reduction by succinate and NADH–NADP transhydrogenase activities of particle. Has tightly bound ADP. Same subunits as F_1. Low inhibitor content.
Factor B	Ammonia–EDTA particle	None	Stimulates energy-linked reactions in the assay particle. M.W. = 29,200. Amino acid composition is known. Two—SH groups. Inhibited by—SH binding reagents. Poorly stained on polyacrylamide gel. Single precipitation band with anti-serum.
OSCP	NaBr extracted OSATPase; A-particles	None	In combination with F_1 and depleted OSATPase, the ATPase activity becomes sensitive to oligomycin. Stimulates energy-linked reactions in the presence of F_1 in A-particles. M.W. = 18,000.
F_6	STA-particle	None	Combined with F_1 and other crude preparations, it stimulates energy-linked reactions in STA particle. Heat stable, trypsin labile. Appears to be a complex of several "isoproteins."

[a] The assay particles are treated in such a way that they respond maximally to the addition specifically of one factor (e.g., ammonia-EDTA particle interacts only with Factor B) or to a combination of factors (e.g., A-particle). The treatments have been devised on an empirical basis. The removal or dissociation of the factors from the membrane is rarely complete, which results in particles with significant basal activity.

interaction between the components than is possible in solution or in a disorganized membrane.

F_1, which was discovered in Racker's laboratory, has been studied most intensively. It is a complex of five subunits of approximately 53,000; 50,000; 25,000; 12,500; and 7,500 daltons (224). Some preparations contain the ATPase inhibitor, referred to as Pullman inhibitor, which is a peptide of 5,700 daltons (225). Added inhibitor produces strong inhibition of the ATPase activity without affecting the coupling factor activity in restoring energy-linked reactions in assay particles (226). There have been recent speculations that the inhibitor might regulate the energy coupling activity of F_1 in oxidative phosphorylation (227).

The *high* ATPase activity of isolated F_1 is a convenient handle for studying the properties of the factor, but it may have no role in ATP production. In fact, this *high* ATPase activity may not be an intrinsic property of the protein when it is in the intact membrane, but may result from a modification during the isolation (228, 145). When the coupled system is operating in the reverse direction, ATP breakdown is linked to

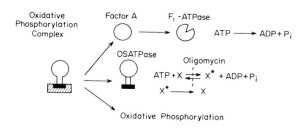

Figure 15. A schematic representation of the functions of the components of a respiratory assembly

energy utilization (*i.e.*, it is an energy utilizing ATPase) and the reaction is inhibited by oligomycin, unlike the ATPase activity of isolated F_1. The ATPase activity of submitochondrial particles may be a result of uncoupling and breakdown of the energized state. Its sensitivity to oligomycin indicates that the energy release or the hydrolysis of ATP is not at the level of F_1 but may involve other components such as X and C (Figure 2) or a proton gradient. The intermediate stage of dissociation of the ATP synthetase (Figure 15) is represented by the oligomycin-sensitive ATPase which contains at least three subunits in addition to F_1 and OSCP (224).

The separation of an OSATPase type of preparation from mitochondria with uncoupler and oligomycin sensitive ATP-P_i exchange activity is an interesting new development (58). An ATP-P_i exchange activity can be induced in the OSATPase of Tzagoloff by promoting vesiculation in the presence of phospholipids (229).

Factor A (or latent ATPase) is probably a more native form of F_1 since it has low ATPase activity which can be stimulated by exposure to heat

(*228, 145*). An antiserum to F_1 inhibits the activities of Factor A and its subunit composition is similar to that of F_1 (*230*). The change produced during the induction of high ATPase activity in Factor A has been ascribed (*231*) to dissociation of the ATPase inhibitor which is present in a small amount in Factor A and in varying amounts in F_1 ATPase or to a conformational change in the enzyme (*228*). Unpublished data from this laboratory support strongly the latter possibility. The relationship between these two forms of ATPase has been discussed more extensively elsewhere (*232*). The latter explanation is further supported by a similar conversion of CF_1 (chloroplast F_1) to an ATPase during heat activation, a process which is not accompanied by release of a peptide (*233*).

Factors A and F_1 have tightly bound adenine nucleotides. Factor A has over 1 mole of ADP per 360,000 g (*145*) and F_1 has 3 moles of ATP and 2 moles of ADP (*234*) bound tightly to its subunits. The latter preparation requires ATP addition to stabilize the activity, and the excess is removed by gel filtration before assay. The preparation of Factor A requires no added nucleotide. These differences might explain the discrepancy in the nucleotide content. These observations are consistent with Boyer's conformational coupling hypothesis.

Factor B is also a well characterized protein whose role is difficult to establish because it has no intrinsic activity (Table II). Recent work (*235*) has shown that it may occur in different forms, one of which has significantly higher specific activity, lower binding affinity to DEAE-cellulose, and a larger molecular weight (47,000), but reacts immunologically with an antiserum to Factor B. The role of coupling factor B in oxidative phosphorylation was suggested to be at the level of the non-phosphorylated energized state (X^* or $X \sim C$) (*236*). Support for the conclusion has come from the observation that the ATP–P_i exchange activity of OSATPase can be enhanced 3 to 4-fold by the addition of purified factor B (*229*).

The oligomycin sensitivity conferring protein (OSCP) is required for the expression of the oligomycin sensitivity of the ATPase system (Figure 15). There have also been speculations that OSCP may be the stalk attaching the 90 A F_1 particles to the mitochondrial membrane (Figure 1).

Racker has succeeded in resolving submitochondrial particles into a heterogenous, hydrophobic protein fraction with a low lipid content and low electron transport activity (*239*). The hydrophobic protein fraction when combined with lipid under controlled conditions yields a vesicular structure which in the presence of F_1 (and sometimes OSCP), is capable of catalyzing ATP–P_i exchange. The activity is low relative to that obtained with intact submitochondrial particle (ETPH) but is uncoupler sensitive. The preparation also appears to associate with purified cytochrome oxidase or NADH dehydrogenase with partial restoration of phosphorylation activity (*238, 44*) and proton transport (*238*) associated with substrate oxidation. Such reconstitution experiments hold promise of yielding further insight into the mechanism of oxidative phosphorylation.

Acknowledgments

This work was supported by grants from the National Institutes of Health (No. PO1 HD 05970 and No. GM 13641) and an institutional grant from the Public Health Service, General Research Support grant No. 5501 RR 05527. H. Wohlrab is an Established Investigator of the American Heart Association.

Literature Cited

1. Engelhardt, W. A., *Biochem. Z.* (1930) **227**, 16.
2. Kalckar, H., "Biological Phosphorylation: Development of Concepts," Prentice-Hall, Inc., New York, 1969.
3. Kalckar, H. M., *Enzym.* (1937) **2**, 47.
4. Belitzer, V. A., Tsibakowa, E. T., *Biokhim.* (1939) **4**, 516.
5. Ochoa, S., *Nature* (1940) **146**, 267.
6. Hogeboom, G. H., Schneider, W. C., Palade, G. H., *J. Biol. Chem.* (1948) **172**, 619.
7. Friedkin, M., Lehninger, A. L., *J. Biol. Chem.* (1948) **174**, 757.
8. Slater, E. C., in "Comprehensive Biochemistry," M. Florkin and E. H. Stotz, Eds., Vol. 14, p. 327, Elsevier, New York, 1966.
9. Lehninger, A. L., "The Mitochondrion," Benjamin Inc., New York, 1965.
10. Ernster, L., Lee, C. P., *Ann. Rev. Biochem.* (1964) **33**, 729.
11. Boyer, P. D., in "Biological Oxidations," T. P. Singer, Ed., p. 193, Interscience, New York, 1968.
12. Waino, W. W., "The Mammalian Mitochondrial Respiratory Chain," Academic Press, New York, 1970.
13. Palmer, G., Brintzinger, H., in "Electron and Coupled Energy Transfer in Biological Systems," T. E. King and M. Klingenberg, Eds., Vol. 113, p. 379, Marcel Dekker, Inc., New York, 1972.
14. Greville, G. D., "Current Topics in Bioenergetics," D. R. Sanadi, Ed., Vol. 3, Academic, New York, 1969.
15. Schnaitman, C., Greenawalt, J. W., *J. Cell Biol.* (1968) **38**, 1968.
16. Tedeschi, H., in "Current Topics in Membranes and Transport," F. Bronner and A. Kleinzeller, Eds., Vol. 2, p. 207, Academic, New York, 1971.
17. Green, D. E., Harris, R. A., *FEBS Lett.* (1969) **5**, 241.
18. Penniston, J. T., Harris, R. A., Asai, J., Green, D. E., *Proc. Nat. Acad. Sci., U.S.* (1968) **59**, 624.
19. Hackenbrock, C. R., *J. Cell Biol.* (1972) **53**, 450.
20. Ernster, L., *Fed. Proc.* (1965) **24**, 1222.
21. Chance, B., *Science* (1954) **120**, 767.
22. Chance, B., *Nature* (1952) **169**, 215.
23. Yang, C. C., Legallais, V., *Rev. Sci. Instruments* (1954) **35**, 801.
24. Wohlrab, H., *Biology Data Book, Fed. Am. Soc. Exptl. Biol.* (1974) **3**, 1588.
25. Griffiths, D. E., Wharton, D. C., *J. Biol. Chem.* (1961) **236**, 1850.
26. Kröger, A., Klingenberg, M., "Current Topics in Bioenergetics," D. R. Sanadi, Ed., Vol. 2, p. 51, Academic, New York, 1969.
27. Wilson, D. F., *Biochim. Biophys. Acta* (1967) **131**, 431.
28. Wilson, D. F., Erecinska, M., Nicholls, P., *FEBS Lett.* (1972) **20**, 61.
29. Orme-Johnson, N. R., Hansen, R. E., Beinert, H., *J. Biol. Chem.* (1974) **249**, 1928.
30. Ohnishi, T., *Biochim. Biophys. Acta* (1973) **301**, 105.
31. Beinert, H., Palmer, G., in "Oxidases and Related Redox Systems," T. E. King, H. S. Mason, and M. Morrison, Eds., Vol. 2, p. 567, Wiley, New York, 1965.

32. Malmstrom, B. G., *Quarterly Rev. Biophys.* (1973) **6**, 389.
33. Beinert, H., Van Gelder, B. F., Hansen, R. E., in "Structure and Function of Cytochromes," K. Okunuki, M. D. Kamen, and I. Sekuzu, Eds., p. 141, University Park Press, Baltimore, 1968.
34. Leigh, J. S., Wilson, D. F., *Biochem. Biophys. Res. Comm.* (1972) **48**, 1266.
35. Wohlrab, H., *Biochem. Biophys. Res. Comm.* (1969) **35**, 560.
36. Chance, B., DeVault, D., Legallais, V., Mela, L., Yonetani, T., in "Fast Reactions and Primary Processes in Chemical Kinetics," S. Claesson, Ed., p. 437, Almqvist and Wiksell, Stockholm, 1967.
37. Chance, B., Wilson, D. F., Dutton, P. L., Erecinska, M., *Proc. Nat. Acad. Sci., U.S.* (1970) **66**, 1175.
38. Orme-Johnson, N. R., Orme-Johnson, W. H., Hansen, R. E., Beinert, H., Hatefi, Y., *Biochem. Biophys. Res. Comm.* (1971) **44**, 446.
39. Ohnishi, T., Hemington, J. S., LaNoue, K. F., Morris, H. P., Williamson, J. R., *Biochem. Biophys. Res. Comm.* (1973) **55**, 372.
40. Dutton, P. L., *Biochim. Biophys. Acta* (1971) **226**, 63.
41. Oshino, R., Oshino, N., Tamura, M., Kobilinsky, L., Chance, B., *Biochim. Biophys. Acta* (1972) **273**, 5.
42. Wilson, D. F., Dutton, P. L., *Arch. Biochem. Biophys.* (1970) **136**, 583.
43. Wilson, D. F., Dutton, P. L., *Biochem. Biophys. Res. Comm.* (1970) **39**, 59.
44. Ragan, C. I., Racker, E., *J. Biol. Chem.* (1973) **248**, 2563.
45. Racker, E., Kandrach, A., *J. Biol. Chem.* (1973) **248**, 5841.
46. Slater, E. C., *Biochim. Biophys. Acta* (1973) **301**, 129.
47. Wikstrom, M. K. F., *Biochim. Biophys. Acta* (1973) **301**, 155.
48. Dutton, P. L., Wilson, D. F., Lee, C. P., *Biochem.* (1970) **9**, 5077.
49. Davis, K. A., Hatefi, Y., Poff, K. L., Butler, W. L., *Biochem. Biophys. Res. Comm.* (1972) **46**, 1984.
50. Boveris, A., Erecinska, M., Wagner, M., *Biochem. Biophys. Acta* (1972) **256**, 223.
51. Sato, N., Wilson, D. F., Chance, B., *FEBS Lett.* (1971) **15**, 209.
52. Jacobs, E. E., Sanadi, D. R., *J. Biol. Chem.* (1960) **235**, 531.
53. Ernster, L., Lee, I. Y., Norling, B., Persson, B., *Eur. J. Biochem.* (1969) **9**, 299.
54. Mitchell, P., *Fed. Europ. Biol. Sci. Proc.* (1972) **28**, 353.
55. Kröger, A., Klingenberg, M., *Eur. J. Biochem.* (1973) **39**, 313.
56. Hatefi, Y., Haavik, A. G., Fowler, L. R., Griffiths, D. E., *J. Biol. Chem.* (1962) **237**, 2661.
57. Wilson, D. F., Erecinska, M., Dutton, P. L., *Ann. Rev. Biophys. Bioengin.* (1974) **3**, 203.
58. Hatefi, Y., Stigall, D. L., Galante, Y., Hanstein, W. G., *Biochem. Biophys. Res. Comm.* (1974) **61**, 313.
59. Rieske, J. S., MacLennan, D. H., Coleman, R., *Biochem. Biophys. Res. Comm.* (1964) **15**, 338.
60. Rieske, J. S., *Arch. Biochem. Biophys.* (1971) **145**, 179.
61. Wilson, D. F., Koppelman, M., Erecinska, M., Dutton, P. L., *Biochem. Biophys. Res. Comm.* (1971) **44**, 759.
62. King, T. E., *Meth. Enzym.* (1967) **10**, 322.
63. Lee, C. P., Johansson, B., King, T. E., *Biochem. Biophys. Res. Comm.* (1969) **35**, 243.
64. Kearney, E. B., Ackrell, B. A. C., Mayr, M., Singer, T. P., *J. Biol. Chem.* (1974) **249**, 2021.
65. Gutman, M., Kearney, E. B., Singer, T. P., *Biochemistry* (1971) **10**, 2726.
66. Kröger, A., Klingenberg, M., *Vitam. Horm.* (1970) **28**, 533.
67. Wohlrab, H., *Biochemistry* (1970) **9**, 474.
68. Wohlrab, H., Ogunmola, G. B., *Biochemistry* (1971) **10**, 1103.
69. Leigh, J. S., Wilson, D. F., Owen, C. S., King, T. E., *Arch. Biochem. Biophys.* (1974) **160**, 476.

70. Yong, F. C., King, T. E., *J. Biol. Chem.* (1972) **247**, 6384.
71. Takano, T., Kallai, O. B., Swanson, R., Dieckerson, R. E., *J. Biol. Chem.* (1973) **248**, 5234.
72. Racker, E., *Essays Biochem.* (1970) **6**, 1.
73. Lee, C. P., in "Probes of Structure and Function of Macromolecules and Membranes," B. Chance, C. P. Lee, and J. K. Blaise, Eds., Vol. 1, p. 417, Academic, New York, 1971.
74. Chance, B., Mela, L., *J. Biol. Chem.* (1967) **242**, 830.
75. Racker, E., Burstein, C., Loyter, A., Christiansen, R. O., in "Electron Transport and Energy Conservation," p. 35, Adriatica Editrice, New York, 1970.
76. Margoliash, E., in "Probes of Structure and Function of Macromolecules and Membranes," B. Chance, C. P. Lee, and J. K. Blaise, Eds., Vol. 1, p. 471, Academic, New York, 1971.
77. Palmieri, F., Klingenberg, M., *Eur. J. Biochem.* (1967) **1**, 439.
78. Wilson, D. F., in "Probes of Structures and Function of Macromolecules and Membranes," B. Chance, C. P. Lee, and J. K. Blaise, Eds., Vol. 2, p. 593, Academic, New York, 1971.
79. Azzi, A., Vainio, H., in "Probes of Structure and Function of Macromolecules and Membranes," B. Chance, C. P. Lee, and J. K. Blaise, Eds., Vol. 1, p. 209, Academic, New York, 1971.
80. Azzi, A., Vainio, H., in "Electron Transport and Energy Conservation," p. 540, Adriatica Editrice, New York, 1970.
81. Heidrich, H. G., *Fed. Europ. Biol. Sci. Lett.* (1971) **17**, 253.
82. Chance, B., Erecinska, M., Lee, C. P., *Proc. Nat. Acad. Sci., U.S.* (1970) **66**, 928.
83. Schneider, D. L., Kagawa, Yl, Racker, E., *J. Biol. Chem.* (1972) **247**, 4074.
84. Maddy, H., *Biochim. Biophys. Acta* (1964) **88**, 390.
85. Berg, H. C., *Biochim. Biophys. Acta* (1969) **183**, 65.
86. Phillips, D. R., Morrison, M., *Biochem.* (1971) **10**, 1766.
87. Carroll, R. C., Eytan, G. D., *Fed. Proc. Am. Soc. Expt. Biol.* (1975) **34**, 579.
88. Chance, B., Williams, G. R., *Adv. Enzym.* (1956) **17**, 65.
89. Kimura, T., Singer, T. P., *Nature* (1959) **184**, 791.
90. Jacobs, E. E., *Biochem. Biophys. Res. Comm.* (1960) **3**, 536.
91. Sanadi, D. R., Fluharty, A. L., *Biochemistry* (1963) **2**, 523.
92. Schatz, G., Racker, E., *J. Biol. Chem.* (1966) **241**, 791.
93. Lee, C. P., Sottocase, G. L., Ernster, L., *Meth. Enzym.* (1967) **10**, 33.
94. Chance, B., Hollinger, G., *Fed. Proc.* (1957) **16**, 163.
95. Wilson, D. F., Dutton, P. L., in "Energy Metabolism and the Regulation of Metabolic Processes in Mitochondria," M. A. Mehlman and R. W. Hanson, Eds., p. 39, Academic, New York, 1972.
96. Storey, B. T., Lee, C. P., in "Mechanisms in Bioenergetics," G. F. Assone, L. Ernster, S. Papa, E. Quagliarlello, and N. Siliprnadi, Eds., p. 127, Academic, New York, 1973.
97. Dutton, P. L., Lindsay, J. G., *Ibid.*, p. 535.
98. Lambowitz, A. M., Bonner, W. D., Wikstrom, M. K. F., *Proc. Nat. Acad. Sci., U.S.* (1974) **71**, 1183.
99. Wikstrom, M. K. F., Lambowitz, A. M., *FEBS Lett.* (1974) **40**, 149.
100. Wikstrom, M. K. F., *Ann. N.Y. Acad. Sci.* (1974) **227**, 146.
101. Slater, E. C., *FEBS Lett.* (1972) **28**, 133.
102. Wilson, D. F., Brocklehurst, E. S., *Arch. Biochem. Biophys.* (1973) **158**, 200.
103. Storey, B. T., *J. Theoret. Biol.* (1971) **31**, 533.
104. DeVault, D., *Biochim. Biophys. Acta* (1971) **225**, 193.
105. Slater, E. C., in "Molecular Basis of Electron Transport," J. Schultz, Ed., p. 95, Academic, New York, 1972.
106. Ting, H. P., Wilson, D. F., Chance, B., *Arch. Biochem. Biophys.* (1970) **141**, 141.

107. Bakker, E. P., Van den Henvel, E. J., Wiechmann, A. H. C. A., Van Dam, K., *Biochim. Biophys. Acta* (1973) **292**, 78.
108. Wilson, D. F., Ting, H. P., Koppelman, M., *Biochem.* (1971) **10**, 2897.
109. Hanstein, W. G., Hatefi, Y., *Proc. Nat. Acad. Sci., U.S.* (1974) **71**, 288.
110. Lee, C. P., *Fed. Proc. Fed. Am. Soc., Expt. Biol.* (1975) **34**, 577.
111. Wilson, D. F., *Biochem.* (1969) **8**, 2475.
112. Sanadi, D. R., *Arch. Biochem. Biophys.* (1968) **128**, 280.
113. Wohlrab, H., *Fed. Proc.* (1971) **30**, 1191.
114. Kaplay, M., Kurup, C. K. R., Lam, K. W., Sanadi, D. R., *Biochem.* (1970) **9**, 3599.
115. Hanstein, W. G., Hatefi, Y., *J. Biol. Chem.* (1974) **249**, 1356.
116. Copeland, L., Deutsch, C. J., Tu, S., Wang, J. H., *Arch. Biochem. Biophys.* (1974) **160**, 451.
117. Cattell, K. J., Lindop, C. R., Knight, I. G., Beechey, R. B., *Biochem. J.* (1971) **125**, 169.
118. Chang, T. M., Penefsky, H. S., *J. Biol. Chem.* (1974) **249**, 1090.
119. Dawson, A. P., Sewlwyn, M. J., *Biochem. J.* (1974) **138**, 349.
120. Bustamante, E., Pedersen, P. L., *Biochem. Biophys. Res. Comm.* (1973) **51**, 292.
121. Howland, J. L., *Biochim. Biophys. Acta* (1963) **73**, 665.
122. Borst, P., Slater, E. C., *Biochim. Biophys. Acta* (1961) **48**, 362.
123. Chappell, J. B., Grevillw, G. D., *Nature* (1961) **190**, 502.
124. Estabrook, R. W., *Biochem. Biophys. Res. Comm.* (1961) **4**, 89.
125. Haijing, F., Slater, E. C., *J. Biochem. (Tokyo)* (1961) **49**, 493.
126. Mitchell, P., *Nature* (1961) **191**, 144.
127. Mitchell, P., *FEBS Lett.* (1974) **43**, 189.
128. Mitchell, P., Moyle, J., *Biochem. J.* (1967) **105**, 1147.
129. Jagendorf, A., Uribe, E., *Proc. Nat. Acad. Sci., U.S.* (1966) **55**, 170.
130. Chance, B., Mela, L., *Proc. Nat. Acad. Sci., U.S.* (1966) **55**, 1246.
131. Chappel, J. B., *Brit. Medical Bull.* (1968) **24**, 150.
132. Mitchell, P., *Fed. Proc.* (1967) **26**, 1370.
133. Chance, B., Lee, C. P., Mela, L., *Fed. Proc.* (1967) **26**, 134.
134. Tedeschi, H., *Proc. Nat. Acad. Sci., U.S.* (1974) **71**, 583.
135. Harris, E. J., Pressman, B. C., *Biochim. Biophys. Acta* (1969) **172**, 66.
136. Mitchell, P., Moyle, J., *Eur. J. Biochem.* (1969) **7**, 471.
137. Schuldiner, S., Rottenberg, H., Avron, M., *Eur. J. Biochem.* (1973) **39**, 455.
138. Vam Dam, K., Engel, G. L., "Mechanisms in Bioenergetics," G. F. Azzone, L. Ernster, S. Papa, E. Quagliariello, and N. Siliprandi, Eds., p. 141, Academic, New York, 1973.
139. Wilson, D. F., Erecinska, M., *FEBS Lett.* (1972) **28**, 119.
140. Hunter, D. R., Komai, H., Haworth, R. A., *Biochem. Biophys. Res. Comm.* (1974) **56**, 647.
141. Boyer, P. D., Cross, R. L., Monsen, W., *Proc. Nat. Acad. Sci., U.S.* (1973) **70**, 2837.
142. Eisenhardt, R. H., Rosenthal, O., *Biochemistry* (1968) **7**, 1327.
143. Chan, P. C., Lehninger, A. L., Enns, T., *J. Biol. Chem.* (1960) **235**, 1790.
144. Boyer, P. D., in "Current Topics in Bioenergetics," D. R. Sanadi, Ed., Vol. 2, p. 99, Academic, New York, 1967.
145. Warshaw, J. B., Lam, K. W., Nagy, B., Sanadi, D. R., *Arch. Biochem. Biophys.* (1968) **123**, 385.
146. Roy, H., Moudrianakis, E. N., *Proc. Nat. Acad. Sci., U.S.* (1971) **68**, 2720.
147. Cross, R. L., Boyer, P. D., *Biochem. Biophys. Res. Comm.* (1973) **51**, 59.
148. Chance, B., Azzi, Lee, I. Y., Lee, C. P., Mela, L., in "Mitochondrial Structure and Function," L. Ernster and Z. Drahota, Eds., pp. 106–114, Academic, New York, 1969.
149. Cohn, M., *J. Biol. Chem.* (1953) **201**, 735.

150. Boyer, P. D., Falcone, A. S., Harrison, W. H., *Nature* (1954) **174**, 401.
151. Swanson, M. A., *Biochim. Biophys. Acta* (1956) **20**, 85.
152. Cohn, M., Drysdale, G. R., *J. Biol. Chem.* (1955) **216**, 851.
153. Boyer, P. D., Luchsinger, W. N., Falcone, A. S., *J. Biol. Chem.* (1956) **223**, 405.
154. Cooper, C., Lehninger, A. L., *J. Biol. Chem.* (1957) **224**, 561.
155. Chiga, M., Plant, G. W. E., *J. Biol. Chem.* (1959) **234**, 3059.
156. Hill, R. D., Boyer, P. D., *J. Biol. Chem.* (1967) **242**, 4320.
157. Lardy, H. A., Johnson, D., McMurray, W. C., *Arch. Biochem. Biophys.* (1958) **78**, 587.
158. Mitchell, R. A., Hill, R. D., Boyer, P. D., *J. Biol. Chem.* (1967) **242**, 1799.
159. Lardy, H. A., Connelly, J. L., Johnson, D., *Biochemistry* (1964) **3**, 1961.
160. Chance, B., Hollenger, B., *Fed. Proc.* (1957) **16**, 163.
161. Klingenberg, M., Schollmayer, P., *Biochem. Z.* (1961) **335**, 243.
162. Löew, H., Kreuger, H., Ziegler, D. M., *Biochem. Biophys. Res. Comm.* (1961) **9**, 307.
163. Lee, C. P., Ernster, L., in "Regulation of Metabolic Processes in Mitochondria," J. M. Tager et al., Eds., Vol. 7, p. 218, Elsevier, New York, 1965.
164. Slater, E. C., Tager, J. M., *Biochem. Biophys. Acta* (1963) **77**, 276.
165. Fluharty, A. L., Sanadi, D. R., *Biochem.* (1963) **2**, 519.
166. Cockrell, R. S., Harris, E. J., Pressman, B. C., *Biochem.* (1966) **5**, 2326.
167. Slater, E. C., Rosing, J., Mol, A., *Biochim. Biophys. Acta* (1973) **292**, 534; Kaplan, N. O., Colowick, S. P., Neufeld, E. E., Ciotti, M. M., *J. Biol. Chem.* (1953) **205**, 17.
168. San Pietro, A., Kaplan, N. O., Colowick, S. P., *J. Biol. Chem.* (1955) **212**, 941.
169. Louie, D. D., Kaplan, N. O., in "Pyridine Nucleotide Dependent Dehydrogenases," H. Sund, Ed., p. 351, Springer-Verlag, Berlin, 1970.
170. Kawasaki, T., Satoh, K., Kaplan, N. O., *Biochem. Biophys. Res. Comm.* (1964) **17**, 648.
171. Kaplan, N. O., Colowick, S. P., Neufeld, E. E., *J. Biol. Chem.* (1952) **195**, 107.
172. Teixeira, Da Cruz, A., Rydstrom, J., Ernster, L., *Eur. J. Biochem.* (1971) **23**, 203.
173. Krebs, H. A., Kornberg, H. L., *Ergbn. Physiol.* (1957) **49**, 212.
174. Klingenberg, M., Slenczka, W., *Biochem. Z.* (1959) **331**, 486.
175. Estabrook, R. W., Nissley, S. P., in "Symp. Funktionelle Morphologische Organization del Zelle," P. Karlson, Ed., pp. 235–248, Springer-Verlag, Heidelberg, 1963.
176. Danielson, L., Ernster, L., *Biochem. Biophys. Res. Comm.* (1963) **10**, 91.
177. Lee, C. P., Ernster, L. B., *Biochim. Biophys. Acta* (1964) **81**, 187.
178. Van de Stadt, R. J., Nieuwenhuis, F. J. R. M., Van Dam, K., *Biochim. Biophys. Acta* (1971) **234**, 173.
179. Moyle, J., Mitchell, P., *Biochem. J.* (1973) **132**, 571.
180. Lee, C. P., Simard-Dusquesne, N., Ernster, L., Hoberman, H. D., *Biochim. Biophys. Acta* (1965) **105**, 397.
181. Orlando, J. A., *Arch. Biochem. Biophys.* (1970) **141**, 111.
182. Rydstrom, J., Teixiera Da Crux, A., Ernster, L., *Eur. J. Biochem.* (1971) **23**, 212.
183. Chance, B., *J. Biol. Chem.* (1965) **240**, 2729.
184. Lehninger, A. L., Carafoli, E., Rossi, C. S., *Adv. in Enzym.* (1967) **29**, 259.
185. Lehninger, A. L., *Proc. Nat. Acad. Sci., U.S.* (1974) **71**, 1520.
186. Chen, C. H., Greenawalt, J. W., Lehninger, A. L., *J. Cell Biol.* (1974) **61**, 301.
187. Vinogradov, A., Scarpa, A., *J. Biol. Chem.* (1973) **248**, 5527.
188. Moore, C., *Biochem. Biophys. Res. Comm.* (1971) **42**, 298.
189. Mela, L., *Biochemistry* (1969) **8**, 2481.

190. Akerman, K. E., Saris, N. E. L., Jarvisalo, J. O., *Biochem. Biophys. Res. Comm.* (1974) **58**, 801.
191. Southard, J. H., Green, D. E., *Biochem. Biophys. Res. Comm.* (1974) **59**, 30.
192. Gomez-Puyou, A., Gomez-Puyou, M. T., Becker, G., Lehninger, A. L., *Biochem. Biophys. Res. Comm.* (1972) **47**, 814.
193. Sottocasa, G., Sandri, G., Panfili, E., Bernard, B., Gazotti, P., Vasington, F. D., Carafoli, E., *Biochem. Biophys. Res. Comm.* (1972) **47**, 808.
194. Reed, P. W., Lardy, H. A., *J. Biol. Chem.* (1972) **247**, 6970.
195. Colonna, R., Massari, S., Azzone, G. F., *Eur. J. Biochem.* (1973) **34**, 577.
196. Case, G. D., *Fed. Proc.* (1974) **33**, 1399.
197. Moore, C., Pressman, B. C., *Biochem. Biophys. Res. Comm.* (1964) **15**, 562.
198. Crofts, A. R., *Ann. N.Y. Acad. Sci.* (1969) **147**, 801.
199. Brierley, G. P., Jurkowitz, M., Scott, K. M., Merola, A. J., *Arch. Biochem. Biophys.* (1971) **147**, 545.
200. Blondin, G. A., DeCastro, A. F., Senior, A. E., *Biochem. Biophys. Res. Comm.* (1971) **43**, 28.
201. Mitchell, P., *Symp. Soc. Gen. Microb.* (1970) **20**, 121.
202. Massari, S., Azzone, G. F., *Eur. J. Biochem.* (1970) **12**, 310.
203. Sluse, F. E., Ranson, M., Lebecq, C., *Eur. J. Biochem.* (1972) **25**, 207.
204. Klingenberg, M., in "Energy-linked Functions of Mitochondria," p. 121, Academic, New York, 1963.
205. Chappell, J. B., Haanhoff, K. N., in "Biochemistry of Mitochondria," p. 75, Academic, New York, 1966.
206. Kraaijenhof, R., Tsou, C. S., Van Dam, K., *Biochim. Biophys. Acta* (1969) **172**, 580.
207. Crompton, M., Chappell, J. B., *Biochemistry J.* (1973) **132**, 35.
208. Pfaff, E., Heldt, H. W., Klingenberg, M., *Eur. J. Biochem.* (1969) **10**, 484.
209. Pfaff, E., Klingenberg, M., *Eur. J. Biochem.* (1968) **6**, 66.
210. Fonyo, A., *Biochem. Biophys. Res. Comm.* (1974) **57**, 1069.
211. Johnson, R. N., Chappell, J. B., *Biochemistry J.* (1974) **138**, 171.
212. Chappell, J. B., Robinson, B. H., in "Metabolic Role of Citrate," p. 123; *Biochem. Soc. Symp.*, Academic, New York, 1968.
213. Robinson, B. H., Williams, G. R., Halperin, M. L., Leznoff, C. C., *J. Biol. Chem.* (1971) **246**, 5280.
214. Halestrap, A. P., Denton, R. M., *Biochemistry J.* (1974) **138**, 313.
215. Meijer, A. J., Brouwer, A., Reingoud, D. J., Hoek, J. B., Tager, J. M., *Biochim. Biophys. Acta* (1972) **283**, 421.
216. LaNoue, K. F., Meijer, A. J., Brouwer, A., *Arch. Biochem. Biophys.* (1974) **161**, 544.
217. Klingenberg, M., *Ess. Biochem.* (1970) **6**, 119.
218. Moyle, J., Mitchell, P., *FEBS Lett.* (1973) **30**, 317.
219. Scherer, B., Klingenberg, M., *Biochem.* (1974) **13**, 161.
220. Vignais, P. V., Vignais, P. M., Defaye, G., *Biochem.* (1973) **12**, 1508.
221. Egan, R. W., Lehninger, A. L., *Biochem. Biophys. Res. Comm.* (1974) **59**, 195.
222. Shertzer, H. G., Racker, E., *J. Biol. Chem.* (1974) **249**, 1320.
223. Beechey, R. B., Cattell, K. J., in "Current Topics in Bioenergetics," D. R. Sanadi and L. Packer, Eds., Vol. 5, p. 306, Academic, New York, 1973.
224. Senior, A. E., *Biochim. Biophys. Acta* (1973) **301**, 249.
225. Knowles, A. F., Penefsky, H. S., *J. Biol. Chem.* (1972) **247**, 6624.
226. Pullman, M. E., Monroy, *J. Biol. Chem.* (1963) **238**, 3762.
227. Van De Stadt, R. J., De Boer, B. L., Van Dam, K., *Biochim. Biophys. Acta* (1973) **292**, 338.
228. Andreoli, T. E., Lam, K. W., Sanadi, D. R., *J. Biol. Chem.* (1965) **240**, 2644.
229. Joshi, S., Shaikh, F., Sanadi, D. R., *Biochem. Biophys. Res. Comm.* (1975) **65**, 1375.

230. Sanadi, D. R., Sani, B. P., Fisher, R. J., Li, O., Taggart, W. V., in "Energy Transduction in Respiration and Photosynthesis," E. Quagliariello, S. Papa, and C. S. Rossi, Eds., p. 89, Adriatica Editrice, Bari, 1971.
231. Penefsky, H. S., Pullman, M. E., Datta, A., Racker, E., *J. Biol. Chem.* (1960) **235,** 3330.
232. Panet, R., Sanadi, D. R., in "Current Topics in Membranes and Transport," A. Kleinzeller and F. Bronner, Eds., Academic, New York, 1975.
233. Farron, F., Racker, E., *Biochemistry* (1970) **9,** 3829.
234. Slater, E. C., Harris, D. A., Rosing, J., Van de Stadt, R., *Biochim. Biophys. Acta* (1973) **314,** 149.
235. Shankaran, R., Sani, B. P., Sanadi, D. R., *Arch. Biochem. Biophys.* (1975) **168,** 394.
236. Sanadi, D. R., Lam, K. W., Kurup, C. K. R., *Proc. Nat. Acad. Sci., U.S.* (1968) **61,** 277.
237. Kagawa, Y., Racker, E., *J. Biol. Chem.* (1971) **246,** 5477.
238. Hinkle, P., Horstman, L., *J. Biol. Chem.* (1971) **246,** 6024.

Chapter

5

Mechanisms in Photosynthesis

Bessel Kok and Richard Radmer, Martin Marietta Corp., Martin Marietta Laboratories, 1450 South Rolling Rd., Baltimore, Md. 21227

DEFINITION:

PHOTOSYNTHESIS IS A BIOLOGICAL process in which light energy is harvested and utilized to drive energy-requiring biochemical reactions. Life is maintained by the continuous influx of sunlight which is harvested by photosynthetic organisms and transformed into chemical energy. All other biological processes utilize the energy obtained by photosynthetic organisms and in this sense might be considered parasitic. Consequently, photosynthesis is the driving force for all metabolic processes in the biosphere.

Under optimal conditions using red light, photosynthetic organisms can convert about 30% of the absorbed energy into chemical energy. For white light (unlight) the efficiency drops to about 20%. Since the functional pigments in green plants absorb all wavelengths shorter than 700 nm, or about one-half of the solar spectrum, this implies that photosynthetic organisms can convert up to 10% of the incident solar radiation.

Under natural conditions only about 1% of the solar energy reaching the earth's surface is channeled through the plant kingdom and supports life; most of the incident energy is re-radiated as heat. However, despite this poor overall yield the total energy conversion by photosynthetic organisms is orders of magnitude higher than the total industrial output of man.

Most of the important qualitative aspects of photosynthesis were deduced over a period of about 100 years centering around 1800. (For a more detailed account of this historical development *see* Refs. *1* and *2*.) The first documented experiments which led to an understanding of the nature of the photosynthetic process were performed by Joseph Priestly, better known for his discovery of oxygen. In 1772 he reported that green plants could "restore" air fouled by respiration or combustion. He did not recognize, however, that light was required for this "restoration," and thus others (and later he himself) had difficulty in repeating his original experiments.

The necessity of light for the production of "dephlogisticated air" (O_2) was recognized a few years later by Jan Ingen-Housz. In 1779 he reported that only the green parts of plants could restore noxious air and only when exposed to sunlight. He also demonstrated that the restoration of air by green plants did not arise merely from the absorption of noxious components, but it was rather a consequence of the production of "vital air" (O_2).

Three years later Jean Senebier demonstrated and reported the accelerating effect of "fixed air" (CO_2) on the production of "pure air" (O_2). This work was elaborated further by Ingen-Housz in 1796; using the terminology of Lavoisier (oxygen, carbon dioxide) he described the nutritive role of air. Thus at that time photosynthesis could be described as:

$$\text{carbon dioxide} + \text{light} \xrightarrow{\text{green plant}} \text{organic matter} + \text{oxygen}$$

In 1804 Nicoles Theodore deSaussure—in addition to confirming the idea of aerial nutrition espoused by Ingen-Housz—conclusively demonstrated the participation of water in the process. The law of the conservation of energy was formulated in 1842 by Julius Robert Mayer. In 1845 Mayer applied this concept to biological processes and, recognizing the process as the conversion of light to chemical energy, gave the first qualitatively correct account of photosynthesis:

$$\text{carbon dioxide} + \text{water} + \text{light} \xrightarrow{\text{light}} \text{organic matter} + \text{oxygen} + \text{chemical energy}$$

A comparison of aerobic (green plant) photosynthesis and bacterial photosynthesis led VanNiel (3) to suggest that the generalized process might be formulated as

$$2H_2A + CO_2 \xrightarrow{h\nu} [CH_2O] + H_2O + 2A$$

(in which A symbolizes any electron acceptor) or for the case of green plant photosynthesis

$$2H_2O + CO_2 \xrightarrow{h\nu} [CH_2O] + H_2O + O_2$$

According to this formulation aerobic photosynthesis is the light-driven reduction of CO_2 by H_2O. This implies that the O_2 evolved is derived from H_2O—a conclusion which has been corroborated by the use of ^{18}O tracers (4) and by the observation that light-driven O_2 evolution occurs in the presence of electron acceptors other than CO_2. The latter observation was made possible when Hill (5) succeeded in isolating chloroplasts which were photochemically active *in vitro*.

Later studies by Calvin and associates (6) elucidated the metabolic pathway by which the light-generated reductant is used to reduce CO_2. The operation of this pathway also requires energy input in the form of ATP. This ATP is generated by photophosphorylation, first observed by Arnon et al. (7) and by Frenkel (8). Later studies by Jagendorf and Hind (9) suggested that this process may utilize energy stored as a concentration gradient across the chloroplast membrane.

The concept of the photosynthetic unit arose from the observation of Emerson and Arnold (10) that the maximum yield of O_2 evolved in a brief flash was orders of magnitude less than the amount of chlorophyll. This concept was supported by the observations of Duysens (11) that energy could migrate through the pigment bed to photochemical conversion centers, such as P_{700} discovered by Kok (12).

The acceptance of a quantum requirement of two photons per equivalent moved from water to NADP, coupled with the observations of chromatic transients by Blinks (13) and of rate enhancement by Emerson (14, 15) led to another important conceptual step forward. About 1960, on the basis of observed light-induced absorption changes (16, 17) and cytochrome potentials (18), three laboratories independently arrived at the concept that green plant photosynthesis is driven by two different photoacts. This hypothesis—in which the photochemical apparatus can be compared with two light-driven batteries connected in series—is often referred to as the "Z" scheme. (Figure 1). It has become one of the cornerstones of modern thinking in aerobic photosynthesis; thus, it is repeatedly referred to throughout this chapter.

The initial process in photosynthesis is the absorption of a quantum of light by a chloroplast pigment molecule—chlorophyll *a* or one of the accessory pigments—which results in the production of an excited state. This excited state is able to migrate from one pigment molecule to another because of the dense packing of the light-harvesting pigments in the lamellae. It eventually encounters a special "trap" chlorophyll *a* molecule which converts the light-generated excited state into chemical energy.

Units of roughly 200 light-harvesting ("antennae") molecules serve each of these "trapping centers" which are specially bound pigment molecules close to an appropriate electron donor and acceptor. Within each unit, an electron of the trapping center chlorophyll *a* molecule is raised to an excited state by a light quantum and reduces its electron acceptor $(A \to A^-)$. The oxidized chlorophyll is returned to its original state by oxidation of its electron donor $(D \to D^+)$. The overall result of this process is the light-driven movement of an electron against the energy gradient. By coupling the back reaction to an energy requiring process, the charge separation within the trapping center can be used to drive energy-requiring chemical reactions. This scheme can be summarized as follows:

$$Chl + h\nu \longrightarrow Chl^*$$
$$Chl^* + A \longrightarrow Chl^+ + A^-$$
$$Chl^+ + D \longrightarrow Chl + D^+$$
$$A^- + D^+ \longrightarrow A + D + energy$$

Although the excited state of chlorophyll might contain all the energy of the photon which produced it, it has a lifetime of nano- to picoseconds (for antenna and trapping center chlorophyll, respectively). On the other hand, although the chemical potential of the products of the photoact is considerably less than that of the original excited state, the stability of these products is

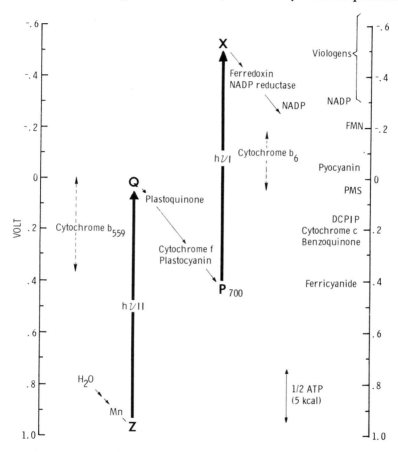

Figure 1. Left: a potential diagram of the two photoacts connected in series (the Z scheme). Right: some redox reagents used in chloroplast reactions, arranged according to their normal potential.

much greater. The difference between the theoretical and observed energy could thus be considered to be the amount of energy spent to obtain products with useful stability.

Green plants are able to produce molecular O_2 and at the same time generate a reductant at least as strong as molecular H_2. Since the potential of the hydrogen electrode (E_m') is -0.42 V and that of the oxygen electrode

+0.81 V, a total chemical potential of about 1.2 V must be created through photosynthesis. To cover the entire 1.2-V span, two light reactions operate in series (using two photons per equivalent moved through the chain) as shown in Figure 1. Photoact I utilizes photons which are mainly absorbed by chlorophyll a and produces a very strong reductant X and a weak oxidant P_{700}^+. Photoact II is sensitized by both chlorophyll a and accesory pigments and produces a weak reductant Q^- and a very strong oxidant Z^+. In both cases a quantum moves one electron equivalent against a gradient of about one eV. Since light quanta of about 700 nm (which are able to initiate this photoconversion) contain about 1.8 eV, the maximum efficiency of each photoact is no more than about 50%.

In the case of photosynthetic bacteria, which do not evolve O_2, it appears that only a single photosystem—somewhat similar to System I of green plants—is operational. These organisms generate only weak oxidizing power and cannot oxidize H_2O to O_2.

Light-driven electron transport is a structure-dependent process. The pigments, grouped in functional units with photochemical conversion centers and other catalysts, are located in lamellae. These structures can be further organized and concentrated in special organelles (discussed later). Much of our present knowledge of photosynthetic electron transport is based on experiments which used chloroplasts isolated from the whole cell. These preparations can perform many photoconversions which are often simpler and more amenable to study than similar processes in the whole cell.

The production of molecular O_2 and reduced organic material by green plants involves the coupling of the light reactions shown in Figure 1 with several different groups of dark reactions. For example, the generation of an O_2 molecule involves the cooperation of four oxidizing equivalents (Z^+) which requires a complex sequence of enzymatic dark reactions. Similarly, the reducing end of System I (X^-) does not reduce NADP directly but proceeds *via* two enzymatic steps.

The turnover of the photosynthetic electron transport system *in vivo* also results in the incorporation of light energy into the pyrophosphate bonds of ATP. This phosphorylation not only occurs coupled to electron transport from water to X^-, but it can also accompany abbreviated electron transport paths (which might involve only one of the photoacts) or cyclic paths which yield no net change of electron donors or acceptors.

In addition to the four primary photoproducts (only one of which has been conclusively identified so far) many other electron carriers might be involved in photosynthetic electron transport. In Figure 1 these components are arranged according to their respective redox potentials, possibly indicative of their sites of operation. There is good evidence that light energy is not only conserved by the oxidation and reduction of specific intermediates but also by the production of membrane potentials and concentration differences of protons (and other ions) between the inner and outer phases of the thylakoid sacs.

The essence of photosynthesis is the conversion of an energy form (light) which is inaccessible to most members of the biosphere into a form (chemical energy) which can be used by other organisms and the photosynthetic organism itself. Photosynthesis can be broken down into a hierarchy of partial processes on the basis of the type of processes occurring—physical vs. chemical—and its order in the overall sequence of events. Our discussions below attempt to adhere to this hierarchy. We discuss aspects related to the harvest of light energy, the processing of this light energy into usable photoproducts, and the chemical utilization of these photoproducts.

Photosynthetic Structures

All photosynthetic organisms contain special intracellular structures which are the loci of the photosynthetic processes. The degree of organization of these structures is related to the organism's position in the phylogenetic tree (*e.g.*, procaryotic *vs.* eucaryotic) and its ability to produce oxygen photosynthetically. Of the four possible combinations of these two criteria, three are encountered in nature;

(1) Photosynthetic procaryotic organisms which do not produce oxygen (photosynthetic bacteria);

(2) Procaryotic organisms which produce oxygen (blue-green algae);

(3) Photosynthetic eucaryotic organisms, all of which produce oxygen (algae other than blue-green and higher plants).

The degree of organization complexity increases in the order listed above. Photosynthetic processes in bacteria are located in membrane structures which reveal no distinct boundaries (except possibly the green sulfur bacteria) and little geometrically arranged structure. Disruption of the bacterial cell wall of purple bacteria produces roughly spherical particles (chromatophores) which arise presumably from the pinching off of vesicular structures during the disruption. There is no organized chloroplast in blue-green algae, but there is a lamellar system surrounded by an embedding matrix.

The most distinct and specialized photosynthetic structures are found in algae (other than blue-green) and higher plants. In these organisms the photosynthetic processes are isolated from the rest of the cell by a double membrane system. In the chloroplast stacks of lamellae (grana) similar to those seen in blue-green algae are apparent. These grana are surrounded by a more mobile phase (the stroma) which can be considered as the cytoplasm of the chloroplast, but it is separated from the true cytoplasm by the double membrane system. These features of the chloroplast are clearly illustrated in Figure 2, which shows a chloroplast in the green alga *Nitella* as viewed by electron microscopy. Eucaryotic organisms seem to have a more organized photosynthetic structure than blue–green algae, but there are no significant differences in the photosynthetic capabilities (*e.g.*, quantum yields, maximum rates, and so forth) of the two groups.

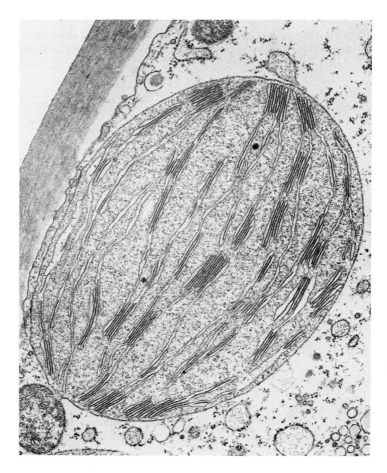

Figure 2. Electron micrograph (magnification ×20,800) of a chloroplast in the green alga Nitella. *(Courtesy of M. C. Ledbetter, Brookhaven National Laboratory.)*

It is possible to separate physically—as well as conceptually—the light-driven formation of reactive intermediates from their subsequent utilization in homogeneous enzyme-mediated chemical reactions. During the past decade much evidence was obtained which suggests that photosynthetic light reactions and the associated electron transport are located in the lamellae, but the dark reactions which utilize the light-generated chemical energy are located in the stroma (chloroplast cytoplasm) (19). Within the lamellar structure itself, the moieties which participate in the primary events appear to be rigidly localized. Those which should interact are located close together while those which should not interact are kept apart. The lamellae appear to be made up of closed double membrane sacs called thylakoids (20). The photosynthetic pigments and reaction chains are arranged on this double membrane in a definite orientation with respect to inside and outside. Consequently, the inside milieu of the thylakoid can be very different from that of the outside.

Early Events of Energy Capture

Pigments; Composition and Absorption Properties. The light-harvesting apparatus of all photosynthetic organisms utilizes pigments derived from two classes of compounds—the tetrapyrroles and the isoprenoids. The general structure of these compounds is given in Figure 3.

The tetrapyrrole pigments fall into two classes. The chlorophylls are closed-chain tetrapyrroles derived (biologically) from and closely akin to porphyrins. Chlorophyll *a* is a "universal pigment" and is present in all oxygen-evolving photosynthetic organisms. Some of the other chlorophylls (*e.g.*, Chl *c*) have a much more limited occurrence.

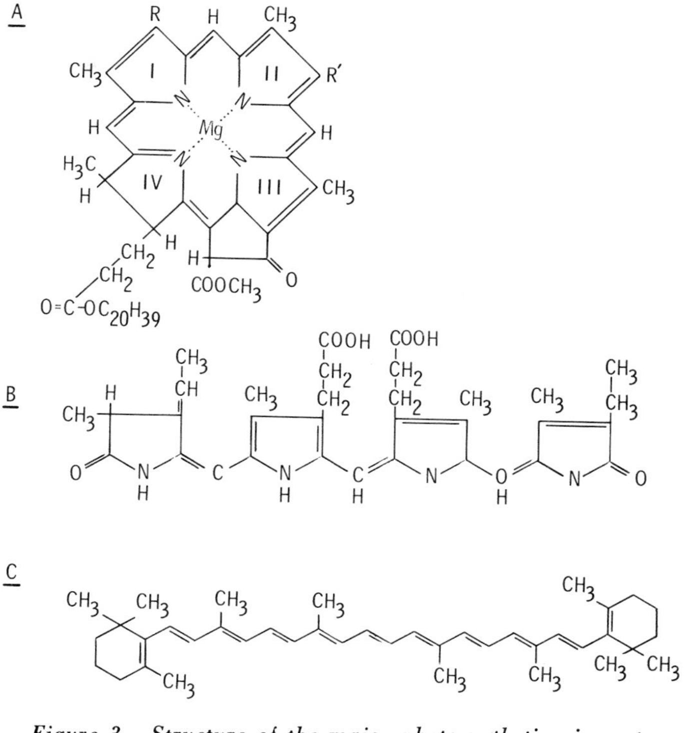

Figure 3. Structure of the major photosynthetic pigments

A: chlorophyll; R = —CH=CH₂; when R′ = —C₂H₅, chlorophyll a*; when R′ = —CHO, chlorophyll* b*. Chlorophyll* c *is not esterified with phytol* (i.e., *(—OC₂₀H₃₉)), and ring IV is not reduced. In bacterochlorophyll, ring II as well as ring IV is reduced; R = —COCH₃, R′ = —C₂H₅. B: Phycocyanobilin, the chromatophore of the biliprotein C-phycocyanin* (21)*. C: β-Carotene.*

Bile pigments—although also apparently derived from porphyrins—are linear (open-chain) tetrapyrroles with physical and chemical properties quite distinct from their biological precursors. These pigments are covalently bound to proteins to form so-called phycobiliproteins, which occur as accessory pigments in certain algae (*e.g.*, blue-green and red algae).

The carotenoids are isoprenoid compounds with chain lengths of 40 carbons (*i.e.*, tetraterpenes). They are usually divided into two classes—the carotenes (which are hydrocarbons) and the xanthophylls, which are oxygen-containing derivatives of the carotenes. Carotenoids are widespread throughout the plant and bacterial kingdoms; one compound, β-carotene, is a universal pigment in the same sense as chlorophyll *a*. In addition to being present in high concentrations in photosynthetic structures they are found in many nonphotosynthetic tissues.

The absorption properties of a pigment are a function of its environment. For example, the location of the red band of chlorophyll *a* in organic solvents varies from 660 to 672 nm, depending on the refractive index and polarity of the solvent. In the condensed crystalline state, the red maximum can be shifted to the far red to 740 nm. Intermediate locations can also arise from variations in the degree of aggregation and the environment. Aggregated chlorophyll tends to be relatively nonfluorescent at room temperature but shows a long wave emission (720 nm) at liquid nitrogen temperatures ($77°K$).

It is not certain in which physical state(s) the chlorophyll occurs in the dense layers of the lamellae. During the greening process in higher plants, the red maximum of chlorophyll *a* shifts from 684 nm to 673 nm. In mature lamellae the red band appears to consist of several overlapping components. Experiments based on analyses of absorption and action spectra, fluorescence emissions at low temperature, and differential extraction all suggest that chlorophyll *a* occurs in several different states of aggregation. French (*22*) showed that many plant spectra could be explained by assuming that they consisted of mixtures of four specific chl *a* bands.

Although chlorophyll *a* appears to be the primary photosynthetic pigment in all plants, other, so-called accessory pigments, contribute to the light-harvesting process. Chlorophyll *b*—present in about one-third the concentration of chlorophyll *a* in green plants, and chlorophyll *c*—a characteristic pigment of brown algae—differ slightly from chlorophyll *a* in their β-substituents (see Figure 3). Their red absorption bands are on the short wave side of the chlorophyll *a* maxima.

The red and blue-green algae contain high concentrations of phycobiliproteins, which are water-soluble protein-like pigment complexes. These pigments are cataloged into three classes—phycocyanins, allophycocyanins, and phycoerythrins—on the basis of their absorption spectra. Like chlorophyll *a*, however, the positions of absorption maxima of the various classes can be influenced by the milieu of the chromatophore. Consequently, different protein carriers or binding modes can cause these pigments to display a variety of band locations.

The bile pigments are found only in aquatic organisms. Since they absorb mainly in the middle of the spectrum (500–600 nm) where chlorophyll absorption is low and transmission by (sea) water is highest—these accessory pigments seem to fill this absorption gap (23).

The role of the carotenoids in light harvesting is not as clear as that of the bile pigments. Some ubiquitous carotenoids—lutein and β-carotene—appear to be relatively inefficient in this respect. There is some evidence that their primary role is to quench triplet excitations of chlorophyll and thus prevent photooxidation of the photosynthetic apparatus.

In all green plants and algae, the two photosystems are sensitized by different pigment assemblies. Chlorophyll a is a major constituent in both photosystems, but at least one long wave absorbing state serves System I exclusively. System II is generally associated with most of the accessory pigments.

Absorption and Emission of Light. Photosynthesis involves the conversion of the energy of light quanta—usually expressed in electron volts, wavelength, or wave number—into chemical energy (usually expressed in kcal/mole). The energy contained in one quantum of light is $E = h\nu$, where h is Planck's constant and ν is the frequency ($\nu = c/\lambda$, where c is the speed of light and λ the wavelength). For wavelengths of light utilized with high efficiency by green plant photosynthesis ($\sim$650–700 nm) this energy is about 1.8 eV/quantum, and a gram–mole of these quanta—1 einstein, 6.02×10^{23} quanta—has an energy of about 40 kcal.

The probability that a pigment molecule will capture a passing quantum is proportional to its extinction coefficient, which is a measure of its effective capture area or cross-section. Figure 4 (solid line) shows the absorption spectrum of chlorophyll a in solution. The strong red and blue absorption bands at 660 and 430 mm—arising from electronic transitions to the first and second singlet excited state, *see* below—have a very high molar extinction coefficients (about 10^5 l mole^{-1} cm^{-1} or 1.7×10^{-16} cm^2/molecule).

The absorption of a visible photon by a photosynthetic pigment molecule results in the transition of an electron to an orbital of higher energy. Since this absorption is quantized, the pigment molecule gains an amount of energy ($h\nu$) equal to that contained in the absorbed light. This high-energy molecule is in an excited state. The excited state is unstable, and it returns to its lower energy ground state with a characteristic decay time—often referred to as its lifetime. The fate of the energy lost during this decay determines whether useful work can be harvested from the light-driven electronic excitation.

The energy obtained by the absorption of light by photosynthetic pigments is either trapped and used to drive energy-requiring processes or is dissipated as light or heat. The loss of the acquired energy as light—the so-called radiative de-excitation—is usually divided into two phenomena—fluorescence and phosphorescence (or luminescence). Fluorescence applies to emissions that accompany allowed radiative transitions from singlet excited states; these emissions usually have a lifetime of less than 10^{-8} sec. Phosphorescence refers to radiative de-excitations from longer-lived metastable

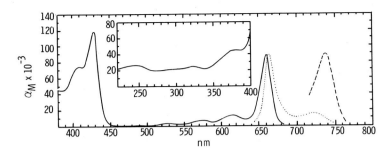

Figure 4. Solid line: absorption spectrum of chlorophyll a dissolved in ether. Dotted line: fluorescence emission spectrum. Dashed line: red absorption band of microcrystals (24).

states such as triplets. Figure 5 is a schematic showing the relationship of the various processes.

In the case of fluorescence from light absorbed in the red by chlorophyll a, these electrons return to the ground state with the emission of a light quantum of somewhat lower energy (longer wavelength) than that which was absorbed. This phenomenon—the Stokes shift—arises from the fact that the excited molecules approach thermal equilibrium among their vibrational states

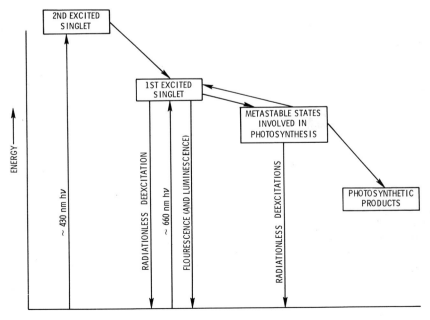

Figure 5. Schematic of excitation and de-excitation pathways

before radiative de-excitation. Consequently, de-excitation occurs from the lowest vibrational levels of the excited state. Since light-induced excitations occur from the ground state to higher vibrational levels of the excited state, most of the quanta absorbed in excitation are of shorter wavelength (higher energy) than those emitted in de-excitation. This is shown in Figure 4; note that the fluorescence emission band (dotted line) is somewhat displaced to the long wave side of the red absorption band at about 660 nm.

If a pigment molecule is promoted to a higher singlet state than the first —e.g., the transition which manifests itself in the blue absorption peak in chl a—it usually decays rapidly (*via* interconnected vibrational levels) to the first singlet excited state before returning to the ground state. This is also illustrated in Figure 4; the fluorescence of chlorphyll a occurs exclusively in the red, and the absorption of blue quanta and red quanta result in exactly the same fluorescence spectrum. The second excited singlet state (430 nm) is extremely unstable (lifetime 10^{-12} second); before fluorescence emission can occur, it converts into the first and lowest singlet. In this radiationless transition the excess energy is dissipated as heat.

The transition from the lowest singlet to the ground state is not always accompanied by the emission of light because of competing thermal processes. In dilute solution in organic solvents the fluorescence yield (quanta emitted/quanta absorbed) is only 0.3 for chl a and 0.1 for chl b (25). The remaining excitations are dissipated as thermal energy. The fluorescence yield can also be greatly affected (quenched) by the interaction of the pigment with other molecules.

In the chloroplast, chlorophyll fluoresces much less ($\sim 1/10$) than in solution. Since the fluorescence yield indicates the lifetime of the excited state, this implies the existence of trapping mechanisms which efficiently drain energy from the pigments. In this case the energy is not dissipated as heat, but it is converted to chemical energy. Since the fluorescence yield should be lower when the photochemical trapping is efficient, the fluorescence yield can be a useful indicator of energy flow. This expectation appears to be fulfilled in bacterial photosynthesis and in Photosystem II of green plants. For unexplained reasons, the fluorescence from the pigment which sensitizes System I is weak and unaffected by the state of the traps.

In addition to the emission of light *via* prompt fluoresence, photosynthetically active materials also re-emit light long after illumination ceases (26). The spectral composition of this delayed light—phosphorescence or luminescence—is that of fluorescence from the same system and, like the fluorescence, appears to originate exclusively from Photosystem II. This delayed light is of very low intensity and decreases continuously with time. Its decay is not homogeneous and appears to consist of several kinetic components with different time constants ranging from microseconds to minutes. The emission is influenced by the state of System II and by high energy states of the thylakoid.

Energy Capture and Distribution; the Photosynthetic Unit. To do chemical work the energy of the excited state must be caught and converted into a stable and manageable form. In gases or solutions an excited molecule can only interact with other molecules through collisions which are infrequent during the brief lifetime of a singlet excited state. Consequently, photochemical conversions in solutions would be rare if it were not for the existence of longer-lived metastable (triplet) excited states.

Although some hypotheses suggested that photosynthesis is initiated by the triplet excited state of chlorophyll, no evidence for this mechanism was found in intact chloroplasts. In fact, it appears that the reactive triplet states are deliberately annihilated by carotenoids, which are ubiquitously dispersed among the chloroplast pigments. Chlorophyll triplets are efficiently transferred to carotenoids and degraded to heat (27). In the absence of carotenoids the chlorophyll is quite susceptible to photobleaching (28).

In the photosynthetic apparatus the need for long-lived excitations is apparently overcome by arranging the pigments so that they can interact efficiently during the brief lifetime of the singlet excited state. A fraction of the photosynthetic pigment has been shown to have a strict spatial orientation (29). Several phenomena have also been observed which are reminiscent of solid state events—e.g., photoconductivity and the trapping of photoproducts at low temperature.

The problem of energy transfer between various photosynthetic pigments arose with the discovery of the photosynthetic unit. In 1932 Emerson and Arnold (10) determined the amounts of O_2 evolved in series of flashes. Each flash was of sufficient intensity to elicit a maximum response (i.e., it was saturating) and short enough (10^{-5} sec) that excitations could not be processed during the flash itself. When the flashes were spaced far enough apart to allow all dark processes to run to completion and restore the trapping centers to the open state, the oxygen yield per flash became maximal and constant. Under these conditions, one O_2 was evolved per 2000 chlorophylls. Since it takes about 10 quanta to produce an O_2 ($\sim$ four in each photosystem), this experiment suggests that in a single flash only one photon per 200 chlorophylls can be processed—i.e., there is only one trap per 200 chl. If we assume that the traps of Photosystems I and II are sensitized by an equal number of chlorophyll molecules, this suggests that there is one conversion center in each system for every 400 chlorophyll molecules. This is indeed the approximate concentration of P_{700}, cytochrome f, plastocyanin, Q, and other components of the electron transport chain (see Figure 1).

The photosynthetic unit derives its efficiency from an energy transport mechanism—which is fast compared with the lifetime of the singlet state—that operates between pigment molecules. Interpigment transfer of some kind does exist. For example, in solutions containing both chlorophyll b (red absorption band at 640 nm) and chlorophyll a (red band at 660 nm) irradiation with a wavelength which mainly excites the b components results in the fluorescence of chlorophyll a rather than chlorophyll b. It appears that the a component, with its slightly lower excitation level, drains the energy from the b component (30).

A similar transfer of energy apparently occurs in photosynthetic tissues.

On the basis of fluorescence measurements, Duysens (31) suggested the transfer of light energy from pigment to pigment. Again this transfer occurs not only between similar molecules but also between unlike molecules, provided the receiving molecule absorbs at the same or slightly longer wavelengths than the emitter.

Since energy transfer in the opposite direction (to the short wavelength) is less efficient, the energy in the lamellar pigment arrays tends to flow towards the longest wave absorption band. The System I pigment array contains a minor long wave chl a fraction ($\sim 5\%$) which absorbs at 700 nm (31) the absorption maximum of P_{700}. P_{700}, the site of the conversion of light into chemical energy, rapidly traps the incoming light energy; (the actual charge separation probably occurs within about 10 psec (32)). The overall result of this pigment arrangement is the funneling of quanta received by many pigment molecules to a single site, much as a lens system concentrates a light beam to its focal point. Thus, the trapping center receives 200 times more photons per unit time than if it operated independently, and the photochemical rate can be that much faster. Each group of pigment molecules which collaborates to enhance the optical crosssection of the trapping center comprises a single photosynthetic unit.

The mechanism of the energy transfer observed in photosynthetic tissue is still not well understood. Energy transport by massive particles (molecules, ions, and radicals) is probably too slow while energy transport *via* conducting states (electrons and "holes" analagous to solid state phenomena) does not seem to be important (33). Current theories ascribe the observed energy transport to the migration of neutral excitation quanta (excitons). However, at present, it is not possible to determine conclusively which of several possible mechanisms for the transfer of neutral excitation energy operates in photosynthetic tissue. It is fairly clear that this process involves the transfer of singlet (rather than triplet) excitation energy (34, 35). The theoretical basis for one type of energy transfer which has achieved a good deal of support is the process of resonance transfer developed by Förster (36).

The Photosystems

Photosystem I. Photosystem I is the "long-wave light" driven reaction system in which an electron is promoted from the redox level of P_{700} (+430 mV) to an as yet unidentified electron acceptor (X) with a potential of about -600 mV. *In vivo*, the donor (P_{700}) is restored to its reduced state by the transfer of an electron from System II, mediated by a series of intermediate electron carriers (*see* below). The reduced electron acceptor is reoxidized by NADP *via* NADP-reductase.

The overall Photosystem I complex appears to contain about 200 light harvesting chlorophylls and about 50 carotenoids. It also contains the specially bound chlorophyll molecule P_{700}, a molecule of cytochrome f and plastocyanin, two molecules of cytochrome b_{563}, and about 10 membrane bound ferredoxin molecules.

P_{700} reacts as a one-electron redox reagent with a pH independent normal potential of $+0.43$ V (12). Either the photochemical or chemical oxidation of P_{700} results in the bleaching of its red absorption band (at about 700 nm) and a partial bleaching of its blue absorption band (at about 430 nm). This is illustrated in Figure 6. In its oxidized state, P_{700} shows a typical electron spin resonance signal (signal I (37, 38)). At room temperature it is rapidly reduced by plastocyanin and/or cytochrome f; at low temperatures the photooxidation is irreversible. The fact that the photooxidation of P_{700} occurs with a high quantum yield (~ 1.0) even at a very low temperature suggests that it is a primary photochemical process.

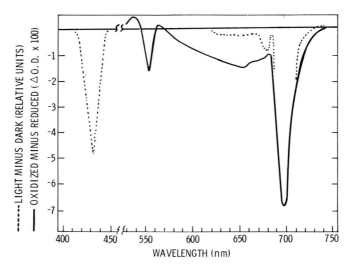

Figure 6. *Difference spectrum measured with chloroplasts from which 80% of the chlorophyll was extracted with aqueous acetone*

Solid line: oxidized minus reduced spectrum showing the α band of cytochrome f (555 nm) and the red band of P700 (698 nm). For clarity the short wave part of this spectrum was omitted because changes caused by the Soret (blue) bands of the two pigments overlap. Dotted line: light minus dark spectrum showing both bands of P700 (12).

Cytochrome f (39) is a high potential (~ 340 mV) c type cytochrome with an absorption band at 554 nm. Several studies suggested that this cytochrome is closely associated with P_{700}, possibly functioning as its primary electron donor.

Plastocyanin (40) is a low molecular weight (21,000) copper–protein which reacts as a single electron redox compound. Its oxidation–reduction potential is constant ($E_m = +0.37$ V) between pH 5.4 and 9.9. Spectroscopically the oxidized form of plastocyanin is blue with a broad absorption band at 597 nm. EPR studies detected a signal which is attributed to the copper component.

At the present time the functions and relative importance of cytochrome f and plastocyanin as electron mediators to P_{700} are not at all clear. Several studies suggested that the reaction sequence is cyt f → plastocyanin → P_{700} (41, 42); others proposed that cyt f and plastocyanin may react in parallel, with both components transferring electrons to P_{700} (43).

Although the primary Photosystem I electron acceptor is not conclusively identified, there is evidence that this acceptor might be the iron–sulfur protein, ferredoxin. Chloroplast ferredoxin is a small none-heme iron protein (molecular weight ~12,000) containing two g atoms of Fe and labile sulfur per mole of protein. It has a midpoint potential of −0.43 V (44). An ESR signal attributable to an iron–sulfur protein has been observed upon excitation of System I at liquid nitrogen temperatures (45). At physiological temperatures a chloroplast component (P430) was identified which behaves kinetically as the primary System I acceptor. This component has a difference spectrum similar to that of an iron–sulfur protein (46).

The role of b-type cytochromes in photosynthetic electron transport is still obscure. Cytochrome b_6—sometimes called b-563—has its α-band at 563 nm, and a midpoint potential ranging from +5 mV under ideal conditions to −140 mV under uncoupling conditions (47).

The photoreduction of NADP requires both soluble ferredoxin and ferredoxin–NADP reductase (48). In addition to the mediation of NADP reduction, the latter enzyme can also function as a transhydrogenase, diaphorase, and NADPH–cyt f reductase and may participate in the cyclic operation of System I (49). The photoreduction of other acceptors (e.g., methaemoglobin and cytochrome c) requires only soluble ferredoxin. Under anaerobic conditions, ferredoxin can transfer electrons to protons via the enzyme hydrogenase, which results in the evolution of molecular H_2 (50).

Numerous artificial electron acceptors can be reduced directly by the primary reductant of System I without mediating enzymes. These include both high potential acceptors—DCPIP, PMS, ferricyanide, and benzoquinone—and low potential acceptors (e.g., FMN and viologen dyes). The photoreduction of low potential autooxidizable acceptors can be observed either directly in chloroplast suspensions from which oxygen was removed or indirectly as a light induced net consumption of O_2. In the complete O_2 evolving system the overall reaction stoichiometry is such that one O_2 is consumed per four equivalents of acceptor (V) reduced by Photosystem I and subsequently reoxidized by molecular O_2; i.e.,

$$2H_2O + 2V \xrightarrow{h\nu} 2VH_2 + O_2$$

$$\frac{2VH_2 + 2O_2 \to 2V + 2H_2O_2}{2H_2O + O_2 \to 2H_2O_2}$$

In an analogous fashion, the primary System I photoreductant itself slowly reacts with oxygen. Consequently, chloroplasts illuminated in the absence of added acceptor show a small net uptake of O_2 (the Mehler reaction) (51).

In the complete system P_{700}^+ is reduced indirectly by Q^-—the electron acceptor of System II—*via* the segment of the electron transport chain which links the two photosystems. If System II is not sensitized (in long wave light) or is inhibited by a specific poison (*e.g.*, DCMU), P_{700} can be maintained in the reduced state by artificial electron donors (*e.g.*, reduced DCPIP and DAD) (52). The sites of entry for these electron donors are not known, and probably vary with the method of chloroplast preparation and the concentration of the donor.

Some reagents (*e.g.*, PMS) are able to function within System I both as a donor and an acceptor and can consequently short-circuit the photoact to produce a cyclic operation. This cycle can be demonstrated by illuminating the system with series of light flashes and spectroscopically observing the redox state of P_{700} which is instantaneously oxidized by each flash and reduced in the subsequent dark period.

System I can also be operated artificially in an open ended (noncyclic) manner. For example, low potential acceptors such as methylviologen or NADP (with ferredoxin and NADP reductase as mediators) can be photoreduced while $DCPIPH_2$ or $DADH_2$ (and ultimately ascorbate, which is usually added to keep the dye reduced) become photooxidized. When an autooxidizable electron acceptor is used, oxygen is consumed in the light with a reaction stoichiometry such that one O_2 is taken up per two equivalents of acceptor (V) photoreduced and subsequently reoxidized. The partial reactions are as follows (DH_2 represents a donor such as $DCPIPH_2$):

$$\begin{array}{c} DH_2 + V \xrightarrow{h\nu} VH_2 + D \\ \underline{VH_2 + O_2 \rightarrow V + H_2O_2} \\ DH_2 + O_2 \rightarrow D + H_2O_2 \end{array}$$

In some instances the operation of System I mediated by artificial donor–acceptor systems (both open and cyclic) can be coupled to the generation of ATP and/or the formation of a proton gradient across the thylakoid membrane. Some of these processes appear to be quite efficient, and the observed rates of energy transformation can be quite high. Some of these reactions are discussed in greater detail later.

System I is relatively stable and apparently can function even in the presence of some detergents. These detergents disrupt membrane structures and cause a loss of soluble proteins, such as plastocyanin and NADP reductase. By the use of differential centrifugation of digitonin treated chloroplasts, Boardman and Anderson (53) were able to obtain fractions which were enriched in System I or II. Similar procedures using other detergents or physical disruption (*e.g.*, by the use of the French press) yield essentially the same results. Preparations enriched in P_{700} have been obtained, but little progress has been made towards purification of the System I reaction center (c.f., bacterial reaction centers discussed later).

Photosystem II. Photosystem II is the photosynthetic light reaction which is normally part of a structure-bound complex intimately connected with the O_2 evolving system. It contains roughly 200 chlorophylls and about 50 carotenoids, which cooperate to channel light energy to a special chlorophyll (aII). It also contains an unidentified primary electron donor (Z), a primary electron acceptor (Q)—which might be plastoquinone—about 10 plastoquinone equivalents, six Mn atoms, and two cytochrome b_{559} molecules.

The photosensitization of System II results in the oxidation of primary donor Z and the reduction of primary acceptor Q. Since the liberation of oxygen from water requires four oxidizing equivalents of an average potential of about $+800$ mV, the primary photooxidant Z probably has a midpoint potential at least this high. The potential of primary reductant Q has been the subject of some dispute but is generally thought to be about 0 mV (54).

In general the overall photochemistry of System II has not been characterized nearly as well as that of System I, and neither the primary donor nor acceptor has been characterized as well as their counterpart in System I. A small flash induced absorption decrease at 680 nm which rapidly returns in subsequent darkness ($t_{1/2} \sim 0.2$ msec) has been interpreted as a bleaching of chlorophyll aII (55). The function of this component may be analogous to that of P_{700}, its oxidized form apparently giving rise to a similar EPR signal (56).

Light-induced absorption changes have also been observed which could be correlated with the photoreductant of System II. A short-lived (0.6 msec) increase in absorption at 325 nm has been reported (57) which may reflect the reduction of the primary electron acceptor Q. However, at present the status of this component in relation to Q is uncertain. It has also been suggested that the observed decrease in absorption at 550 nm (C550) (58) may reflect the redox state of Q.

The redox state of Q to a large extent determines the fluorescence yield of the chlorophyll a associated with System II. Even with all traps open—i.e., Q oxidized and Z reduced—a residual fluorescence yield (F_o) is observed, which suggests that under optimal conditions not all excitations are trapped. With all traps closed—i.e., either Z or Q in the wrong redox state—the yield reaches a maximum (F_{max}). Between these two limits—F_{max} and F_o—the fluorescence yield (F) varies as a function of the degree of openness of the traps. To a first approximation the magnitude of (F_{max} minus F) is proportional to the efficiency (quantum yield) of O_2 evolution. Since under most conditions the primary donor Z is fully reduced, the fluorescence yield reflects the redox state of primary acceptor Q (59). The fluorescence reaches F_{max} (Q reduced) under reducing conditions or in strong, rate-saturating light. It becomes minimal (F_o) upon oxidation of Q in dark or in far red light which preferentially sensitizes System I.

In contrast to System I, System II can be affected by a number of specific inhibitors. The herbicide DCMU is frequently used to block System II and thus chemically isolate System I. In the presence of DCMU, a brief flash (10^{-5} sec) raises the fluorescence from F_o to F_{max}, indicating that a single charge separation can still take place. However, the return to F_o in a subsequent dark period is slow. In the absence of DCMU but the presence of an electron acceptor the fluorescence is also raised by a flash, but it returns rapidly to F_o in the subsequent dark period ($\sim$0.2 msec). In this case, Z^+ and Q^- are restored quickly, and a series of flashes will result in the evolution of O_2.

In the absence of both DCMU and added electron acceptor the fluorescence return in a series of flashes is normal initially but becomes incomplete after 10 or 15 flashes. This behavior is also reflected in the O_2 evolved per flash, which is high initially and becomes negligible after a series of flashes

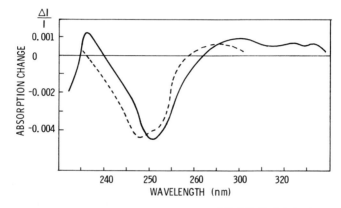

Zeitschrift für Naturforschung

Figure 7. Solid line: absorption changes in the UV induced in isolated chloroplasts by alternating strong red light (660 nm $\sim$0.1 sec) with 720 nm light ($\sim$0.4 sec). Dashed line: difference of the absorption spectra of reduced and oxidized plastoquinone A dissolved in ethanol (63).

(see Figure 12). These two phenomena—both related to the so-called gush of O_2 evolution observed at the onset of illumination (60)—arise from the imbalance of System II–System I turnover in the absence of an electron acceptor. The time course of the gush reflects the reduction of the intially oxidized electron carrier pools by System II. During the gush approximately one O_2 is evolved per $\sim$200 chlorophylls—roughly 10 times more than that evolved by a single flash (which induces a single turnover of the system) (61)—suggesting that there is a pool of about 10 equivalents of electron carrier located between the two photoacts. Kinetic and spectral studies suggest that at least part of this pool is plastoquinone (62).

The absorption spectrum of plastoquinone A is shown as a dashed line in Figure 7. The solid line illustrates the absorption spectrum after a 0.1-sec illumination by strong red light ($\sim$660 nm). The changes in absorption presumably reflect the complete reduction of the quinone pool. A brief flash, which converts each System II trap only once, reduces only about one-tenth of the pool. In very bright light, the quinone pool is reduced with a half-time of about 10 msec, which corresponds to a reaction time of about 1 msec per equivalent transferred from water to the quinone pool. The quinone pool is slowly oxidized in the dark and is rapidly oxidized by long wave (System I) light. Since the rate of its oxidation by System I (about 10 msecs per equivalent) is lower than its reduction by System II in strong saturating light, the quinone pool is almost completely reduced under these conditions (discussed later).

The role of cytochrome b_{559} (64) is unclear although it appears to be closely associated with System II (65). Depending on conditions, this cytochrome can be photochemically reduced or oxidized by System II (66). Since no rapid turnover of this cytochrome is observed, it is probably located on a side path. Its normally high midpoint potential ($+0.370$ mV) irreversibly changes to $+60$ mV in chloroplasts treated with certain uncouplers (CCCP, antimycin A) or subjected to disruptive treatments with detergents (67, 68). It appears that membrane phenomena determine its redox state and modify its light induced absorption changes.

Manganese has been well documented as an essential component of Photosystem II. It appears to function primarily as part of the O_2-evolving mechanism; hence, it is considered in greater detail later.

Several high potential electron acceptors—e.g., benzoquinone, DCPIP, DAD_{ox} and ferricyanide—can be reduced by System II. However, under most circumstances, reduction of these acceptors by System I is more rapid and is consequently the predominant reaction path. Since there is no specific System I poison available comparable with DCMU for System II, these acceptors are specific for System II only where communication with System I is destroyed (e.g., in some mutants, in System II particles, or in the presence of certain inhibitors.)

The strong oxidizing power generated by System II appears to be rather well shielded from external reductants. Few if any compounds are known which readily donate electrons to Z without damaging the O_2-evolving system. Hydroxylamine, for instance, can serve as an electron donor, but in the process it removes the Mn of the O_2 enzyme.

The O_2 evolution system—as discussed below—is quite fragile in the presence of detergents and is destroyed by a few minutes incubation at 50°C. Consequently, the isolation of O_2-evolving System II particles has proved quite difficult. However, the photosystem itself is relatively stable and can participate in many partial (and possibly spurious) reactions after treatment with detergents. System II particles isolated in this manner still show the 680-nm absorption changes and are able to photooxidize cytochrome b_{559}. Such preparations also photooxidize artificial donors—e.g., ascorbate, hydroquinone, and hydroxylamine (69).

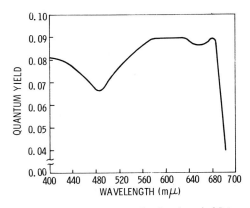

Figure 8. Dependence of the quantum yield of Chlorella photosynthesis on wavelength (70)

Interactions between the Photosystems. The interactions between photosystems I and II take place on several different levels. Probably the most important of these interactions involve (1) the transfer of electrons and (2) the distribution of light energy. These interactions are not independent of each other; consequently, factors which affect one interaction (*e.g.*, electron transfer) can also influence the other (*e.g.*, distribution of light energy).

In green algae, the quantum yield of O_2 evolution varies with the wavelength of the actinic light. This suggests that not all pigments are equally efficient in channeling their absorbed energy to O_2 evolution. A plot of the quantum yield of O_2 evolution as a function of wavelength—an effectiveness spectrum—is shown in Figure 8. Note that there is a dip of about 25% at 480 nm, which seems to indicate that energy absorbed by carotenoids is not fully utilized. A somewhat smaller dip at 650 nm might imply that chlorophyll *b* excitation is also inefficiently utilized. Beyond 690 nm, the quantum yield drops severely, as if an inactive long wave pigment were present.

A similar phenomenon is also observed in other aerobic photosynthetic organisms. Figure 9 shows the absorption spectrum and the action spectrum of O_2 evolution (the rate as a function of wavelength) observed with the red alga, *Porphyra nereocystis*. A plot of the ratio between the two curves yields a relative effectiveness spectrum. This ratio is much higher in green light ($\sim$550 nm)—where the accessory pigment phycoerythrin is the main absorber—than in red and blue light, where chlorophyll *a* and carotenoids are the predominant absorbing pigments. This again suggests that the light energy absorbed by some of the photosynthetic pigments is inefficiently utilized when assayed *via* the rate of O_2 evolution.

Although long-wave quanta are used inefficiently to produce O_2, they can be harvested efficiently to drive photoreactions related to System I. For example, the quantum yield of NADP reduction observed with isolated chloroplasts decreases precipitously at wavelengths greater than 690 nm. However, if reduced DCPIP is added to donate electrons to P_{700} and DCMU added to inhibit O_2 evolution, the efficiency of NADP reduction rises at wavelengths greater than 690 nm and approaches 1 eq/$h\nu$.

Two observations which led to the concept of two series-connected photosystems were (1) chromatic rate transients, first reported by Blinks (13) and (2) enhancement, reported by Emerson (14, 15). Chromatic rate transients can be observed when two monochromatic light beams—one absorbed mainly by accessory pigments, the other by chlorophyll a—are alternately used as the actinic light. If the intensities are initially adjusted to yield identical steady state rates of O_2 evolution, transitory changes of the rate—undershoots or overshoots—can be observed when the beams are alternated. The phenomenon of photosynthetic enhancement is observed when two such wavelengths are given separately or together. In this case, it is observed that the two beams together are more effective than the sum of their individual effects. If V_I and V_{II} are the rates of O_2 evolution in lights I and II separately and V_{I+II} is the rate in the two beams together, then in the case of enhancement, $V_{I+II} > V_I + V_{II}$.

We now recognize these two phenomena as being the transient and steady state manifestations of what is called push and pull. The observations are readily explained by the hypothesis of two cooperating photosystems, each equipped with different pigment assemblies. According to this model, the

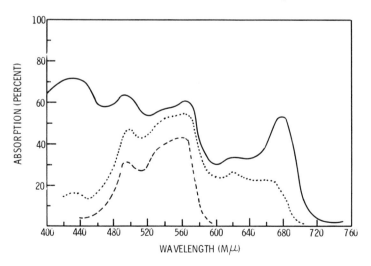

Journal of General Physiology

Figure 9. Absorption and action spectra measured with the red alga Porphyra nereocystis

Solid line: absorption spectrum of the thallus; Dotted line: action spectrum of O_2 evolution; Dashed line: absorption spectrum of phycoerythrin extracted from the thallus. The effectiveness spectrum is the ratio between the two upper curves at each wavelength (71).

quantum yield of the overall process will be optimal when the two systems absorb an equal fraction of the incident light, and low at wavelengths in which the light is preferentially absorbed by one of the two systems. Thus, two light beams—one mostly absorbed by System II and the other mostly by System I—complement each other, resulting in enhancement.

More direct evidence for a scheme of two series connected photoacts (c.f. Figure 1) came from spectroscopic observations of intermediates of the electron transport chain. For example, in the blue-green alga, *Anacystis*, the accessory pigment phycocyanin sensitizes the reduction of P_{700}, while light absorbed by chlorophyll *a* and carotenoids causes its oxidation (72). Similar push and pull effects were reported for cytochrome f, plastoquinone, the rate of O_2 evolution, and the yield of fluorescence (which reflects the redox state of Q). In excess System I light the pool of redox intermediates located between the two photoacts and the primary reductant Q become oxidized since the rate of their photoreduction by System II is low. If System II light is now given, it initially operates with maximum efficiency until the intermediates in the chain (including Q) are reduced because of a shortage of System I photons. This is reflected by a gush of O_2 and a rise in the fluorescence yield, which reflects the conversion of Q to Q^-. In isolated chloroplasts the converse experiment is also possible. After pre-illumination with System II light—which reduces the pool—the efficiency of System I light is temporarily high, and a gush is observed in the reduction of NADP or viologen.

In all oxygen-evolving photosynthetic organisms the pigment assemblies which sensitize the two photosystems are different spectrally as well as chemically. For example, in the red and blue-green algae System I is sensitized mainly by chlorophyll *a* while System II is sensitized mainly by bile pigments. However in white (sun) light the two photosystems absorb approximately equal amounts of light, and the distribution of light energy between the two systems is balanced.

In contrast to those of blue-green algae, the action spectra of the two photosystems in green plants are rather similar. Figure 10 shows the action spectra of the two photosystems. Note that the maxima are slightly displaced; most of the long wave components are associated with System I while the pigment system of System II consists of more short wave pigments (*e.g.*, chlorophyll *b*.)

The mode and mechanism of quantum distribution between the photosystems are not well understood. There are currently two main hypotheses which attempt to explain this phenomenon. The concept of separate packages assumes that the two photosystems have their own independent pigment beds. This hypothesis is consistent with many observations but does not explain, without further assumptions, the flatness of the effectiveness spectrum in green cells below 690 nm. Joliot *et al.* (73) assume that the reaction by which the traps are restored $[P^+ + Q^- \rightleftharpoons P + Q]$ has a low equilibrium constant ($K_{eq} \leqslant 10$). Theoretical curves based on this modification are consistent with the observed quantum yield spectrum. However, this assumption does not agree with the difference in midpoint potentials ($\Delta E \sim 0.4$ V) between P and Q.

The alternate, spill-over concept of quantum distribution (74) assumes that except for the long wave System I component, most pigments, and thus

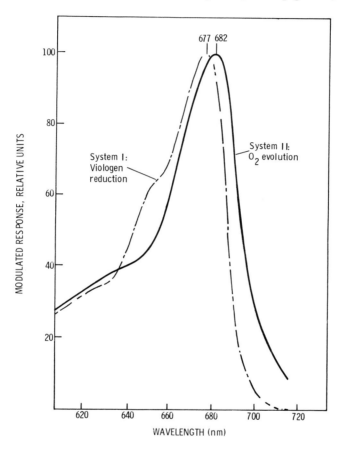

Figure 10. Action spectra of the two photosystems observed in isolated chloroplasts

Dashed line: rate of O_2 evolution using NADP as the terminal electron acceptor. Solid line: viologen reduction in the presence of DCMU, using DAD as the System I electron donor (73).

light quanta, are shared by the two systems. A quantum preferentially flows to a System II trap, but finding it closed can move to a System I trap. This hypothesis predicts a flat quantum yield spectrum of O_2 evolution for $\lambda < 690$ but fails to explain the poor quantum yield of NADP reduction at wavelengths shorter than 690 nm observed in the presence of DCMU and $DCPIPH_2$ (75).

In red and blue-green algae the relative abundance of the pigments can vary considerably, depending upon the color of light in which the organisms are grown (76). It also appears that in green algae some pigments can shift from one system to the other within a few minutes in response to changes in the wavelength of illumination (77, 78). In both cases the observed shifts are in the direction of optimal quantum distribution. In isolated chloroplasts the ion content of the medium can alter the distribution of quanta between the photosystems, suggesting that the observed shifts *in vivo* might be related to conformational changes of the thylakoid.

A chloroplast thylakoid contains many photosynthetic units. There is evidence that in this superstructure interactions occur between the pigment units and between the electron transport chains. For example, the pigment units of System II are not rigidly separated, and photons absorbed by a unit in which the trapping center is closed can migrate to a neighboring unit (79). No comparable interunit transfer seems to occur in System I.

There is also interaction between units on the reducing side of System II. The pools of redox equivalents (the so-called A pools) communicate and thus allow each System II unit to interact with several other System I and System II centers (80). In contrast, the O_2-generating enzymes—one per System II center—are strictly independent of each other, and there is no migration of positive charges (81) (see below).

The Bacterial Photosystem

The photosynthetic apparatus of bacteria differs from that of algae and higher plants in that bacteria are unable to use water as an electron donor and evolve O_2 in the light. Instead, donors such as H_2S or organic compounds are oxidized in a light-driven process. In aerobic (*i.e.*, green plant or algal) photosynthesis the electrons derived from H_2O are promoted to a redox level at least as low as -400 mV (and thus are able to reduce NADP). The analogous process—although implicated by some workers—has not been conclusively demonstrated in a bacterial system.

Photosynthetic bacteria are generally divided into different groups: the purple sulfur bacteria (*Thiorhodaceae*), purple non-sulfur bacteria (*Athiorhodaceae*), and green sulfur bacteria (*Chlorobacteriaceae*). The sulfur bacteria use sulfide or thiosulfate as electron donors while the non-sulfur bacteria use organic compounds. Generally in bacteria the pigment system is spectrally much more variable from species to species than in aerobic photosynthetic organisms. The chlorophyll component in most purple bacteria is bacteriochlorophyll *a*, which has its red absorption maximum at 770 nm in ether solution. *In vivo*, its absorption maximum varies with the species, occurring in bands at about 800, 850, and 880 nm. *R. virdis* contains bacteriochlorophyll *b*, which has its major absorption peak at 1012 nm (82). The green bacteria contain *Chlorobium* chlorophyll as their primary pigment; its major absorption peak *in vivo* is at about 740 nm. The spectral changes attributable to reaction center bacteriochlorophyll molecules—analogous to P_{700} in green plants—range from 870 nm in *Rhodospseudomonas spheroides* (*83*) to 985 in *R. viridis* (*84*).

Our knowledge of photosynthetic reactions at the biophysical level is more detailed in bacteria than in algae and green plants. This is partially the result of the smaller photosynthetic unit of purple bacteria (about 50 Bchl per reaction chain compared with 400 chl in green plants); the smaller amount of bulk pigment has facilitated the observation of spectral changes in the reaction center. The isolation of reaction centers in which most of the bulk pigment is readily bleached or removed (*85*) has also been instrumental in elucidating the primary events occurring at the photosynthetic trap.

In every bacterial photosynthetic system studied to date a specialized bacteriochlorophyll molecule—designated $P(n)$, where n is the wavelength of maximum absorption change—undergoes a reversible photooxidation (Figure 11). This phenomenon was first reported by Duysens (*11*), who observed

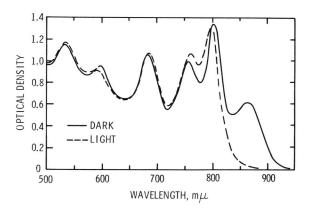

Photochemistry and Photobiology

Figure 11. Light and dark absorption spectra of R. spheroides *chromatophores treated with K_2IrCl_6 to bleach the light-harvesting bacteriochlorophyll* (*85*)

reversible spectral changes in *Rhodospirillum rubrum* and *Chromatium*. These spectral changes could be mimicked by chemical oxidation (*86*), which suggested that the reaction center bacteriocholophyll *a* molecule was undergoing a reversible photoinduced oxidation. Clayton (*85*) was able to show that all of the observed spectral changes at 870 nm in *R. spheroides* were caused by a specialized Bchl *a* molecule present in about one part in 50.

In addition to the reaction center molecule (P870 in *R. spheroides*), spectral changes at 800 nm occur (*see* Figure 11) which are attributed to two additional Bchl *a* molecules called P800 (*87*). It appears that the reaction center bacteriochlorophyll molecules are closely coupled and thus able to exchange energy (*88*).

Evidence to date supports a reaction center structure in which the primary photochemical event is the photooxidation of $P(n)$, followed by the re-reduction of $P(n)$ by a c-type cytochrome. The net result is thus the promotion of an electron from the cytochrome to the acceptor X, *i.e.*,

$$\text{cyt } P(n) \text{ X} \xrightarrow{h\nu} \text{cyt } P(n)^* \text{ X} \rightarrow \text{cyt } P(n)^+ X^- \rightarrow \text{cyt}^+ \, P(n) \text{ X}^-$$

For example, in *Chromatium* chromatophores, Parson (*89*) [using short (30 nsec) flashes from a Q-switched ruby laser] observed that the photooxidation of P870 (sometimes called P890, the reported band locations vary) preceded that of the associated cytochrome (C422, also called C555 if the α rather than the Soret band is used for notation). The re-reduction of P870 occurred in 2 μsec and corresponded kinetically to the oxidation of C422. Subsequently, P870 could only be reoxidized again after a much longer time (50 μsec to 1 $m$sec), which evidently reflected the reoxidation of X by a secondary electron acceptor (*90*).

The identity of the primary electron acceptor X is still debated. The literature (*91*) reveals recurrent suggestions—but no convincing evidence—that the primary electron acceptor in bacteria is ferredoxin, a compound which does occur in these organisms. Clayton (*92*) interpreted a light-induced spectral change as the reduction of ubiquinone in a primary photochemical reaction. Other workers, however, consider this compound to be of secondary importance, perhaps located on a back flow path from the unknown primary acceptor. Its high midpoint potential ($+100$ mV at pH 7.4) makes it unattractive from another standpoint. Since P890 has a midpoint potential of $+490$ mV, a light-driven transfer of an electron from P890 to X would only cover a potential span of about 400 mV, which is less than one-half the span presumably traversed in both green plant photosystems. Such a reaction appears to be somewhat inefficient since only $\sim 30\%$ of the energy of the 890-nm photon would be conserved.

Photosynthetic bacteria perform an efficient light-induced reduction of NAD although the mechanism of this process *in vivo* is unresolved. If ferredoxin were the primary electron acceptor, the major route could be direct (like System I in green plants). According to some views (*93, 94*) the primary purpose of the bacterial photosystem is the production of ATP by a cyclic light-driven electron transport system. The reduction of NAD is then accomplished at the expense of this ATP or a high energy intermediate generated by the cycle in so-called reverse electron flow.

To date there is no report of photosynthetic enhancement in bacterial systems, and traditionally bacteria have been considered to have only one photosystem similar to Photosystem I in green plants. Some reports, however, suggest that bacteria may contain two photosystems. Presumably, these operate in parallel or at least do not interact directly to yield any enhancement. For example, in *Chromatium* the action spectra for the photooxidation of the two cytochromes C555 and C552 were found to be distinctly different (*95*); similar observations were reported in *R. rubrum* (*96*). The overall significance of the two reactions is not clear. At present no electron transport scheme is available for the bacteria which incorporates available data to the extent that the Z scheme does for green plants.

Oxygen Evolution

The process of O_2 evolution in green plants involves the utilization of oxidizing equivalents produced by Photosystem II. The O_2-evolving enzyme system is very fragile, and its operation depends on the preservation of its structural integrity. To date, efforts to fractionate and reconstruct the total system have not been successful.

Manganese appears to be a specific constituent of the O_2-evolving system (97). The work of Cheniae et al. showed that the O_2-evolving Mn catalyst is formed in a low-yield multiquantum process requiring System II light (98). The Mn occurs with an abundance of about six protein-bound atoms per trapping center. Four of these Mn atoms can be removed by mild heating or extraction with a parallel loss of O_2-evolving capacity. However, the loss of this Mn fraction does not impair the photooxidation of electron donors other than water (99). In addition to Mn, the oxidation of water—but not of artificial donors—also requires the presence of chloride or other monovalent anion of a strong acid (100) and bicarbonate (101).

The oxidation of water to molecular oxygen involves the accumulation and collaboration of four oxidizing equivalents. Without the intervention of a special catalytic system, the four successive oxidation steps would involve widely different free energies and unstable intermediates of very different potentials. Within the framework of the Z-scheme the reported quantum yield of $\sim 10\ hv/O_2$ suggests that photosynthetic O_2 evolution is a four-quantum process. Thus the overall picture of O_2 evolution probably involves a stoichiometric quantum-per-electron oxidation of the O_2-evolving enzyme.

Early observations showed that after a dark period the amount of O_2 evolved by single short flash of light is much lower than that obtained from subsequent flashes (102). With improvements in the technique of O_2 measurement Joliot (103) obtained more precise and informative kinetic data. He demonstrated conclusively that after a period of darkness no O_2 is evolved after a single short saturating flash. More importantly, however, he found that a sequence of these flashes results in a periodic oscillatory behavior of the O_2 yields. This showed that the lack of O_2 on the first flash is inherent in the O_2-evolving process itself and not merely an activation of the enzyme system.

The process of O_2 evolution is most readily described by recourse to the O_2 "clock" (81). According to this picture each O_2-evolving reaction center independently accumulates and processes four positive charges (Z^+) to evolve an O_2 molecule from water: $4Z^+ + 2H_2O \rightarrow 4Z + O_2 + 4H^+$. Thus, O_2 evolution is a linear process in which each O_2 center (S) goes through four oxidizing states:

$$S_0 \xrightarrow{h\nu} S_0' \xrightarrow{k_0} S_1 \xrightarrow{h\nu} S_1' \xrightarrow{k_1} S_2 \xrightarrow{h\nu} S_2' \xrightarrow{k_2} S_3 \xrightarrow{h\nu} S_3' \xrightarrow{k_3} S_0 + O_2$$

with back reactions k_{-2} (from S_2 to S_1) and k_{-3} (from S_3 to S_2).

S_3 and S_2 are unstable (back reactions k_{-3} and k_{-2}) and return to the S_1 state which, like the S_0 state, is stable in darkness. This picture is consistent with most of the kinetic observations reported to date. For example, in very weak light the observed efficiency of O_2 evolution decreases, and a plot of O_2 rate *vs.* light intensity is S-shaped. This is consistent with the above scheme since in weak light the rate of the forward reactions ($h\nu$) is the same order of magnitude as the rate of the back reactions. The observed lag in O_2 evoltuion after a few minutes of dark is also consistent with this hypothesis.

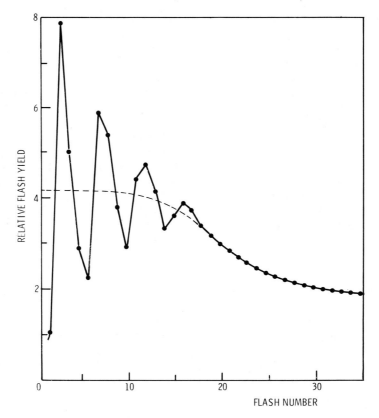

Figure 12. Flash yield patterns obtained with chloroplasts after 10 min dark in the absence of electron acceptor

Dashed line shows the computed state of Q and, except for the lack of an initial lag, is equivalent to the O_2 gush which would be observed under these conditions (106)

The cooperation of oxidizing equivalents in the light and their loss in the dark is most easily analyzed by measuring the O_2 yields of flash sequences spaced about 0.1 to 1 sec apart (104). Each flash produces one plus charge in each trapping center, and photochemical events thus proceed in discrete steps. After a long dark period, the O_2 flash yield oscillates with a period of four, which is consistent with the occurrence of a four-step process. The first two yields are negligible, and the third flash yield is maximal. These observations are also consistent with the above scheme. The fact that the O_2 system must be primed and does not yield significant O_2 on the first two flashes suggests that O_2 centers cannot exchange holes and that each trapping center and its O_2-evolving enzyme acts as a self-sufficient unit.

Similar kinetic experiments in which fast pH changes were monitored showed that the protons liberated in the process $2H_2O \rightarrow O_2 + 4H^+$ are released in synchrony with the oxygen. Thus, the actual decomposition of water appears to occur in a terminal concerted reaction by each reaction center which has accumulated four plus charges (105).

Even under optimal conditions, in which Q is (presumably) completely oxidized, imperfections (both biological and technical) result in "misses" and "double hits," and thus an eventual damping of the O_2 oscillation. The phenomenon becomes even more complex when the redox state of Q changes during the sequence of flashes (c.f. O_2 gush discussed previously). In general, O_2 evolution is a function of both the donor and acceptor sides of System II; consequently, the O_2 flash yield will be proportional to the product $[S_3] \cdot [Q_{ox}]$ if the Q's of different units interact (106). In the absence of an electron acceptor, the O_2 yields change from strongly oscillating—while Q is oxidized—to quickly damped, and finally to a lower steady-state yield, presumably limited by the Mehler reaction (Figure 12).

The various reaction steps involved in O_2 evolution are relatively rapid. For example, the delay between photochemical excitation of the S_3 state and the appearance of molecular oxygen (k_3) is about 0.8 msec (103); the relaxation times of the other steps are somewhat faster—(0.2–0.4 msec). Consequently, the O_2 evolution system does not limit the rate of photosynthesis in strong light.

As described above, however, in very weak light, where photons arrive infrequently at the trapping centers the deactivation reactions $S_3 \xrightarrow{k_{-3}} S_2 \xrightarrow{k_{-2}} S_1$ can be significant. The rate of this deactivation varies considerably with the material—chloroplasts or algae—and experimental conditions and appears to correlate well with the redox state of the A pools. For example, in chloroplasts with added electron acceptor the half-life of the S_3 state is at least 10 sec (81, 107) while under conditions in which the A pools are largely reduced it is about 0.6 sec (106).

Photophosphorylation

Energy Conservation Sites. In green plants the sum of the energy generated by the two photoreactions is higher than that required to oxidize water and to reduce ferredoxin. This additional energy can be conserved as ATP. The electron transport chain includes one or more sites where this conversion can take place (7, 8).

Although the standard free energy of the conversion of ADP and P_i to ATP is about 10 kcal/mole, illuminated chloroplasts can establish ATP/ADP ratios which are equivalent to a phosphate potential of about 15 kcal/mole (108). This corresponds to the conservation of up to 660 mV for a one-equivalent reaction. If more than one equivalent were to move through this site per ATP formed, this value would be lower (e.g., 330 mV for two equivalents).

We distinguish two different types of electron transfer which are associated with the formation of ATP. In noncyclic photophosphorylation electrons are transferred from water to a low potential acceptor with the concomitant formation of ATP and O_2. For example, using NADP as the terminal electron acceptor one would observe the overall reaction

$$(1-x)H_2O + NADP + ADP + xP_i \xrightarrow{h\nu} xATP + NADPH + H^+ + 1/2\ O_2$$

where x is the ATP/2e ratio.

As in other ATP forming reactions, the addition of Mg^{2+} or Mn^{2+} is required for photophosphorylation. The most reasonable scheme for this process involves the transfer of electrons through the entire electron transport system (c.f., Figure 1) with the excess energy alluded to above drained off enroute (for example between Q and P). In the absence of an exogenous electron acceptor or with an autooxidizable electron acceptor—e.g., viologen—the O_2 produced by System II can be consumed by System I so that in the presence of catalase ATP is formed without net O_2 exchange. These processes require the participation of both photosystems and are inhibited by DCMU, which inhibits System II.

In the presence of DCMU, ATP production can be partially restored by the addition of a System I electron donor—e.g., $DCPIPH_2$—which allows electron flow through an energy conserving site. Another mode of System I electron flow coupled to ATP formation is truly cyclic—i.e., it does not involve a net oxidation–reduction reaction. In this process electrons are transferred from X^- to P^+ in a cyclic DCMU-insensitive operation of System I, which evidently passes through an energy conservation site. In vitro such a cycle can be mediated by electron carriers (e.g., PMS and pyocyanine).

The precise loci for the conservation of energy (e.g., ATP) are still debated. Several possible sites have been suggested:

(1) The reaction chain between the photoacts; the reaction $Q^- \rightarrow P^+_{700}$ pumps protons into the thylakoids, forming a proton gradient.

(2) Within the O_2 evolution process itself; this reaction occurs inside the thylakoid so that protons can accumulate and form a proton gradient.

(3) The reaction chain between X^- and P^+_{700} via cytochrome b_{563}; this transition entails a total potential change of 0.8 V, and thus could contain two sites associated with cyclic phosphorylation.

(4) Within an oxidative path—*i.e.*, reactions associated with the light-enhanced uptake of O_2 which is observed in whole cells.

The mechanism(s) by which electron flow is coupled to the production of ATP are still controversial (*see* Chapter 4). The main hypotheses are:

(1) The chemiosmotic hypothesis (*109, 110*) which postulates that electron transfer generates an electrochemical gradient across the thylakoid membrane. The energy stored in this gradient is used to drive the synthesis of ATP—*i.e.*,

$$e^- \text{ transfer} \rightarrow H^+ \text{ gradient} \rightarrow ATP$$

(2) The chemical hypothesis (*111*), which postulates that electron transfer results in the formation of a chemical high-energy intermediate (denoted [∼]). According to this scheme proton transfer is not on a direct route to ATP synthesis; rather it is on a side path competing with phosphorylation for the high-energy intermediate [∼], *i.e.*,

$$e^- \text{ transfer} \rightarrow [\sim] \rightarrow ATP$$
$$\downarrow \uparrow$$
$$H^+ \text{ gradient}$$

(3) The conformational hypothesis (*112*), which is formally equivalent to the chemical hypothesis but postulates that energy is stored in the conformational change of a protein rather than a covalent bond.

Regardless of the precise mode of energy conservation, it is clear that the energy of electron transport reactions can be conserved (∼I in Equation 1). This energy can either be consumed in the formation of ATP (Equation 2) or degraded (Equation 3). If neither reaction occurs, ∼I accumulates, and electron transport (Equation 1) is severely inhibited.

$$P^+_{700} + Q^- + I \rightleftharpoons P_{700} Q + \sim I \qquad (1)$$

$$\sim I + ADP + P_i \rightleftharpoons ATP + I \qquad (2)$$

$$\sim I \longrightarrow I \qquad (3)$$

Although many early measurements of phosphorylation stoichiometry suggested that one ATP was formed per pair of electrons transferred to NADP, more recent reports indicate ratios which are consistently greater than one (*113*). Moreover when NADP is replaced by electron acceptors such as benzoquinone and oxidized *p*-phenylenediamine—which are presumably reduced directly by System II—this ratio is decreased by one-half (*114*). These observations suggest that two sites of ATP formation are operative between O_2 evolution and NADP reduction—one within the O_2 evolution process, the other within the electron transport chain between the photoacts. Since the overall ATP/2e ratio depends not only upon the number of sites but also upon the mechanism of ATP formation, a determination of the amount of energy conserved at the sites and the efficiency with which this energy is converted to ATP is intimately related to the type of energy conservation scheme assumed (*see* above)

Photophosphorylation is a structure dependent process; consequently, its rate and efficiency *in vitro* are influenced by the conditions and procedures of chloroplast isolation. Even in chloroplasts exhibiting high phosphorylation efficiency, a residual (basal) rate of electron transport is observed in the absence of ADP and P_i. This basal rate probably reflects the decay of $\sim I$ (Reaction 3), which under coupled non-phosphorylation conditions limits the overall rate of the system. The addition of ADP and P_i removes the rate limitation imposed by $\sim I$. In the absence of an ATP consuming process, most of the ADP is phosphorylated, a high "phosphate potential" and high $\sim I/I$ ratio are reestablished, and the rate of electron transfer again returns to the low, basal value.

Numerous agents—loosely termed uncouplers—are known which somehow can prevent ATP formation without impairing the associated rate of electron transport. The stimulation of the rate of electron flow by these uncouplers—or for that matter, the phosphorylating agents ADP, Mg^{2+}, and P_i—is observed only when the rate of photon absorption exceeds the rate limitation imposed by the breakdown of $\sim I$. In weak light, the electron transport rate is a function of intensity only—*i.e.*, the rate of photon absorption is much lower than the rate of Reaction 3.

Figure 13 illustrates the effect of the uncoupler methylamine on the rate of ferricyanide reduction in saturating light. Note that without uncoupler the rate of ferricyanide reduction is approximately doubled by the addition of ADP and P_i, and one ATP is formed per two equivalents of ferricyanide reduced. As the amine concentration is increased up to about $10^{-2} M$, ATP formation is inhibited, and the rate of electron flow stimulated. This probably

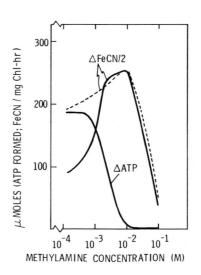

National Academy of Science and National Research Council Publication

Figure 13. Effect of the uncoupler methylamine on the rate of ferricyanide reduction in the presence (dashed line) and absence (solid line) of phosphorylation reagents. The other solid curve shows the rate of ATP formation accompanying the reduction of ferricyanide in the latter experiment (115).

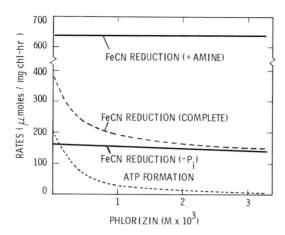

Figure 14. Effect of phlorizin on the rates of ferricyanide reduction and ATP formation in illuminated spinach chloroplasts

FeCN is ferricyanide. The rate of FeCN reduction in the absence of P_i is a measure of the basal rate of electron transport. The two dashed curves show the rates of FeCN reduction and ATP formation at different concentrations of inhibitor (117).

reflects the acceleration of Reaction 3—*i.e.*, the loss of a high energy precursor. (Note that further increases in the concentration of uncoupler severely inhibit the uncoupled rate of electron transport, possibly because of inhibition of the O_2-evolving system). Ammonia and amines also dissipate the pH gradient which, according to one hypothesis, is an intermediate in the energy conversion process. A number of agents which allow the free exchange of ions through membranes—"ionophores," *e.g.*, gramicidin—also uncouple phosphorylation.

The rates of electron transport and phosphorylation are strongly affected by the pH of the reaction mixture and even the type of buffer used (*116*). Basal electron transport has a maximal rate at pH 8.5–9.0. The rate coupled to ATP formation has a similar pH profile but drops less rapidly with decreasing pH. In the presence of methylamine the rate shows a broader optimum at a lower pH value. Consequently uncoupling can greatly enhance the rate ($\sim 50 \times$) at low pH while at high pH its effect on the rate becomes small.

Another type of phosphorylation inhibitor interferes with the final ADP phosphorylating step (Equation 2); thus, it annihilates the rate stimulation observed upon the addition of ADP and P_i. In the presence of these inhibitors, the rate of basal or amine uncoupled electron transport is unaffected. Figure 14 shows the effect of one such inhibitor—phlorizin—on electron transport and ATP formation. In this experiment the ratio ATP/2e is 2 and constant if calculated on the basis of the extra ADP-stimulated electron transport.

These data are consistent with the idea of two phosphorylation "sites" provided one makes the *ad hoc* assumptions that (1) a "site" yields one ATP per two electron passages, and (2) the basal rate is the result of an independent, non-phosphorylating pathway (or an already uncoupled chloroplast fraction).

Uncoupling phenomena (*i.e.*, stimulation of electron transport rate and loss of phosphorylation capacity) can also be induced by EDTA (*118*). Chloroplasts washed with EDTA or suspended in a medium of low salt content release a protein which mediates the phosphorylation of ADP. This so-called "coupling factor" has been isolated and can restore the phosphorylating capacity of EDTA treated chloroplasts (*119*).

Certain treatments of the coupling factor can cause it to act as an ATPase. In general, chloroplasts, in contrast to mitochondria, show extremely little ATPase activity in darkness (*120*). However, ATPase activity is rapidly induced by light, presumably as a consequence of the formation of some intermediate (*121*).

Ion Movements. Photosynthetic phosphorylation can be separated experimentally into "light" and "dark" steps so that energy harvested in the light can be utilized for ATP synthesis in a subsequent dark period. This has been demonstrated by experiments in which a chloroplast suspension was illuminated in the absence of ADP and P_i. If these reagents were then added to the suspension within a second or so after darkening, ATP was formed (*122, 123*). Jagendorf and Hind (*9*) identified this light-induced ATP precursor as a pH gradient across the thylakoid membrane.

Upon illumination, an unbuffered chloroplast suspension medium becomes alkaline, which implies that protons are transported into the thylakoid vesicles. A change of 2.5–3 pH units can be established in this manner, and at a low initial pH as much as one proton per chlorophyll can be transported. In the absence of light the system re-equilibrates with a halftime of several seconds. This equilibration is faster at high pH and is accelerated by the presence of uncouplers or detergents, which seem to make the membrane permeable to protons.

The generation of ATP in the absence of light has been observed by equilibrating a chloroplast suspension at low pH ($\sim$4) and then quickly raising it to pH 8 in the presence of ADP and P_i. If the low pH was obtained by equilibration with weak organic acids, as much as 0.2 ATP/chlorophyll was formed (*124*). (Apparently these acids penetrate the thylakoids and enhance the proton storage capacity of the vesicles by their buffering effect). Since no electron carrier in the chloroplast occurs in such high a concentration, the source of energy must be attributed to the pH gradient itself. These observations have given a strong impetus to the chemiosmotic view of ATP formation, which ascribes the driving potential of phosphorylation to a charge and/or concentration difference across an anisotropic membrane (*see* above and Chapter 4). Noncyclic electron transport results in the translocation of two protons for each equivalent that moves through the electron transport chain (*125*). A consumption of three protons from the gradient for each ATP generated (*126*) thus corresponds to an overall stoichiometry of 1.33 ATP/2e. Overall ATP/2e ratios somewhat higher (~1.5) (*127*) have been reported, and one could speculate that under ideal conditions the H^+/ATP ratio, and consequently the ATP/2e ratio, is equal to 2. Some ramifications of this stoichiometry are considered later.

Figure 15 is a schematic of a mechanism for energy conservation proposed by Mitchell (*110*) which has gained considerable experimental support, particularly in the laboratory of Witt (*128*). In this scheme it is assumed that both photoacts move an electron from the inside to the outside of the thylakoid membrane. The concerted action of many units can then establish a significant membrane potential. (This electrical field presumably causes absorption changes—electrochromic shifts—of some of the pigments in the

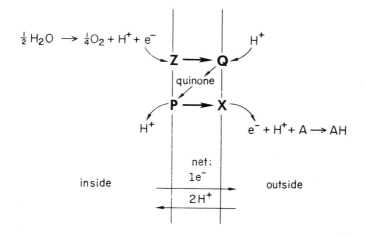

Glynn Research Ltd.

Figure 15. Hypothetical scheme for the accumulation of protons inside the chloroplast vesicle arising from O_2 evolution and electron transport via plastoquinone (110)

membrane (*129*). The most pronounced change—a red shift from 480 → 515 nm which was originally observed by Duysens (*130*)—probably arises from carotenoid.) Subsequently, the reduction of plastoquinone, which is assumed to be arranged across the membrane, consumes protons from the outer phase while its oxidation releases protons in the inner phase. With continued illumination, proton transfer is electrically balanced by flows of K^+ and Cl^- ions. The efflux of protons occurs only through specific sites in the thylakoid membrane. At these sites the coupling enzyme in some as yet unclarified way can translate the stored energy into ATP.

Although there is considerable evidence for a cause-and-effect relationship between proton flow and phosphorylation, the concept is by no means proved. Several experiments have been reported which suggest that a proton gradient is not—or not always—an intermediate of phosphorylation (*131, 132*). For example, in subchoroplast particles NH$_4$Cl can abolish proton uptake without affecting ATP formation (*133*). In addition, some chloroplast preparations which are rich in System I can show high rates of cyclic phosphorylation but no appreciable proton translocation (*134*).

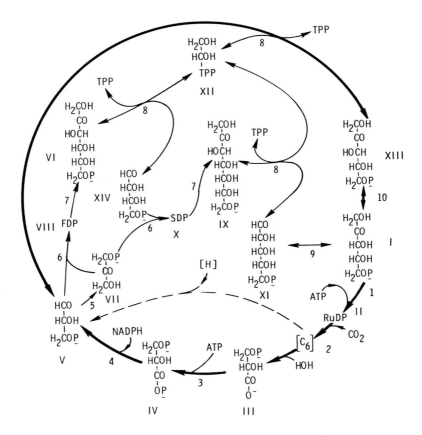

"Harvesting the Sun"

Figure 16. The pentose (Calvin–Benson) cycle

I. Ribulose-5-phosphate (Ru5P); II. ribulose-1,5-diphosphate (RuDP); III. 3-phosphoglyceric acid (PGA); IV. phosphoryl-3-phosphoglyceric acid; V. 3-phosphoglyceraldehyde (GA3P); VI. fructose-6-phosphate (F6P); VII. dihydroxyacetone phosphate (DHAP); VIII. fructose-1,6-diphosphate (FDP); IX. sedoheptulose-7-phosphate (S7P); X. sedoheptulose-1,7-diphosphate (SDP); XI. ribose-5-phosphate (R5P); XII. thiamine pyrophosphate glycoaldehyde addition compound (TPP—CHOH—CH$_2$OH), thiamine pyrophosphate (TPP); XIII. xylulose-5-phosphate. Enzymes: 1. ribulose-5-phosphate kinase; 2. ribulose diphosphate carboxylase (carboxydismutase); 3. phosphoglyceryl kinase; 4. triose phosphate dehydrogenase; 5. triose phosphate isomerase; 6. aldolase; 7. diphosphatase; 8. transketolase; 9. phosphoribose isomerase; 10. ribulose phosphate-xylulose phosphate isomerase; dashed line, hypothetical reduction of PGA moiety (*135*).

Utilization of Reduced Photoproducts

Production of Reduced Carbon Compounds. The primary function of the photosynthetic apparatus is the production of reduced carbon compounds from CO_2 and light. As described previously, System I produces a reductant at least as strong as H_2 (*i.e.*, -420 mV) which is probably a special molecule of ferredoxin. Reduced ferredoxin is able to interact with many diverse biological systems (*see below*). The most important reaction from the standpoint of carbon metabolism is its interaction with NADP, mediated by the enzyme NADP reductase. NADPH, in turn, is utilized to reduce CO_2 to the redox level of an aldehyde. In this process, it is assisted by ATP, generated (at least in part) by the process of photophosphorylation.

The CO_2 fixation process takes place *via* the series of intraconversion reactions—the pentose or Calvin-Benson cycle—shown in Figure 16. This reaction scheme is generally accepted as the primary CO_2 fixation pathway in photosynthetic organisms. It is beyond the scope of this article to describe this scheme in detail; for a thorough description of this metabolic pathway the reader is referred to Ref. *136*. We discuss a few salient aspects of the scheme and some additional features of carbon metabolism in plants which appear physiologically significant.

(1) The cycle may be said to start with the transfer of a phosphate group from ATP to ribulose-5-phosphate (RuP) by the enzyme phosphoribulokinase. This reaction activates or primes the pentose phosphate, and the resulting ribulose-1,5-diphosphate (RuDP) is the substrate in the succeeding carboxylation reaction:

(2) The carboxylation step itself, catalyzed by the enzyme ribulose diphosphate carboxylase—sometimes called carboxydismutase—takes place according to the reaction:

$$RuDP + CO_2 \to 2 \text{ phosphoglyceric acid (PGA)}$$

In this reaction a 5-carbon compound, RuDP, is carboxylated and cleaved to yield two 3-carbon PGA molecules. This is the only reaction by which the overall system realizes a net gain of carbon from CO_2.

(3) The reduction of PGA to the level of an aldehyde requires another priming reaction. In this reaction—mediated by the enzyme phosphoglcerylkinase—a phosphate group is transferred from ATP to PGA. The resulting compound—phosphoryl-3-phosphoglyceric acid (P3GA)—is then reduced by NADPH to 3-phosphoglyceraldehyde (GA3P). The reactions are:

$$PGA + ATP \to P3GA + ADP$$
$$P3GA + NADPH + H^+ \to GA3P + NADP^+$$

This last reaction is the sole reductive step in the pentose cycle.

(4) Although the total cycle is quite complex and is catalyzed by at least 10 different enzymes, it consists of only one carboxylating step, one reductive step, and two ATP-utilizing primary steps. All other reactions are merely

rearrangements by which the starting material (RuDP) is regenerated and products are eliminated. The overall stoichiometry is such that three moles of ATP and two of NADPH are required to reduce one mole of CO_2 to the level of a hexose and regenerate the components of the cycle.

In addition to the pentose cycle, many plants—notably tropical grasses such as sugar cane—have a special CO_2 pathway [the Hatch-Slack-Kortschak pathway (*137, 138*)] which appears to act as a CO_2 pump, providing the carboxylating enzyme of the pentose cycle with CO_2 even when the external CO_2 pressure is vanishingly low. In this pathway oxaloacetate is produced by the ATP-driven carboxylation of pyruvate; the overall process may be written as

$$CO_2 + \text{pyruvate} + 2\ ATP \rightarrow \text{oxaloacetate} + 2\ ADP + 2\ P_i$$

The oxaloacetate—or a compound derived from it (*e.g.*, malate)—then acts as a CO_2 donor for RuDP, regenerating the pyruvate.

Plants which possess this cycle—usually called C_4 plants in contrast to C_3 plants which lack it—require about 5 moles of ATP per mole of CO_2 reduced to the level of carbohydrate. Presumably, at the additional expense of 2 ATP per CO_2 fixed C_4 plants can use low CO_2 concentrations more effectively.

The main rationale for proposing this pathway as a CO_2 pump was that RuDP carboxylase seemed to have a low affinity for CO_2. However, recent reports (*139*) have shown that the CO_2 affinity of this enzyme is sufficient to account for the observed rates *in vivo*. Consequently, the proposed function of this pathway may require reconsideration.

We should also mention a presumably aberrant pathway in which glycolate is formed in appreciable quantities. The physiological function for this glycolate is unknown, and its metabolism poorly understood. According to some workers (*140*) glycolate formation is the result of competition of O_2 with CO_2 on the RuDP carboxylase—instead of a CO_2 being added to the C_5 piece ($C_5 \rightarrow C_6$), an oxidized C_2 piece is generated ($C_5 \rightarrow C_3 + C_2$). Photorespiration—an extra light-induced uptake of O_2 and evolution of CO_2 observed mainly in C_3 plants—is usually ascribed to the oxidation of this glycolate.

Other Reactions Utilizing Reduced Photoproduct. The light-produced strong reductant functions mainly to reduce CO_2. However, under some physiological conditions this reductant can apparently be channeled toward other reactions. For example, since the pioneering work of Gaffron (*141*), it has been recognized that some algae could be adapted to metabolize molecular hydrogen by incubating the cells anaerobically in the dark. Upon exposure to light these cells were able to photoevolve H_2. The source of the electrons which are promoted by light to the level of H_2 apparently is not H_2O but internal reductants; photosynthetic enhancement of H_2 evolution is not observed, and System II is not required for H_2 photoevolution (*142*).

H_2 photoevolution by algae is not a usual metabolic process; it occurs only under conditions of extreme anaerobiosis. There is evidence that the photoevolution of H_2 serves as a priming reaction for the photosynthetic apparatus (143). Under anaerobic conditions all components of the photosynthetic electron transport chain become reduced and thus Q and X become inoperative. However, in the presence of light and an active hydrogenase the electrons from X^- can be transferred to protons and ejected as H_2. This continued noncyclic operation would result in the oxidation of the photosynthetic electron transport chain by P^+_{700}.

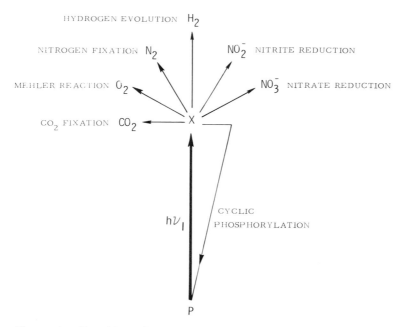

Figure 17. *Possible pathways for the utilization of the strong reductant of System I*

X may also play a role in nitrogen metabolism. Many blue–green algae are able to fix N_2 in the light (*i.e.*, mediate the reaction $N_2 \rightarrow NH_3$). These organisms appear to utilize the reduced System I acceptor X^- as the source of the strong reductant necessary to drive this reaction (144). There is some evidence that the biological reduction of nitrite (and possibly nitrate) by photosynthetic organisms may also be coupled to X (145). A scheme representing possible pathways for the utilization of the strong reductant of System I is shown in Figure 17. At present it is difficult to assess the relative importance of some of these pathways *in vivo*.

Energetics and Kinetics

Optimal Efficiency of Photosynthesis and Plant Growth. In photochemical reactions a single photon interacts with a single molecule, inducing the transition of a single electron. Therefore, it is convenient to express photochemical efficiency in terms of the quantum yield ϕ—equivalents converted per quantum absorbed—or its inverse, the quantum requirement ($h\nu$/equiv). However, this equivalence law pertains only to the primary conversions; in whole cell processes the quantum yield may not be stoichiometrically related to the ultimate products.

In green plant photosynthesis the two photoacts collaborate to split water into its elements. This process results in a gain of chemical potential of 1.2 V or 120 kcal/mole of oxygen. Considerable additional energy is required to stabilize reaction intermediates and generate ATP. Since a mole of red (680–700 nm) quanta represents about 40 kcal/mole, the minimum conceivable requirement of 1 $h\nu$ per equivalent or 4 $h\nu/O_2$ would correspond to an efficiency of about 75%.

The true optimal quantum yield of photosynthesis has been a hotly controversial issue (*see* Ref. *146* for review). In whole cells this seemingly simple determination is complicated by the peculiar transient behavior of CO_2 at the onset of illumination (*147*) and by interactions of photosynthetic and respiratory pathways in weak light (*148*). Presently, a minimum requirement of ~ 10 $h\nu/O_2$—corresponding to $\geqslant 2$ $h\nu$/equiv—is generally accepted. This requirement corresponds to an energy conversion efficiency of about 30% in red light and about 20% in white (solar) radiation (because the extra energy in quanta of wavelengths < 680 nm is dissipated). Since plant pigments do not absorb wavelengths beyond ~ 700 nm (about half the solar radiation), the best possible efficiency of photosynthesis and plant growth in sunlight is about 10%.

The minimum requirement of 10 $h\nu/O_2$ which corresponds to $\geqslant 2$ $h\nu$ per equivalent moved from water to NADP, underlies the two-photoact scheme of Figure 1. This efficiency is not only approached in isolated chloroplasts but also over long time periods with growing algae (*149*). Growth experiments with heterotrophic organisms (*150*) indicate that three ATP's are required to convert carbohydrate to cellstuff. This, coupled with the three to five ATP's required to reduce a CO_2 molecule to the level of carbohydrate suggests that under growing conditions at least six to eight ATP's are needed per O_2 evolved or per carbon incorporated—*i.e.*, the ATP/2e ratio must be at least 3 *in vivo*. It is doubtful whether the ATP generation coupled to the noncyclic operation of the two photoacts (probably less than 2 ATP/2e) is sufficient to meet the overall ATP requirements of the cell.

The source of the extra ATP required for CO_2 assimilation and growth is still unresolved. In whole algae, long wave light can suppress respiratory O_2 uptake (*148, 151*), and maintain other ATP requiring conversions (*152*). In these instances ATP is probably generated in a cyclic operation of System I. However, the rate of this cycle appears to be low. Other cycles can be conceived—*e.g.*, photochemically produced NADPH might enter directly into mitochondrial oxidative phosphorylation. In any case, the extra ATP would be generated at the expense of extra photons which would yield no net O_2

and reducing power, and thus the overall efficiency of autotrophic growth would be decreased (153).

Factors Affecting the Rate and Efficiency of Photosynthesis.
Photosynthesis involves an interplay between two photochemical reactions and many dark reactions. The rate of the overall process depends upon the availability of various substrates—e.g., water, CO_2, and light—and the turnover rate of the enzymatic machinery. The CO_2 concentration that supports half-maximal rate *in vivo* (called the apparent affinity or Michaelis constant K_m), is the resultant of a complex set of diffusion and reaction rates. This constant does not appear to vary greatly among land plants, despite considerable variation of V_{max} and different pathways of CO_2 fixation (*i.e.*, C_3 vs. C_4 plants) (154). Its value is typically about 0.03%; consequently, the normal CO_2 content of the air is insufficient to sustain the maximum photosynthetic rate.

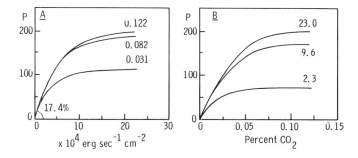

Mededelingen van de Landbouwhogeschool, Wageninger

Figure 18. Rate of photosynthesis observed with a turnip leaf (P, in μl $O_2/cm^2/hr$) as a function of light intensity and CO_2 concentration

A: rate as a function of light intensity at three different CO_2 concentrations (units, % CO_2); B: rate as a function of CO_2 concentration at three different light intensities (units, 10^4 erg sec/cm^2). Magnitude of the slope in A (17.4%) can be used as an index of efficiency since $CO_2/quantum = (CO_2/sec)/(quantum/sec)$. Thus in this experiment 17.4 cal were fixed per 100 cal of absorbed light. This value was derived assuming (1) the assimilatory quotient—O_2 evolved/CO_2 fixed—was about 1.0, and (2) all visible light falling on the leaf was absorbed (155).

Figure 18 illustrates the effect of light intensity and CO_2 concentration on the photosynthetic rate of a C_3 plant. In these experiments, the rate of CO_2 uptake was measured at three CO_2 concentrations and three light intensities. At the weakest intensity, 0.03% CO_2 (the normal concentration in air) was sufficient to maintain the rate, which in this instance was limited by the slow influx of photons. However, when the light intensity was increased tenfold, at least 0.1% CO_2 was required to maintain the increased rate. The maximal rate seen in this experiment was presumably determined by the maximum turnover rate of one or more steps in the enzymatic reaction chain, since all substrates were present in excess.

The picture can be further complicated—especially in C_3 plants—by the phenomenon of photorespiration in which light enhances O_2 uptake and simultaneous CO_2 release. (For a detailed account of photorespiration *see* Ref. *156*.) The rate of photorespiration varies widely; it is highest at elevated temperatures, high O_2 and low CO_2. This process can result in a severe loss of photosynthetic product and a significant decrease in net production under "field" conditions (*see* Ref. *157*). However, under optimal conditions (weak light, ample CO_2 and low O_2), these losses are minor.

In contrast to the rather evenly distributed CO_2, natural illumination can fluctuate widely with time and location. Consequently, the "dynamic range" of the photosynthetic apparatus with respect to light intensity is of prime importance. In whole cells in weak light, the half life of the S_3 state is on the order of 3 sec. Thus, the efficiency of light conversion is half maximal at an intensity in which each System II reaction center receives a quantum every 3 sec. For a unit of 200 chlorophylls per trap this corresponds to each chlorophyll absorbing a quantum once per 10 min or about 10^{-4} times the intensity of full sunlight.

In full sunlight at noon, a single chlorophyll molecule absorbs about 10 photons per second. Consequently, a trapping center receiving excitation energy from the 200 chlorophyll molecules of the unit will receive 2000 quanta per sec. Flashing light experiments—like those described earlier—showed that the amount of O_2 made in each flash is half maximal—the coinversion of the photoproducts is half completed—if the dark time between the flashes is 5–10 msec. This suggests that the maximum attainable rate is about 100 equivalents/sec/reaction chain. Consequently, in most algae and crop plants photosynthesis is half saturated at about one-tenth the intensity of full sunlight. This early light saturation of the photosynthetic apparatus, in conjunction with the low CO_2 content of the atmosphere, causes the efficiency of photosynthetic energy conversion to be suboptimal at high intensities. Plant leaves have an arsenal of structural adaptations to spread (dilute) the light and the CO_2 uptake over a large leaf area. Still, these saturation phenomena cause most crop yields to remain well below the 10% efficiency which can be observed in weak light.

Depending upon habitat, there can be considerable differences in chlorophyll content and rate between shade plants and sun plants and between C_3 and C_4 species. In general the photosynthetic apparatus seems designed to operate under low-light conditions. This is graphically illustrated by the various parameters plotted in Figure 19. At less than one incident quantum per trap per 3 sec the system is limited by the stability of the O_2 precursors (the S states), while on the other end of the intensity scale the process saturates when the centers are hit every 5–10 msec. This means that photosynthetic light conversion proceeds efficiently over approximately a 1000-fold intensity range, all well below the intensity of bright sunlight.

The 10 msec maximal turnover rate observed with whole leaves and algae is also observed in isolated chloroplasts using acceptors other than CO_2. Consequently, the endogenous rate-limiting step is probably located within the light-driven electron transport chain (cf. Figure 1). In strong, rate-saturating

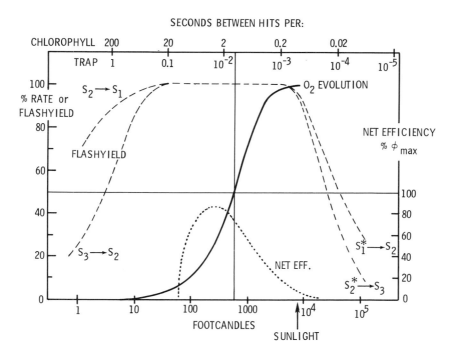

Figure 19. *Illustration of the working range of photosynthesis and some of the parameters which affect this range*

Rate limitations of O_2 system (dotted lines) refer to the scheme

$$S_0 \xrightarrow{h\nu} S_0^* \xrightarrow{k_0} S_1 \xrightarrow{h\nu} S_1^* \xrightarrow{k_1} S_2 \xrightarrow{h\nu} S_2^* \xrightarrow{k_2} S_3 \xrightarrow{h\nu} S_3^* \xrightarrow{k_3} S_0 + O_2$$

with k_{d2} and k_{d3} back-reactions.

The rate limitation of O_2 evolution (solid line) at about 1000 footcandles probably arises from the transfer of electrons from Q to P (see text) (153).

light in the presence of an electron acceptor for System I and an uncoupler of phosphorylation, P_{700} and cytochrome f become oxidized, while Q and most of the plastoquinone pool become reduced. This suggests that the rate limitation probably arises from the dark reactions which interconnect the two photoacts.

Nomenclature

ADP, adenosine diphosphate
ATP, adenosine triphosphate
CCCP, carbonyl cyanide-m-chlorophenyl hydrozone
Cyt, C555, C552, and so forth, cytochrome
DAD, DAD_{ox}, diaminodurol and its oxidized form
DCMU, 3-(3,4-dichlorophenyl)-1,1-dimethylurea
DCPIP, $DCPIPH_2$, 2,6-dichlorophenolindophenol and its reduced form
EDTA, ethylenediaminetetraacetic acid
ESR (EPR), electron spin resonance
FMN, $FMNH_2$, flavine mononucleotide and its reduced form
NAD^+, NADH, nicotinamide adenine dinucleotide and its reduced form
$NADP^+$, NADPH, nicotinamide adenine dinucleotide phosphate and its reduced form
P_i, orthophosphate
PMS, phenazine methosulfate.
All potential values quoted are E_0 at pH 7.

Acknowledgments

The authors thank G. Cheniae, C. F. Fowler, and K. L. Zankel for many helpful discussions.

Literature Cited

1. Rabinowitch, E. I., "Photosynthesis and Related Processes," Interscience Pub., Inc., New York, 1945.
2. Loomis, W. E., "Handbuch der Pflanzenphysiologie," W. Ruhland, Ed., Vol. V, Part 1, 85, Springer, Berlin, 1960.
3. van Niel, C. B., *Cold Spring Harbor Symp. Quant. Biol., 1935,* **3,** 138.
4. Reuben, S., Randall, M., Kamen, M. D., Hyde, J. L., *J. Amer. Chem. Soc.* (1941) **63,** 877.
5. Hill, R., *Nature* (1937) **139,** 881.
6. Bassham, J. A., Calvin, M., "The Path of Carbon in Photosynthesis," Prentice-Hall, Englewood Cliffs, 1957.
7. Arnon, D. I., Allen, M. B., Whatley, F. R., *Nature* (1954) **174,** 394.
8. Frenkel, A., *J. Amer. Chem. Soc.* (1954) **76,** 5568.
9. Jagendorf, A. T., Hind, G., *Nat. Acad. Sci. Publ.* (1963) **1145,** 599, Washington, D. C.
10. Emerson, R., Arnold, W., *J. Gen. Physiol.* (1932) **15,** 391.
11. Duysens, L. N. M., Transfer of Excitation Energy in Photosynthesis, Ph.D. Thesis, Utrecht, 1952.
12. Kok, B., *Biochim. Biophys. Acta* (1961) **48,** 527.
13. Blinks, L. R., "Research in Photosynthesis," H. Gaffron, Ed., 444, Wiley (Interscience), New York, 1957.
14. Emerson, R., Chalmers, R., Cederstrand, C., *Proc. Nat. Acad. Sci., U.S.A.* (1957) **43,** 133.
15. Emerson, R., *Ann. Rev. Plant Physiol.* (1958) **9,** 1.
16. Kok, B., Hoch, G., "Light and Life," W. D. McElroy, B. Glass, Eds., 397, Johns Hopkins Press, Baltimore, 1961.

17. Duysens, L. N. M., Amesz, J., Kamp, B. M., *Nature* (1961) **190,** 510.
18. Hill, R., Bendall, F., *Nature* (1960) **186,** 136.
19. Park, R. B., Pon, N. G., *J. Mol. Biol.* (1961) **3,** 1.
20. Menke, W., *Ann. Rev. Plant Physiol.* (1962) **13,** 27.
21. Cole, W. J., Chapman, D. J., Siegelman, H. W., *J. Amer. Chem. Soc.* (1967) **89,** 3643.
22. Brown, J. S., *Ann. Rev. Plant Physiol.* (1972) **23,** 73.
23. Engelmann, T. W., *Bot. Z.* (1884) **41,** 1.
24. Holt, A. S., Jacobs, E. E., *Amer. J. Botany* (1954) **41,** 710.
25. Livingston, R., *Quart. Rev. No. 2* (1960) **14,** 174.
26. Strehler, B. L., Arnold, W., *J. Gen. Physiol.* (1951) **34,** 809.
27. Chessin, M., Livingston, R., Truscott, T. G., *Trans. Farad. Soc.* (1966) **62,** 1519.
28. Griffith, M., Sistrom, W. R., Cohen-Bazire, G., Stainer, R. Y., *Nature* (1955) **176,** 1211.
29. Olson, R. A., *Nat. Acad. Sci. Res. Council Publ.* (1963) **1145,** 545.
30. Watson, W. F., Livingston, R., *J. Chem. Phys.* (1950) **18,** 802.
31. Butler, W. L., *Biochim. Biophys. Acta* (1962) **69,** 309.
32. Zankel, K. L., Reed, D. W., Clayton, R. K., *Nat. Acad. Sci.* (1968) **61,** 1243.
33. Pearlstein, R. M., *Brookhaven Symp. Biol.* (1967) **19,** 16.
35. Clayton, R. K., "Molecular Physics of Photosynthesis," 181, Blaisdell Pub. Co., New York, 1965.
36. Förster, T., *Discuss. Faraday Soc.* (1959) **27,** 7.
37. Commoner, B., *In* "Light and Life," W. D. McElroy, B. Glass, Eds., 356, Johns Hopkins Press, Baltimore, 1961.
38. Beinert, H., Kok, B., *Nat. Acad. Sci. Nat. Res. Council, Publ.* (1963) **1145,** 131.
39. Davenport, H. E., Hill, R., *Proc. Roy. Soc.* (1952) **B139,** 327.
40. Katoh, S., *Nature* (1969) **186,** 553.
41. Gorman, D. S., Levine, R. P., *Plant Physiol.* (1966) **41,** 1648.
42. Curtis, V. A., Siedow, J. N., San Pietro, A., *Arch. Biochem. Biophys.* (1973) **158,** 898.
43. Kok, B., In "Photosynthetic Mechanisms of Green Plants," *Nat. Acad. Sci.-Nat. Res. Council Publ.* (1963) **1145,** 66.
44. Tagawa, K., Arnon, D. I., *Nature* (1962) **195,** 537.
45. Malkin, R., Bearden, J., *Proc. Nat. Acad. Sci.* (1971) **68,** 16.
46. Hiyama, T., Ke, B., *Proc. Nat. Acad. Sci.* (1971) **68.**
47. Böhme, H., Cramer, W. A., *Biochim. Biophys. Acta* (1973) **325,** 275.
48. San Pietro, A., Lang, H. M., *J. Biol. Chem.* (1958) **231,** 211.
49. Avron, M., Jagendorf, A. T., *Arch. Biochem. Biophys.* (1956) **65,** 475.
50. Arnon, D. I., *Nat. Acad. Sci.-Nat. Res. Council, Publ.* (1963) **1145,** 195.
51. Mehler, A. H., *Arch. Biochem. Biophys.* (1951) **34,** 339.
52. Vernon, L. P., Zaugg, W. S., *J. Biol. Chem.* (1960) **265,** 2728.
53. Boardman, N. K., Anderson, J. M., *Nature* (1964) **203,** 166.
54. Cramer, W. A., Butler, W. L., *Biochim. Biophys. Acta* (1969) **172,** 503.
55. Doring, G., Stiehl, H. H., Witt, H. T., *Z. Naturf.* (1967) **22b,** 639.
56. Malkin, R., Bearden, A. J., *Proc. Nat. Acad. Sci.* (1973) **70,** 294.
57. Stiehl, H. H., Witt, H. T., *Z. Naturf.* (1969) **24b,** 1588.
58. Knaff, D. B., Arnon, D. I., *Proc. Nat. Acad. Sci., U.S.A.* (1969) **63,** 963.
59. Duysens, L. N. M., Sweers, H. E., *In* "Studies on Microalgae and Photosynthetic Bacteria," 353, Univ. Tokyo Press, Tokyo, 1963.
60. Blinks, L. R., Skow, R. K., *Proc. Nat. Acad. Sci. U.S.A.* (1938) **24,** 420.
61. Joliot, P., *J. Chem. Phys.* (1961) **58,** 584.
62. Amesz, J., *Biochim. Biophys. Acta* (1964) **79,** 257.
63. Stiehl, H. H., Witt, H. T., *Z. Naturf.* (1968) **23b,** 220.

64. Lundegarth, H., *Physiol. Plantarum* (1952) **15**, 390.
65. Boardman, N. K., Anderson, J. M., *Biochim. Biophys. Acta* (1967) **143**, 187.
66. Hind, G., Olson, J. M., *Ann. Rev. Plant Physiol.* (1968) **19**, 249.
67. Bendall, D. S., Sofrova, D., *Biochim. Biophys. Acta* (1971) **234**, 371.
68. Cramer, W. A., Fan, H. N., Bohme, H., *Bioenergetics* (1971) **2**, 289.
69. Yamashita, T., Butler, W. L., *Plant Physiol.* (1968) **43**, 1978.
70. Emerson, R., Lewis, C. M., *Amer. J. Botany* (1943) **30**, 165.
71. Haxo, F. T., Blinks, L. R., *J. Gen. Physiol.* (1950) **33**, 389.
72. Kok, B., Gott, W., *Plant Physiol.* (1960) **35**, 802.
73. Joliot, P., Joliot, A., Kok, B., *Biochim. Biophys. Acta* (1968) **153**, 635.
74. Myers, J., *Nat. Acad. Sci.-Nat. Res. Council Publ.* (1963) **1145**, 301.
75. Hoch, G. E., Martin, I., *Arch. Biochem. Biophys.* (1963) **102**, 430.
76. Jones, L. W., Myers, J., *Plant Physiol.* (1964) **39**, 938.
77. Bonaventura, C., Myers, J., *Biochim. Biophys. Acta* (1969) **189**, 366.
78. Murata, N., *Biochim. Biophys. Acta* (1969) **172**, 242.
79. Joliot, A., Joliot, P., *C. R. Acad. Sci.* (1964) **258**, 4622.
80. Siggel, U., Renger, G., Rumberg, B., *2nd Int. Cong. Photosynthesis, 1971*, 753, Stresa, Italy.
81. Kok, B., Forbush, B., McGloin, M., *Photochem. Photobiol.* (1970) **11**, 457.
82. Eimhjellen, K. E., Aasmundrud, O., Jensen, A., *Biochem. Biophys. Res. Commun.* (1963) **10**, 232.
83. Clayton, R. K., *Ann. Rev. Plant. Physiol.* (1963) **14**, 159.
84. Holt, A. S., Clayton, R. K., *Photochem. Photobiol.* (1965) **4**, 829.
85. Clayton, R. K., *Photochem. Photobiol.* (1966) **5**, 669.
86. Goedheer, J. C., *Biochim. Biophys. Acta* (1960) **38**, 389.
87. Clayton, R. K., Sistrom, W. R., *Photochem. Photobiol.* (1966) **5**, 661.
88. Sauer, K., Dratz, E. A., Coyne, L., *Proc. Nat. Acad. Sci. U.S.A.* (1968) **61**, 17.
89. Parson, W. W., *Biochim. Biophys. Acta* (1968) **153**, 248.
90. Parson, W. W., *Biochim. Biophys. Acta* (1969) **189**, 384.
91. Sybesma, C., *in* "Photobiology of Microorganisms," P. Halldal, Ed., 57, Wiley, New York, 1969.
92. Clayton, R. K., *Biochem. Biophys. Res. Commun.* (1962) **9**, 49.
93. Chance, B., *Nature* (1961) **189**, 719.
94. Gest, H., "Bacterial Photosynthesis," H. Gest, A. San Pietro, and L. P. Vernon, Eds., 129, Antioch Press, Yellow Springs, 1963.
95. Morita, S., *Biochim. Biophys. Acta* (1968) **153**, 241.
96. Sybesma, C., Fowler, C. F., *Proc. Natl. Acad. Sci. U.S.* (1968) **61**, 1343.
97. Kessler, E., *Planta* (1957) **49**, 435.
98. Cheniae, G. M., Martin, I. F., *Biochem. Biophys. Res. Commun.* (1967) **28**, 89.
99. Cheniae, G. M., *Ann. Rev. Plant Physiol.* (1970) **21**, 467.
100. Izawa, S., Heath, R., Hind, G., *Biochim. Biophys. Acta* (1969) **180**, 338.
101. Stemler, A., Govindjee, *Plant Physiol.* (1973) **52**, 119.
102. Allen, F. L., Franck, J., *Arch. Biochem. Biophys.* (1955) **58**, 510.
103. Joliot, P., *Brookhaven Symp. Biol.* (1966) **19**, 418.
104. Joliot, P., *Photochem. Photobiol.* (1968) **8**, 451.
105. Fowler, C. F., Kok, B., *Proc. 3rd Internat. Congress on Photosynthesis Research, 1973*, Bochum.
106. Radmer, R., Kok, B., *Biochim. Biophys. Acta* (1973) **314**, 28.
107. Forbush, B., Kok, B., McGloin, M., *Photochem. Photobiol.* (1971) **14**, 307.
108. Kraayenhof, R., *Biochim. Biophys. Acta* (1969) **180**, 213.
109. Mitchell, P., *Nature* (1961) **191**, 144.
110. Mitchell, P., Glynn. Res. Ltd. (1966) Bodmin., Cornwall.
111. Slater, E. C., *Nature* (1953) **172**, 975.
112. Boyer, P. D., "Oxidases and Related Redox Systems," T. E. King, H. S. Mason, M. Morrison, Eds., Vol. **2**, 994, John Wiley, New York, 1965.

113. Izawa, S., Good, N. E., *Biochim. Biophys. Acta* (1968) **162,** 380.
114. Saha, S., Ouitrakul, R., Izawa, S., Good, N. E., *J. Biol. Chem.* (1971) **246,** 3204.
115. Avron, M., Shavit, N., *Nat. Acad. Sci.-Nat. Res. Council Publ.* (1963) **1143,** 611.
116. Avron, M., *Proc. 2nd Int. Congr. Photosynthesis, 1972,* 861, Stresa, Italy, Junk, The Hague.
117. Izawa, S., Winget, G. D., Good, N. E., *Biochem. Biophys. Res. Commun.* (1966) **22,** 223.
118. Avron, M., *Biochim. Biophys. Acta* (1963) **77,** 699.
119. McCarty, R. E., Racker, E., *Brookhaven Symp. Biol.* (1966) **19,** 202.
120. Arnon, D. I., Allen, M. B., Whatley, F. R., *Biochim. Biophys. Acta* (1956) **20,** 449.
121. Avron, M., *J. Biol. Chem.* (1962) **237,** 2011.
122. Shen, Y. K., Shen, G. M., *Sci. Sinica* (1962) **11,** 109.
123. Hind, G., Jagendorf, A. T., *Proc. Nat. Acad. Sci.* (1963) **49,** 715.
124. Jagendorf, A. T., Uribe, *Brookhaven Symp. Biol.* (1967) **19,** 215.
125. Izawa, S., Hind, G., *Biochim. Biophys. Acta* (1967) **143,** 377.
126. Schroder, H., Muhle, H., Rumberg, B., *Proc. 2nd Int. Photosynthesis Congr., 1972,* 919, Stresa, Italy.
127. Reeves, S. G., Hall, D. O., *Biochim. Biophys. Acta* (1973) **314,** 66.
128. Witt, H. T., *Quart. Rev. Biophys.* (1971) **4,** 365.
129. Junge, W., Witt, H. T., *Z. Naturf.* (1968) **23b,** 244.
130. Duysens, L. N. M., *Science* (1954) **120,** 353.
131. Walker, D. A., Crofts, A. R., *Ann. Rev. Biochem.* (1970) **38,** 389.
132. Dilley, R., "Current Topics in Bioenergetics," D. R. Sanadi, Ed., Vol. **4,** 237, Academic Press, New York, 1971.
133. McCarty, R. E., *J. Biol. Chem.* (1969) **244,** 4292.
134. Arntzen, C. J., Dilley, R. A., Neumann, *Biochim. Biophys. Acta* (1971) **245,** 409.
135. Bassham, J. A., Jensen, R. G., "Harvesting the Sun," A. San Pietro, F. A. Greer, T. J. Army, Eds., 74, Academic Press, New York, 1967.
136. Bassham, J. A., "Plant Biochemistry," J. Bonner, J. E. Varner, Eds., 875, Academic Press, New York, 1965.
137. Hatch, M. D., Slack, C. R., *Biochem. J.* (1966) **101,** 103.
138. Hatch, M. D., Slack, C. R., *Ann. Rev. Plant Physiol.* (1970) **21,** 141.
139. Bahr, J. T., Jensen, R. G., *Biochem. Biophys. Res. Commun.* (1974) **57,** 1180.
140. Andrews, T. J., Lorimer, G. H., Tolbert, N. E., *Biochem.* (1973) **12,** 11.
141. Gaffron, H., *Nature* (1939) **143,** 204.
142. Healey, F. P., *Plant Physiol.* (1970) **45,** 153.
143. Kessler, E., *Arch. Mikrobiol.* (1973) **93,** 91.
144. Bothe, H., *Proc. 2nd Intl. Congress on Photosynthesis, 1972,* 2169, Stresa, Italy, Junk, The Hague.
145. San Pietro, A., Black, C. C., *Ann. Rev. Plant Physiol.* (1965) **16,** 155.
146. Kok, B., "Handbuch der Pflanzenphysiologie," W. Ruhland, Ed., Vol. V, Part 1, 566, Springer, Berlin, 1960.
147. Emerson, R., Lewis, C. M., *Amer. J. Bot.* (1941) **28,** 789.
148. Kok, B., *Biochem. Biophys. Acta* (1949) **3,** 625.
149. Kok, B., *Acta. Bot. Nee'rl* (1952) **1,** 445.
150. Bauchop, T., Elsden, S. R., *J. Gen. Microbiol.* (1960) **23,** 457.
151. Hoch, G., Owens, O. v. H., Kok, B., *Arch. Biochem. Biophys.* (1963) **101,** 171.
152. Tanner, W., Loos, E., Kandler, O., "Currents in Photosynthesis," J. B. Thomas, J. C. Goedheer, Eds., 243, Donker, Rotterdam, 1966.
153. Kok, B., "Horizons in Bioenergetics," A. San Pietro, H. Gest, Eds., 153, Academic Press, New York, 1972.
154. Goldworthy, A., *Nature* (1968) **217,** 62.

155. Gaastra, P., *Mededel. Landbouwhogeschool, Wageningen* (1959) **59**, 1.
156. Zelitch, I., "Photosynthesis, Photorespiration, and Plant Productivity," Academic Press, New York, 1971.
157. Tolbert, N. E., Nelson, E. B., Bruin, W. J., "Photosynthesis and Respiration," M. D. Hatch *et al.*, Eds., 506, Wiley, New York, 1971.

Work supported in part by the Energy Research and Development Administration Contract E(11-1)3326. This chapter was prepared with the support of the National Science Foundation Grant BMS 74-20736. Any opinions, findings, conclusions, or recommendations expressed herein are those of the authors and do not necessarily reflect the views of NSF.

Chapter

6

The Mechanism of Muscle Contraction

John Gergely, Department of Muscle Research, Boston Biomedical Research Institute; Department of Neurology, Massachusetts General Hospital; and Department of Biological Chemistry, Harvard Medical School, Boston, Mass.

Historical

MUSCLE, AS IT HAS been made clear in Dorothy Needham's recent monumental monograph (1), has been studied by biologists, physicians, physicists, and chemists for over 2000 years. Although it is only within the last two or three decades that the current picture has essentially fully emerged, many of the contemporary themes can be traced back for as much as 100 years. Muscle contraction has been a particularly fascinating field of research because it affords relatively easy access to measurements of mechanical work, the accompanying chemical changes, and the determination of concomitant heat liberation or absorption. Thus even before much was known about the detailed chemical processes that take place in muscle, the consideration of muscle contraction in terms of conservation of energy became possible and provided an important example of extending physical laws to living organisms.

Moreover, the highly ordered structure of striated muscle has provided opportunities for relating macromolecular properties and structure to the organization of structures of increasing complexity. Some of these structural features, fully resolved only with the electron microscope, were observed in outline with much more primitive techniques in earlier times (see Huxley (2) for interesting historical material).

Studies on the metabolism of active and resting muscle have occupied a central place in efforts to determine the precise sequence of events that leads to the production of energy available for the performance of work by muscle. The path to final results has been tortuous since cause and effect were not always clearly distinguishable. Although many important facts emerged fairly early—e.g., oxygen consumption, lactic acid production, changes in pH, etc.— the causal and temporal relation of these processes to each other and to muscle contraction itself remained long clouded. This introduction deals only briefly with a few selected aspects that shed some light on the emergence of currently held views.

Lactic Acid Era. By the middle of the nineteenth century with the advent of the law of energy conservation the need arose to look for some source of heat and work energy in muscle. The fact that Liebig and others had shown that muscle contraction can take place without oxygen dispelled the theory of direct combination of oxygen with constituents of muscle as a source of heat and energy and initiated a search for a substance whose transformation would serve as a source of energy. It was during this search that the term inogen was coined by Hermann denoting an unknown "complicated N-containing substance," whose breakdown would furnish the required energy. Eventually attention was focused on lactic acid whose formation from glycogen in rigor and concomitantly with muscle activity became well established as a key reaction in providing energy for contraction. The names of Fletcher and Hopkins, Hill, and Meyerhof stand out in the period 1907–1930 in what might be described as the lactic acid era of muscle contraction [for historical background *see* Needham (*1*), Kalckar (*3, 4*), Cori (*5*)]. This period was characterized by careful studies on the relation of mechanical work, heat production, and lactic acid formation which led to the discovery of the finer details of glycogen breakdown and also began to show the importance of phosphate compounds in glycogen metabolism. An important line of research during this period was concerned with the connection between oxygen consumption, on the one hand, and the removal of lactic acid and restoration of glycogen during the so-called recovery stage of muscle contraction, on the other. The discovery in muscle of a new phosphate compound, phosphocreatine (*6, 7, 8, 9*) in the late 1920s, the demonstration of its decrease in a series of contractions, and the discovery by Lundsgaard (*10*) that muscle poisoned with iodoacetic acid, in which the formation of lactic acid had been prevented, contracts in the absence of oxygen, ushered in the end of this era and led to revised views concerning the source of energy of muscle contraction.

ATP and Muscle Contraction. Shortly before Lundsgaard's crucial discovery there appeared reports on another phosphate compound (*11, 12*), adenyl pyrophosphate or adenosine triphosphate (ATP), its present name. Soon the breakdown of ATP in a series of contractions was shown, and the connection between creatine phosphate breakdown and ATP breakdown was established when Lohmann discovered that the addition of ATP to a muscle extract in which most of the creatine phosphate had broken down resulted in its resynthesis. The fact that no enzyme catalyzing the hydrolysis of creatine phosphate in the absence of adenine nucleotides is found in muscle and that muscle extracts and muscle homogenates were able to catalyze the hydrolyis of ATP led to idea that ATP breakdown must preclude phosphocreatine breakdown (*13*). A new epoch in muscle biochemistry was ushered in when Engelhardt and Ljubimova (*14*) discovered that the protein long known as myosin is the carrier of the catalytic activity leading to the hydrolysis of ATP. Their work and the subsequent work of the Needham group in England (*15*) showed that ATP was not only hydrolyzed by myosin but interacted with it while producing profound changes in its physicochemical properties. Thus a new avenue to an understanding of the relation between the chemical processes underlying the contraction mechanism and the molecular machinery of contraction appeared. At this point a brief consideration of the history of the proteins of the contractile machinery is in order.

Myosin and Actomyosin. The term myosin was coined by Kühne (*15a*) to denote a water-insoluble protein extractable from muscle in strong salt solution—e.g., 10% NaCl discovered five years earlier. Although several investigators had carried out studies on myosin, its relation to muscle structure and function remained obscure for a long time.

Among the attempts to localize myosin within the structural framework of muscle the observation by T. H. Huxley (*16*) that treatment of muscle fibers with reagents that at that time were thought to dissolve myosin abolished the birefringence of what we today would describe as the A band is of particular interest in the light of our current knowledge. An explanation of both the existence of fibrils in muscle and the birefringence of muscle has been sought in terms of the existence of hypothetical entities (Disdiaklasten), minute bodies of higher refractive index, longitudinally aligned and embedded in an isotropic medium. These and similar structures play an important role in nineteenth century views about the mechanism of muscle contraction (*1*). Although early attempts were made to identify these hypothetical entities with myosin, it was the study by von Muralt and Edsall (*17, 18*) leading to a detailed characterization of the birefringence of flow properties of what in 1930 seemed to be the best available myosin preparation obtained according to the technique of Weber that laid a firm foundation for such a view. The birefringence of flow of myosin was explained in terms of the existence of anisotropic particles, and von Muralt and Edsall concluded that the particles, revealed in their physicochemical measurements, play an important structural and functional role in muscle.

Reference has been made above to the discovery of Engelhardt and Ljubimova linking myosin with ATPase activity and the work of Needham *et al.* demonstrating changes in flow birefringence and viscosity of myosin on the addition of ATP. Both the Russian and the British workers recognized the importance of the existence of a protein which is both catalytically active and is part of the contractile mechanism of the muscle fiber. Needham *et al.* (*15*), believing that they had a definite single entity in hand, interpreted the changes in the physicochemical properties of myosin on addition of ATP in terms of changes in protein configuration. The work of Albert Szent-Györgyi and his colleague, F. B. Straub, in Szeged, Hungary, not known in the West until the end of World War II, established that what had been known as myosin is actually a complex of two proteins—myosin proper and actin (*19*). This complex became known as actomyosin, or in its natural form myosin B, and viscosity changes upon addition of ATP were eventually ascribed to the dissociation of actomyosin into actin and myosin. As we shall see later, this dissociating effect of ATP on the actomyosin complex is a key reaction in living muscle.

Ultrastructural Aspects

General. [For background details *see* Refs. *20, 21*]. Light microscopists have long known that the physiological unit of muscle—the cell or fiber—contains typical repeating structures along its length. These repeating units are known as sarcomeres and are separated from each other by Z discs. Within each sarcomere one distinguishes the A and I band, the former occupying the center of each sarcomere; the A band is highly birefringent, a fact referred to in the preceding section. Within the A band a central lighter zone, the H zone, is seen, and in the center of the H zone there is the darker M band. The contractile material is subdivided into smaller units—myofibrils separated by mitochondria and sarcoplasmic reticulum (*see* Chapter 3). The muscle cell is surrounded by a plasma membrane which together with the various connective tissue elements and collagen filaments forms the sarcolemma. The plasma membrane is involved in maintaining the interior of the plasma membrane of the resting cell at an electrical potential of about 100 mV more negative than the exterior; upon stimulation of the muscle by its nerve, activation of the contractile machinery results in contraction and tension development. As discussed elsewhere in this volume, one aspect of the excitation–contraction coupling involves certain elements of the muscle cell that are continuous with the plasma membrane, the so-called transverse or T tubules. These tubules are seen as openings on the surface of the muscle cell disposed either at the level of the Z bands or at the junction of the A and I bands, depending on the species, and the depolarization of the membrane spreads activity along the tubules to the interior of the fiber (*22*).

This rough outline applies to most muscles, but morphological differences exist between muscles of differing functional properties (cardiac muscle, fast and slow skeletal twitch muscles, slow tonic muscles). The discussion of the details is outside the scope of this chapter, and the reader is referred to special reviews and monographs (*23, 24*).

It is only during the past 20 years that sufficiently detailed information on the molecular level became available to account for most features of the light microscopic appearance of the muscle cell. During the same period, concomitant with the elucidation of the ultrastructure of muscle, a profound change took place in our understanding of the contractile mechanism itself. The discoveries leading to what one might properly refer to as a scientific revolution (*cf. 25*) are linked to the names of H. E. Huxley and the late Jean Hanson (*26, 27*), A. F. Huxley and R. Niedergerke (*28*), and W. Hasselbach (*29*), and resulted in (a) the abandonment of the view that muscle contains continuous filaments running from one end of the cell to the other and the acceptance of the existence of two sets of discrete discontinuous filaments that become linked only upon excitation; and (b) the recognition that muscle contraction does not depend on shortening on an ultramicroscopic scale but rather on the relative motion of the two sets of filaments (sliding filament mechanism).

Electron micrographs of muscle, both in longitudinal and cross section, reveal these two sets of filaments (Figure 1). The so-called thick filaments, about 11 nm in diameter disposed in a hexagonal array 40 nm apart occupy the A band. Their length is about 1.6 μm in vertebrates. Longer sarcomeres and longer thick filaments are found in other species. Thin filaments, about 50 nm in diameter, originate in the Z band and extend into the A band to a varying degree depending, as it is now clear, on the length of the sarcomere. The length of the thin filaments also varies from species to species; in frog muscle, on which much of the physiological work has been done, their length is about 1 μm on each scale of the disk. The H zone is the area between the ends of the thin filaments. In cross section the thick filaments form a hexagonal array, and in vertebrate muscles each triangle formed by three thick filaments contains at its center a thin filament. Thus the number of thin

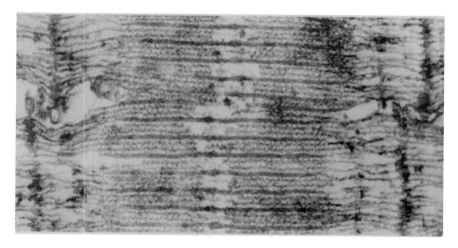

"The Structure and Function of Muscle"

Figure 1. Electron micrograph of a longitudinal section through the sarcomeres of two adjacent myofibrils. The Z lines bounding the sarcomeres run vertically at the right and left edge of the picture. Two kinds of filaments are visible; thick ones (about 110 Å) in an array confined to the A band and thin ones (about 50 Å) in two arrays which terminate at the borders of the H zone in the middle of the picture. The two kinds of filaments interdigitate in the A band (except the H zone). Crosslinks between thick and thin filaments are visible (×266,400) (20).

filaments is twice that of the thick filaments. Electron micrographs of longitudinal sections show that the I band contains exclusively thin filaments, the central part of the A band only thick filaments while the peripheral portions of the A band contain both thick and thin filaments. Projections, or crossbridges, extending over the whole length of the thick filaments except a central zone of about 0.22 μm are an important aspect of these filaments. These projections, as we shall see, can be attributed to the so-called head portion of the myosin molecules.

X-ray diffraction studies of living muscles complement electron microscopic data on the thick filaments and, in particular, the crossbridges. The original evidence seemed to support the view that crossbridges emerge, sepa-

Table I. Composition of Myofibrils in Rabbit Skeletal Muscle (35)

Protein	Quantities, % of myofibrillar protein	Role	Location
Myosin	60	contraction	thick filaments
Actin	20	contraction	thin filaments
Tropomyosin	4	regulation	thin filaments
Troponin	3	regulation	thin filaments
α-Actinin	1.5	structure?	Z-band
C-protein		?	thick filaments
M-protein	11.5	structure?	M-zone
Other		?	?

Quarterly Review of Biophysics

rated by a distance of 14.3 nm along the filament axis, that on each level there are a pair of diametrically opposed crossbridges, and that crossbridges are rotated by 60° at successive levels resulting in a helical repeat of 42.9 nm, with a total of six crossbridges per repeat (30). Recent evidence has been adduced to support two different types of arrangement of crossbridges. Tregear and Squire (31) favor a "three-stranded" model (32) for the vertebrate-thick filament while Morimoto and Harrington's data (33) support the presence of either four crossbridges, each containing one myosin molecule, or two crossbridges, each containing two myosin molecules with an axial repeat. It is important to note that the thick filaments are lined up in register and thus constitute sharp A band. The M lines are attributable to material—protein in nature—attached to the thick filaments.

Subsequent sections of this chapter deal in detail with the constituents of these filaments and their interaction. At this point it may be helpful to summarize the current state of knowledge concerning the localization of the various proteins in relation to the ultrastructure of muscle and to indicate the functional or structural role attributed to the various proteins. This information is summarized in Table I. While the association of myosin and actin with thick and thin filaments, respectively, was established in the mid-1950s, it is only during the last decade that information regarding the other proteins has developed. Of greatest importance is the tropomyosin–troponin system associated with the thin filaments. This system, as discovered by Ebashi and his colleagues (34, 35), plays a crucial role in the regulation of contraction and relaxation by the intracellular Ca^{2+} level, which in turn is regulated by the sarcoplasmic reticulum (see Chapter 3). A combination of Ca^{2+} with troponin produces conditions that permit the association of actin and myosin.

Sliding Filaments. The counterpart of a modern ultrastructural view of muscle in terms of function can be best summed up in what has become known as the sliding filament theory or mechanism. Although the process of sliding is well established, the precise details of how sliding is produced are still at the center of much active research.

Concomitant with the discovery that the ultrastructurally identifiable filaments in muscle that determine its microscopic structure are not continuous, it also became clear that muscle contraction does not depend on the shortening of microscopic or ultramicroscopic elements but rather on the relative motion of the two distinct sets of filaments (Figure 2). Thus while muscle contraction is accompanied by a shortening of sarcomeres, there is no significant shortening of either the thick or thin filaments although, according to quite recent results, activation of muscle is accompanied by a very small ($\sim 1\%$) shortening of the A band thick filaments (36). While the length of the thick and thin filaments remains essentially constant, the overlap region increases as the sarcomere shortens and the width of the I band decreases. This view was clearly stated in the early 1950s by Huxley and Hanson and by Huxley and Niedergerke, but it took many years before it became generally accepted. Another important physiological counterpart of the sliding filament theory is the relation of filament overlap to tension development. Precise measurements on single muscle fibers carried out in A. F. Huxley's laboratory (37) have shown that as the overlap increases, there is a linear increase in tension. At the point corresponding to the thin filaments' reaching the central portion of the thick filaments there is a small span ($\sim .2$ μm) over which shortening of the sarcomere does not produce a change in tension. This is consistent with the penetration of the thin filaments into that zone of the thick filaments where there are no projections. On further shortening, tension declines, at first gradually, and then as the Z bands reach the end of the thick filaments, more precipitously. This dependence of the tension on sarcomere length is entirely consistent with the view that tension development is proportional to the number of projections present in the overlap zone and indeed is attributed, as we shall see in greater detail, to the formation of connections—the so called crossbridges—between thick and thin filaments.

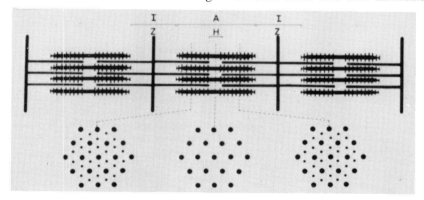

"The Structure and Function of Muscle"

Figure 2. Structure of striated muscle, showing overlapping arrays of actin- and myosin-containing filaments, the latter with projecting crossbridges on them. The bottom half of the picture shows cross-sections through the center of the A-band (middle), and through the two lateral halves (20).

The variation of tension with overlap is best explained by considering each crossbridge as an independent force generator operating in a cyclic fashion (for a review evaluating other views *see* Ref. *38*).

Myofibrillar Proteins

Myosin. SUBUNITS, PHYSICAL CHEMISTRY. For an understanding of the structural aspects of the thick filaments and of their interaction with thin filaments it is necessary to turn to a detailed discussion of the myosin molecule, which is the chief constituent of the thick filaments (for background *see* Refs. *39, 40, 41*). The anatomy of the myosin molecule can be conveniently described in terms of a rigid rod to which two globular regions are attached. This structure is attributed to two polypeptide chains, each having a mass of about 200,000 daltons. The two chains combine over a considerable portion of their length into a coiled-coil α-helical structure, but each chain ends in a distinct globular region forming what have become known as the two heads of the myosin molecule. The two-headed structure is well supported by several types of evidence including physicochemical measurements and, most directly, electron microscopy. The fact that by means of limited proteolytic digestion myosin can be broken up into various fragments preserving some of the original functional and structural aspects of the molecule has greatly helped in developing our present knowledge. While at first myosin was thought to consist of a single polypeptide chain, the next stage of development proposed a three-stranded model before the current two-chain model was accepted (*see e.g., 42, 43*). Actually myosin is not merely a simple complex of the two heavy chains but, in addition, each myosin molecule contains, according to the best evidence available, four so-called light chains, each head region being associated with two of these. The relation of the various fragments obtained by proteolysis to the overall structure of myosin is shown in Figure 3. In current nomenclature heavy meromyosin (HMM) and light meromyosin (LMM) represent the two duplexes formed by mild tryptic or chymotryptic digestion. Heavy meromyosin subfragment 1 (HMM S-1), or sometimes only subfragment 1, is essentially identical with a globular head portion although finer differences exist between preparations obtained with trypsin or papain, for example (*44, 45, 46*).

Some confusion may arise from the established nomenclature in that the adjectives heavy and light are used both for the polypeptide chains that are the subunits of myosin and for the fragments that are obtained by proteolysis. The fact that myosin is insoluble at ionic strengths ≤ 0.1 has been helpful in purifying it and in analyzing the functional role of various fragments. It appears that the solubility of myosin is determined by the properties of the rod portion, or light meromyosin, which is also insoluble at low ionic strength. The ATPase and actin-combining properties of myosin (*see below*) reside in the globular head portions (HMM S-1) as discussed in greater detail below. Note that the light chains are associated with the globular portions and that the removal of the light chains by such methods as heating, high pH, or use of denaturing agents leads to the loss of ATPase activity.

The differences among muscles of different functional characteristics referred to above in connection with structure are also reflected in some molecular constituents. Among these are the light chains (for differences in ATPase *see below*). The fast twitch muscle myosin contains three types of light chains, designated as LC_1, LC_2, and LC_3 in order of increasing speed of migration in SDS polyacrylamide gel electrophoresis. LC_2 is also known as the DTNB light chain since DTNB treatment removes it, and LC_1 and LC_3 are also referred to as alkali 1 and 2 (A1 and A2) light chains since after removal of LC_2 by DTNB the remaining light chains can be removed by exposure to mildly alkaline conditions. Cardiac and slow twitch muscle

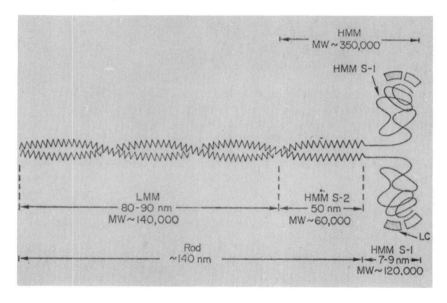

"Disorders of Voluntary Muscle"

Figure 3. Schematic structure of the myosin molecule, based chiefly on Lowey et al. (39). The rod portion of the molecule has a coiled coil α-helical structure. Hinge regions postulated in the mechanism of contraction (212) are at the junction of HMM S-1 and HMM S-2 and of HMM S-2 and LMM. Note that HMM S-1 has one chief polypeptide chain while the other fragments have two. Note the light chains (LC) in the head region. The scheme suggests the presence of two different subunits in each HMM S-1 (41).

myosin contains only two types of light chains whose mobility is similar to but different from that of LC_1 and LC_2, respectively, of fast muscle type myosin (*46a, 47, 48, 49, 50*). Cardiac and slow twitch muscle LC_2 cannot be removed by DTNB treatment.

The stoichiometric relation of the various light chains seems to indicate that each molecule contains two LC_2s. In fast muscle myosin the sum of $LC_1 + LC_3$ is two per molecule. At present one can only speculate on the role which the light chains play in the functional properties of myosin (*51*).

Weeds and Taylor (*52*) have just produced evidence for the existence of two isoenzymes of HMM S-1 (head region of myosin) of the fast muscle type. They contain different light chains (LC_1 and LC_2, or A1 and A2 in their terminology), and they differ in the parameters of the actin-activated ATPase activity. It is not clear whether a given myosin molecule contains different types of alkali light chains or whether LC_1 is always present with LC_1 and LC_3 with LC_3. In slow and cardiac muscle myosin it appears that there are a pair of LC_1s and a pair of LC_2s per molecule. The light chains are useful markers of the type of myosin present and have been used to study the transformation of myosin from one type to another under experimentally changed conditions of innervation or activity (*53, 54*).

CHEMICAL ASPECTS. Our knowledge of the detailed chemical structure of myosin is rather incomplete. Only limited information is available on sequences scattered over various parts of the molecule. Thus the amino acid sequences of two of the light chains, LC_1 and LC_3, distinguishable by their mobilities on SDS-polyacrylamide gel electrophoresis, have been determined, and considerable similarities between the two have been found (*50*). Work on LC_2 is in progress (*55*). The N terminal region of the heavy chain has been located in the HMM S-1 region of the molecule, and amino acid sequence studies have indicated the existence of at least two types of heavy chains (*56*). The existence of various types of molecules containing either LC_1 or LC_3, discussed above, may also be reflected in the two different kinds of heavy chains.

An interesting feature of the chemistry of myosin is the presence of unusual methylated amino acids. One of these is 3-methylhistidine, which is found only in fast adult muscle. It is absent in embryonic muscle and in cardiac and slow twitch muscle myosin (*57, 58*). In view of the differences in ATPase activity and in the speed of contraction between the different types of muscles (*59, 60*) it would be tempting to suggest association between methylation of this particular histidine residue also located in the subfragment 1 region of myosin and the biological properties of myosin. This, however, at this point cannot be considered more than a tentative hypothesis. There are also lysines methylated to various degrees (*61, 62*), but so far no differences have been found among the various types of myosin in this respect.

While nothing can be said about the relation of these methylated residues to either the actin-binding site or to the ATP-binding site of the myosin head, there are two sulfhydryl groups out of about 20 per heavy chain whose closeness to or presence in the ATP binding site has been suggested. These two sulfhydryl groups are generally referred to as SH_1 and SH_2 (*63, 64*). The SH_1 group is characterized by its rapid reaction with sulfhydryl reagents (*65*) although the selectivity depends somewhat on the reagent chosen. What is interesting is that the reaction of these —SH groups with thiol reagents results in changes in ATPase activity (for more on this point *see* p. 239).

Since myosin is an ATPase, the search for its phosphorylated form has been going on for a long time with the idea that phosphorylated myosin might be an intermediate in the enzymatic reaction and play a role in the mechanism of contraction (*see* Ref. *66*). Some students of muscle contraction consider phosphorylation and dephosphorylation of myosin to be a means of altering the charges on the myosin or actomyosin filaments which in turn would result in length changes (*e.g.*, Riseman and Kirkwood, Ref. 67). While phosphorylation of this dynamic type has not been established (68), and in fact in the light of current studies on the kinetics of ATPase activity phosphorylation of this type is unlikely (*see e.g.*, Taylor, Refs. 69, 70), recent evidence has been obtained for the existence of a phosphorylated form of one of the light chains of myosin *viz.*, LC_2—the light chain that can be removed with DTNB treatment (71). The existence of a phosphorylated form is revealed in polyacrylamide gel electrophoresis in the presence of urea and in the absence of SDS as a more rapidly moving peptide band. Phosphorylation of LC_2 appears to be the result of an enzyme which is specific for this protein (72). Since enzymatic activity of myosin is not impaired by the removal of this light chain, it is unlikely that its phosphorylation plays a direct role in the mechanism of ATP hydrolysis. However, the possible role the phosphorylation of myosin plays in modulating the activity of myosin is an exciting prospect for future research. It is particularly intriguing that the light chain whose phosphorylation has been demonstrated appears to be related to a light chain found in molluscan myosin that is known to play a regulatory role (51, 73).

RELATION OF MOLECULAR STRUCTURE OF MYOSIN TO THICK FILAMENTS. The electron microscope picture of a myofibril shows that myosin filaments have a diameter of about 100 A. A consideration of the dimensions of the myosin molecule and the number of myosin molecules present in a sarcomere (20) would suggest that a given myosin filament is a bundle of many myosin molecules per cross section; the rigid rod-like part of the myosin molecule makes up the body of these filaments, and the enzymatically active globular end would correspond to the bridges seen in electron micrographs between myosin and actin filaments. Huxley (74) was able to demonstrate in myosin solutions at low ionic strength the formation of filamentous aggregates possessing central symmetry resulting from the sidewise apposition of the rod-like part of the myosin molecules and the globular heads appearing as the lateral projections. The formation of the aggregates proceeded in two directions, leaving a central zone free of lateral projections, quite reminscent of the appearance of the thick filaments seen in electron micrographs of muscle itself. The significance of the bipolar orientation of myosin molecules in the thick filaments has been pointed out by Huxley in relation to the sliding filament mechanism. This arrangement makes it possible for the myosin molecules in the two halves of the sarcomere to act in the same directional sense with regard to the thin filaments originating at the two Z bands, thereby moving the thin filaments toward the center of the sarcomere.

Electron microscopic investigation of negatively stained aggregates of LMM or of the whole rod portion of myosin (75, 76, 77) may also furnish useful insight into the forces responsible for the self-assembly of myosin molecules into thick filaments. Similarly, investigations of the aggregation properties of myosin in solution (78) appear to throw new light on the mechanism of filament formation. Rather detailed analysis of the myosin filaments involving the combined use of electron microscopic techniques and antibody staining techniques has been made (*e.g., see* Refs. *79, 80*), but the precise arrangement of the myosin molecules in the thick filaments, particularly in relation to the number of crossbridges and the number of myosin molecules per crossbridge, has not yet been settled.

Actin. The discovery of actin by Straub (*81*) in Szent-Györgyi's laboratory has opened up a new epoch in the biochemistry of muscle. What in the early literature was described as myosin turns out to be the complex of two proteins: myosin and actin. Actin itself can be extracted from acetone-dried muscle powder with distilled water. It is present in the watery extract as a globular protein : G-actin. Estimates of its molecular weight have ranged from 46,000 to 60,000. Probably the most reliable value is that derived from the recently determined amino acid sequence—*viz.*, 41,785 (*82*).

On addition of various salts in a concentration of about $0.1M$ for monovalent cations and about 1 mM for divalent cations, a G-actin solution undergoes a drastic change; its viscosity increases, and it exhibits birefringence of flow, suggesting the presence of large, asymmetric molecules. These are so-called F-actin (for fibrous actin) particles produced through polymerization of the globular units. The average molecular weight of F-actin *in vitro* is of the order of 3 million, and its length is 10 μm or longer, in contrast with actin filaments *in vivo* having a length of 1 μm per half sarcomere. The differences have been attributed to a protein similar in amino acid composition to actin, β-actinin (*83*). The polymerization reaction is probably important in the process by which actin filaments are laid down *in vivo*, but it is unlikely (*84*) that the polymerization–depolymerization cycle plays a significant role in the mechanism of muscle contraction.

G-actin contains tightly bound ATP and Ca^{2+} or Mg^{2+} (*85, 86, 87*), one mole of each per mole of actin; the bound divalent cation *in vivo* is likely to be Mg^{2+} (*88*). Both can undergo rapid exchange with nucleotides and divalent metals, respectively (*84, 89*); for a recent review *see* Tonomura and Oosawa (*90*). The polymerization process is accompanied by the dephosphorylation of ATP to ADP (*85, 91*), which remains firmly bound to actin, but in contrast to the firmly bound ATP of G-actin the bound nucleotide of F-actin is not available for rapid exchange reaction. Rapid exchange of ADP bound to F-actin with ATP and other nucleoside triphosphates, accompanied by the liberation of inorganic phosphate, has also been obtained by ultrasonic treatment (*92, 93, 94*). Slower exchange of ADP bound to F-actin with either ADP or ATP has been demonstrated (*95, 96, 97*). The splitting of ATP by actin during polymerization is not a typical enzymatic process in the sense that a small amount of protein can split an unlimited amount of substrate, but there is a stoichiometric relationship between the amount of actin present and the amount of phosphate liberated. Globular actin preparations that contain bound ADP and can undergo polymerization without the splitting of ATP (*98, 99, 100*) have also been obtained. Since under certain conditions actin that is free of either tightly bound Ca or ATP with the ability to polymerize and to react with myosin can be obtained, it seems likely that the chief function of bound metal and nucleotide is to stabilize the structure of the actin molecule rather than to participate directly in interactions with myosin or with other actin molecules (101, 102).

Recent work by Cooke shows that some interactions in the actin–myosin system depend on the nature of the nucleotide bound to actin (103).

With the availability of the amino acid sequence of actin it will be possible to localize those residues that may play a functional role in the structure and function of actin. SH groups are essential both for ATP binding and polymerization and for combination with myosin (104, 105, 106). Of the five –SH groups (107) three are reactive in the native molecule (108) without preventing polymerization or combination with myosin. The cysteine adjacent to the C-terminal phenylalanine (107) contains the most reactive –SH group. This group reacts selectively with Cu^{2+} (109), and a spin label attached to it can indicate conformational change in actin (110, 111, 112). As suggested by Elzinga and Collins (107) the region containing the single 3-methylhistidine residue and a tyrosine residue readily methylated by the reagent tetranitromethane may be involved in the polymerization process. It appears so far (107) that actins isolated from various sources (cf., Ref. 113) including non-muscular contractile systems have similar chemical structures. It is also clear that the 3-methylhistidine-containing peptides in actin differ in their sequence from the 3-methylhistidine-containing peptides in myosin, thereby rendering a previously suggested evolutionary relation between actin and myosin unlikely (114).

Electron microscopic studies by Hanson and Lowy (115) have shown that F-actin consists of a double helix of globular actin units. Actin filaments directly isolated from muscle also have the same structure. The detailed structure of these filaments is discussed below after two of their other components—troponin and tropomyosin—are discussed. It should be added that conventional actin preparations were usually contaminated with these two proteins; this contamination can be avoided by extracting actin at 0°C (116) and using small amounts of divalent cations in the purification process (117).

Regulatory Proteins. In this section we consider briefly those proteins that participate in regulating the interaction of actin and myosin *in vivo* and reconstituted systems *in vitro*—viz., tropomyosin and troponin. The other crucially important ingredient in this regulatory process is Ca^{2+}, which exerts its control function by combining with one of the subunits of troponin. We first review the regulatory proteins as proteins and discuss their localization in the contractile apparatus. We return to the functional aspects after having dealt with myosin ATPase and some details of the actin–myosin interaction.

TROPOMYOSIN. Tropomyosin, discovered by Bailey in the late 1940s, can be extracted from an ethanol-treated muscle (118). It consists of two subunits, each having a molecular weight of about 35,000 (119). Tropomyosin is notable for its high α-helical content and in this respect is very similar to the light meromyosin fragment of myosin (120); it is thought that the two polypeptide subunits are arranged in a coiled-coil fashion (121). Determination of its amino acid sequence has recently been completed, and it will now become possible to relate chemical structure to functional properties. The pattern of hydrophobic residues forming two ridges on the surface of each α-helical subunit has been invoked (*122*) to explain the coiled-coil arrangement of the two subunits.

Evidence has been obtained for the existence of two kinds of subunits (α and β) in tropomyosin preparations (*123, 124*). It seems that $\alpha\beta$ and $\alpha\alpha$ forms may exist, and different types of muscle contain the two subunits in different proportions.

Aggregation of tropomyosin at low ionic strength can occur both in the form of true three-dimensional crystals and so called paracrystalline aggregates (*125, 126*). Various patterns are observed in the electron microscope that greatly contribute to our understanding of the factors responsible for the interaction among the tropomyosin molecules and also throw some light on their interaction with troponin (*see below*).

The participation of tropomyosin in the Ca^{2+}-dependent regulation of actin–myosin interaction was discovered by Ebashi and his colleagues. Highly purified actin produces EGTA-insensitive actomyosin (*127*), and these authors initially assumed that there existed a "native tropomyosin" responsible for the sensitization. They, however, soon established that another moiety—troponin—is also involved (*34, 128, 129*) in the Ca^{2+} sensitivity of the actin–myosin interaction.

Tropomyosin by itself can either inhibit or activate the ATPase activity of actomyosin or acto-HMM (*130, 131*). Both effects require a high Mg^{2+} (~ 5 mM) concentration and low ATP concentrations for activation. The relation of these observations to the role played by tropomyosin in the Ca^{2+}-dependent processes is discussed below.

TROPONIN. Troponin was at first considered a homogeneous protein, and various molecular weights have been suggested (*see e.g.*, Ref. *132*). The work of Hartshorne and his colleagues (*133, 134*) and Schaub and Perry (*135*) has shown that troponin can be separated into at least two components. Starting with the work of Greaser and Gergely (*136*), evidence has begun to accumulate (*129, 137*) that three components are necessary for reconstituting troponin activity, defined as the conferring of Ca^{2+} sensitivity on actomyosin in the presence of tropomyosin. These troponin components are being designated as TnC, TnT, and TnI (*138, 139*), and the following scheme lists some of their characteristics.

Subunit	Mol. Wt.[a]	Characteristics
TnC	18,000	Ca^{2+} binding
TnT	37,000	Binding to tropomyosin
TnI	24,000	Inhibition of actomyosin ATPase with and without Ca^{2+}

[a] Determined from SDS-gel electrophoresis. The molecular weight of TnC is also in agreement with its recently determined amino acid sequence; the amino acid sequence of TnI suggests a M_r of 21,000 (*140*).

The three subunits of troponin occur in a 1:1:1 ratio in myofibrils (*141*), there being one troponin complex for each tropomyosin molecule.

Apparent reconstitution of troponin activity without TnT can be obtained at higher than physiological ratios of TnC and TnI to actin (142).

Perhaps the functionally most important property of the subunits, or at least the property of the subunits that can be most readily related to the function of troponin, is Ca^{2+} binding to TnC. Earlier studies on Ca^{2+} binding were hampered by various technical difficulties in controlling the Ca^{2+} concentration and by the lack of well-defined preparations. Recent measurements of Potter and Gergely that have tried to circumvent these difficulties by using highly purified TnC, a Ca^{2+}–EGTA system permitting binding studies at very low free Ca^{2+}, and a computerized non-linear curve fitting procedure, are summarized in Table II.

Table II. Calcium Binding: Number of Sites (n) and Binding Constants (K) (220) [a]

	n_1 [b]	K_1 [c]	n_2 [b]	K_2 [c]
TnC	1.8	2.5×10^7	2.6	3.7×10^5
TnC + Mg^{2+}	1.7	3.0×10^6	2.4	1.1×10^5
CI	2.0	2.4×10^8	2.4	3.0×10^6
CI + Mg^{2+}	4.0	5.0×10^6		
Tn	1.9	5.2×10^8	2.0	5.2×10^6
Tn + Mg^{2+}	4.0	4.0×10^6		

[a] TnC and TnI were mixed in a 1:1 molar ratio. The TnC content of Tn was determined by densitometry of fast green stained gels.
[b] Binding sites expressed as moles Ca^{2+}/mole of TnC present.
[c] M^{-1}.

"Calcium Binding Proteins"

There are two classes of binding sites, and a good fit to the experimental data can be obtained without assuming any cooperative behavior in the binding process. The sites of higher affinity also bind Mg^{2+}, and the effect of Mg^{2+} on Ca^{2+} binding can be accounted for by assuming competition between Ca^{2+} and Mg^{2+} for the same site. The sites having a lower affinity for Ca^{2+} do not appear influenced by Mg^{2+}. We return to the relation between Ca^{2+} and Mg^{2+} in discussing the regulatory process itself.

Ca^{2+} binding studies on TnC show an enhancement of its affinity for Ca^{2+} upon interaction with TnI, either in the TnC–TnI complex or in whole troponin. Effects of this interaction (143, 144, 145) are also reflected in other physicochemical properties that have recently been studied in various laboratories suggesting conformational changes induced in TnC by other subunits and probably vice-versa.

An important relationship between the family of proteins first found in fish muscle that are able to bind Ca^{2+} (146)—parvalbumins or fish muscle Ca^{2+} binding proteins—and more recently discovered in muscles of higher species (147, 148), has emerged. The functional role of the parvalbumins is not clear, and they differ from the myofibril-bound TnC in that they appear free in the cytoplasm; their molecular weight is about half that of TnC. X-ray and amino acid sequence studies have suggested certain repeating features within the parvalbumin molecules attributable to gene duplication or triplication (*149*). Two Ca^{2+} binding sites have been identified, each involving a "loop" flanked by a pair of α-helices (*150*). Kretsinger and his colleagues predicted that there would be similarities between parvalbumins and TnC, and Collins and his colleagues (*151*) in our group have clearly shown that extensive homologies exist between the amino acid sequences of TnC and parvalbumin. By comparison with the parvalbumin sequence there are four regions that would be likely candidates for the four Ca^{2+} binding sites in TnC referred to above. This is an area in which much work is going on, and homologies between the light chains of myosin and parvalbumin and TnC have also been found (*152, 153*). Discussion of the evolutionary implications of these findings is beyond the scope of this chapter, but evidence is accumulating concerning the role of myosin light chains in Ca^{2+} interactions; in molluscan muscle this process plays a key role (*51*).

TnT combines with tropomyosin, which can be directly demonstrated by ultracentrifugal experiments (*139*) or electron microscopy by virtue of

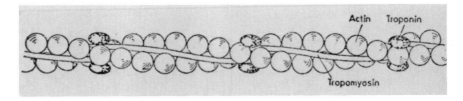

Quarterly Review of Biophysics

Figure 4. A model for the fine structure of the thin filament (35). *Ebashi and his colleagues merely had evidence for the periodic distribution of the troponin complex along the thin filaments; the relation of troponin to tropomyosin is hypothetical.*

the changes brought about in the electron microscopic staining pattern of tropomyosin aggregates (*123, 138, 154*). Work on its amino acid sequence is rapidly proceeding. TnI obtained its name from the fact that it inhibits actomyosin ATPase with or without Ca^{2+}; this inhibitory effect is enhanced by tropomyosin (*155*). Both TnI and TnT interact with TnC, and while it is generally assumed that the ability of TnT to combine with tropomyosin plays an important role in the functioning of the intact Tn complex *in vivo*, whether the "inhibitory" effect of isolated TnI is actually exercised when the Tn complex inhibits in the absence of Ca^{2+} (*see below*) is not quite settled. The amino acid sequence of TnI (*140*) is now available, and studies with fragments of known sequence will greatly aid in analyzing the functional significance of different parts of the molecule.

We return to the question of how the troponin subunits interact when we take up the molecular mechanism of the regulation of actin–myosin interaction. The possible regulatory role of the recently discovered phosphorylation of TnI and TnT is also discussed there. Another aspect of the troponin–

tropomyosin system which also receives more attention below is its localization in the myofibrillar structure.

Hanson and Lowy (*115*) were the first to suggest that tropomyosin is situated as two long strands in the two grooves of the F-actin double helix. This has since been borne out by various electron microscopic and x-ray studies (*156, 157*). Ebashi, Endo, and Ohtsuki (*35*) proposed that troponin would be attached to the tropomyosin strands, separated by periods (Figure 4). This periodicity has been revealed before in electron microscopic studies (*see e.g.,* Ref. *157*). The precise localization of troponin in relation to the tropomyosin molecule—whether situated in the middle or near the end of these molecules—is still not settled (*138, 158*).

Additional Myofibrillar Proteins. During the last few years a number of myofibrillar proteins have been described whose function and location have not been completely settled. It is also not clear in every case whether one is really dealing with a new protein or whether previously known proteins are appearing in modified form.

α-ACTININ AND β-ACTININ. Both these proteins have amino acid compositions similar to that of actin (*127, 159, 160*). Although α-actinin has been described as promoting superprecipitation of actomyosin (*159*), this effect has been questioned by Briskey and his colleagues (*161, 162, 163*). Evidence is accumulating for the localization of α-actinin in the Z band (*164*). The ability of α-actinin to crosslink F-actin (*162, 165*) may play a role in anchoring the actin filaments in the Z band. β-actinin, as mentioned above, has been implicated in limiting the length of actin filaments *in vivo* and in producing—*in vitro*—a homogeneous population of short ($\sim 1\,\mu$m) actin filaments (*83, 166, 167*). The role of demonstrated effects of β-actinin on actin–myosin systems (*168*) is less clear.

M PROTEIN. Recent efforts directed at identifying the proteins involved in the structure known as the M band, occupying the center of the sarcomere and probably playing a role in holding together the thick filaments, are still in progress (*169, 170*). Several components ranging in molecular weight from 40,000 to 190,000 have been reported. Some of these have been found identical with known enzymes—a 40,000-dalton chain (*171*) is creatine kinase (*172*), the 100,000-dalton component, phosphorylase (*169*)—and are unlikely to be the structural components. The protein with a chain weight of about 160,000 (*173, 174*) may be a better candidate since antibodies directed against it stain the M band.

C PROTEIN. The most recently discovered myofibrillar protein is the C protein which was found on SDS–gel electrophoresis of myosin preparations. It has been suggested that the C protein is a constituent of the thick filaments and may play a structural role in the architecture of these filaments (*175*).

Myosin ATPase and Interactions with Actin

Since the historic discovery of Engelhardt and Ljubimova (*14*) it has become a central dogma of muscle biochemistry that the hydrolysis of ATP by myosin is the key chemical reaction in muscle contraction. The availability of highly purified myosin preparations as well as of enzymatically active fragments has greatly contributed to our current understanding of many of the details of the reaction, of the effect of various modifiers, and, most importantly and more recently, of the intermediate steps in the ATP hydrolysis process.

Metal Effects. Early work on myosin, at first usually containing actin (myosin B), has dealt with the effect of various metals (*19*). Note that many papers bearing the word myosin in the title were actually dealing with actomyosin without the authors' recognizing that fact; many studies even in the late 1950s require careful scrutiny since it is important to distinguish between the ATPase activity of myosin and, as we shall see, the ATPase activity of actomyosin. The actomyosin complex in the presence of ATP exists only at low ionic strength—in 0.6M KCl, ATP causes essentially complete dissociation of actomyosin (*176, 177*). Variation in the ionic strength produces changes that are attributable to the dissociation of the actomyosin complex to myosin. An early recognized feature of pure myosin was its high ATPase activity in the presence of calcium and its low activity in the presence of magnesium. This is often expressed in terms of inhibition of myosin activity by Mg^{2+}. Two points are to be discussed in this connection. The first concerns the level of Ca^{2+} required for activating myosin ATPase. This is usually of the order of several millimolar in contrast to the low concentration of Ca^{2+}—1 μmole—required to regulate actomyosin in the presence of the troponin–tropomyosin complex as discussed below. The second point is that strictly speaking Mg^{2+} does not inhibit since in the absence of divalent cations ATPase activity of myosin is even lower. Nihei and Tonomura (*178*) have shown that the effect of divalent cations on the ATPase activity of myosin when plotted against an ionic radius is a bell-shaped curve. Ca^{2+} is very close to the peak with a radius of slightly over 0.1 nm, and Mg^{2+}, having a smaller radius, is close to 0.05 nm at one of the inhibitory tails of the curve. As shown by Morales and his colleagues (*179*) the ionic strength plays an important role in determining the effect of divalent cations; the activation by Ca^{2+} is largest at the lowest ionic strength and declines with increasing ionic strength. Typical values for the steady state ATPase rate of myosin are of the order of 0.02 mole per second per mole in the presence of Mg^{2+} and about 10 moles per second per mole with Ca^{2+}. The physiological importance of the low ATPase rate in the presence of Mg^{2+} will become clear when we turn to the activation of myosin ATPase by actin.

The discovery that EDTA activates myosin ATPase in the presence of high K^+ concentration remained puzzling for some time. A long series of papers deals with this phenomenon in terms of EDTA activation of ATPase. It has, however, become clear through the work of several investigators that the real activator in this case is K^+ and that the effect of EDTA is attributable to the removal by chelation of Mg^{2+} ions that would in the presence of high K^+ concentrations reduce the ATPase activity (*180, 181*). Detailed studies have shown that one can properly speak of divalent cation-activated myosin ATPase and monovalent cation-activated ATPase since it is not only potassium but lithium and ammonium ions that can activate whereas sodium ions inhibit (see e.g., Ref. 182). From a physiological point of view the significance of the monovalent cation-activated ATPase is not likely to be important.

Thiol and Other Groups. From the earliest times of enzymatic studies on myosin it was clear that SH groups have some role in ATPase activity. Myosin ATPase activity was shown to be lost when thiol reagents combined with myosin and the reducing reagents known to protect SH groups preserved the myosin ATPase. More detailed information concerning the thiol groups that are in some way closely tied to the ATPase activity of myosin became available when Kielley and Bradley (65) showed that the effect was diphasic: the ATPase activity increased on adding increasing amounts of reagent up to a point, followed by inhibition on adding more reagent. There seems to be good evidence that there is one rapidly reacting sulfhydryl group in each heavy chain of myosin whose blocking leads to activation of the Ca^{2+}–ATPase and inhibition of K^+–ATPase (183). There is a second sulfhydryl group in each heavy chain whose reaction with the thiol reagent produces inhibition of both types of activities. The reactivity of this group, the so called SH_2 thiol group, is enhanced by ATP, ADP, or pyrophosphate (63). The fact that the two SH groups can under some conditions play identical roles in the ATPase activity has been shown by Seidel (182) by first blocking SH_1 with DTNB followed by the reaction of NEM with SH_2. Removal of the DTNB with reducing agents such as dithiothreitol led to reactivation of Ca^{2+} and inhibition of K^+ activated ATPase—a property typical of SH_1-blocked myosin. Recent work by Harrington and his colleagues (184) suggests that the two sulfhydryl groups are close to the ATP and metal-binding site of myosin; they propose a model in which the ATP and Mg^{2+} would form a complex with both sulfhydryl groups. If this proposal is borne out by further experimental evidence, it would be an important step in our obtaining more detailed information about the active site of myosin. The fact in itself that modification of sulfhydryl groups affects ATPase, or even that the mobility of the spin label attached to SH_1 changes on adding ATP (185), does not in itself prove that the sulfhydryl groups are at the same site where the ATP binds since the ATP-induced conformation of the protein could spread over larger distances and produce changes in the reactivity of sulfhydryl groups or in the behavior of a group attached to a sulfhydryl group. Conversely, modification of a sulfhydryl group remote from the active site could, by reciprocally induced conformational change, alter the ATPase activity.

The work of Yount and his colleagues (see e.g., Ref. 186) shows that ATP analogs capable of forming disulfide bonds with myosin react at sites other than the substrate-binding site and produce inhibition of ATP hydrolysis and of actin binding. This work also bears on the question, which requires further clarification, of the existence of "regulatory," as distinct from hydrolytic, nucleotide binding sites. Changes in ATPase after reaction with other chemical groups has been reported although the precise relation of these groups at the active site is even more difficult to assess than that of thiol groups (for reviews see Ref. 66).

Distance measurements deduced from energy transfer between fluorescent donor/acceptor molecules covalently attached to side chains in the HMM S-1 region or energy transfer between fluorescent ATP analogs and tryptophan residues or attached acceptors, may play an important role in mapping the topography of the active site (*187, 188*). Histidine groups have been implicated either as being in the active center or as being involved in producing conformational changes in myosin that have an effect on ATPase activity. Modification by trinitrophenylation of a lysine located in the subfragment 1 portion leads to increased ATPase activity in the presence of Mg^{2+}. The stoichiometry of this reaction is of interest since it throws light on the question of whether or not the two heads of the myosin molecule are identical (for an extensive discussion *see* Refs. *66* and *90*). A recent paper by Muhlrad et al. (*189*) suggests that there are two lysines per myosin molecule, one in each chain, that react preferentially with dinitrobenzene, but without showing a difference between the two chains.

Intermediate Steps in the Hydrolysis of ATP. Until a few years ago the general scheme in which myosin ATPase was discussed implied the Michaelis-Menten type reaction with the ATP enzyme complex rapidly formed and the hydrolytic step rate limiting, presumably followed by rapid dissociation of the products. Many years ago Blum (*190*) suggested that the dissociation of the product might play a role in determining the relative rates of hydrolysis catalyzed by myosin of various nucleotides although this suggestion has not been extensively followed up. As mentioned before, the question of a phosphorylated intermediate has been recurring. This concept has been supported by the work of Tonomura and his colleagues (*66, 191*) although a careful search by Sartorelli et al. (*68*) failed to produce evidence for a covalent phosphorylated intermediate. An interesting feature of the ATPase reaction is the exchange reaction observed with $H_2^{18}O$. Two kinds of exchanges were distinguished: one referred to as the "intermediate exchange" required the hydrolysis of ATP and resulted in the incorporation of three O_2 atoms per phosphate released in addition to the O_2 atom attributable to the hydrolytic step itself. The "medium exchange" occurs without hydrolysis of ATP between [^{18}O] P_i and $H_2^{16}O$ (*192*); it is catalyzed by ADP and divalent cations. Finally, the so-called burst phenomenon, first described by Weber and Hasselbach (*193*) refers to the rapid liberation of phosphate at the beginning of the ATPase reaction. This burst, as originally described, amounted to many moles of inorganic phosphate released in the early phase of the reaction per mole of enzyme. It was Tonomura and his colleagues who first showed that in the presence of Mg^{2+} the burst was nearly stoichiometric with the amount of enzyme present (*66, 191*), and, although this problem is still not quite resolved, it seems that the amount of phosphate rapidly liberated at the beginning of the reaction is between 1.5 and 2 moles of myosin (*69, 70*). There is also rapid proton liberation, and, according to Tokiwa and Tonomura (*194*), proton absorption accompanies the early stage of ATP hydrolysis.

During the past few years a fairly simple and consistent scheme accounting for most of these facts has been developed through the efforts of many investigators as an outgrowth of the work initiated by Taylor and his colleagues. They have shown that the initial burst in P_i liberation is caused by rapid hydrolysis of ATP with the products remaining bound to myosin. This initial hydrolysis takes place with a rate constant of about 200 sec^{-1} while the overall hydrolysis in the presence of Mg^{2+} is of the order of 0.02 sec^{-1}. While Lymn and Taylor (*195*) at first suggested that the rate-limiting step in the hydrolysis of ATP is the dissociation of the bound product, it soon became clear that the complex formed by the products of hydrolysis is different from the complex that is formed upon adding ADP and P_i to myosin or its fragments. The first recognition of this fact stemmed from the interpretation of the experiments showing that the mobility of a spin label analog of iodoacetamide attached to the SH_2 thiol group of myosin was greatly increased during the steady state of ATP hydrolysis, and that when ATP was completely hydrolyzed, the spectrum changed to one identical with that obtained on adding ADP to myosin (*196*). Previous studies with a fluorescent myosin dye complex (*197*) and on the ultraviolet difference spectra associated with the interaction of myosin and HMM with ATP (*198*) are also subject to the same interpretation if one recognizes that the steady state complex does not contain ATP but tightly bound products. The demonstration that the tryptophan fluorescence in myosin was enhanced during the hydrolysis of ATP (*199*) and a rapid kinetic approach showing that the intermediate present during the steady state interacts differently with an NADH coupled enzyme system than does the complex formed from added ADP (*200*) also support the view that the steady state product complex is not identical with the complex formed with added ADP.

Another aspect of the ATPase reaction that has already been mentioned is rapid proton liberation. It now appears well established that the initial rapid proton liberation is not related to the hydrolysis of ATP but is the result of changes in conformation of the protein following the binding of ATP (*201*). The same proton burst occurs on binding of ADP or unhydrolyzable ATP analogs—*e.g.*, AMP.PNP. On the basis of all the evidence available to date, including rapid kinetic measurements involving protein fluorescence and fluorescent substrates (*202, 203*), the minimal scheme that accounts for all the known observations on the hydrolysis of ATP by myosin involves at least seven intermediate steps. These are shown in the following scheme:

$$\underset{(1)}{M + ATP \rightleftharpoons} \underset{(2)}{MATP \rightleftharpoons} \underset{(3)}{M^*ATP \rightleftharpoons} \underset{(4)}{M^{**}ADP.P_i \rightleftharpoons} M^*ADP\,P_i$$

$$\underset{(5)}{\rightleftharpoons} \underset{+P_i}{M^*ADP} \underset{(6)}{\rightleftharpoons} MADP \underset{(7)}{\rightleftharpoons} M + ADP$$

Asterisks indicate species of increased fluorescence. Those marked with two asterisks have a higher fluorescence than those marked with one. It appears that, at least at room temperature, the rate limiting step is Step 4 rather than the actual dissociation of the product; and while according to several reports (*202, 204*), the dissociation is rate limiting at low temperature, this has recently been called into question (*205*).

An important, and perhaps surprising, fact that has emerged from these kinetic studies is the value of the equilibrium constant between bound ATP and the bound products of its hydrolysis (Step 3). In contrast to the value of the equilibrium constant in solution which is $10^6 M^{-1}$ or greater, the equilibrium constant for the bound species is about $9 M^{-1}$ (*206*), indicating that ATP hydrolysis in the bound form is characterized by a much smaller negative free energy than in solution.

More than 20 years ago Morales and his colleagues (*207*) concluded that the main part of the net $\Delta G°$ of ATP hydrolysis is released as ATP binds to myosin—or actomyosin—and that the free energy of hydrolysis in the bound state is small. In retrospect it seems that the essentially correct insights became clouded by the fact that they were based on somewhat indirect evidence and that the correct thermodynamic conclusions were imbedded in a theory of muscle contraction that was based on a polyelectrolyte-charge neutralization mechanism which would be considered by most, including the original proponents, untenable today.

The decrease in the equilibrium constant for ATP hydrolysis in the bound form has made it possible to show the direct formation of ATP from inorganic phosphate and ADP. There is some controversy concerning the precise value of the dissociation rate of bound ATP. Mannherz and his colleagues (*208*) suggest a value of about 10^{-7} sec^{-1}, from studies involving the net synthesis of enzyme-bound (HMM S-1) ATP from ADP and ^{32}P, for the back rate of Step 2 which would, for all practical purposes, make the once unbound ATP undissociable. However, Wolcott and Boyer (*209*) deduced a higher value from their studies on the incorporation of radioactive ^{32}P into free ATP. The incorporation of ^{18}O from water into the P_i from ATP can be explained by repeated cleavage and reformation of ATP about six times for each P_i released by net hydrolysis. The so-called medium exchange between P_i and H_2O which depends on nucleotides and divalent metal ion cofactors is also presumably a manifestation of the same mechanism (*210*).

Interaction of Myosin with Actin and Activation of Its ATPase Activity by Actin. PHYSICOCHEMICAL EFFECTS. Ever since the discovery of myosin and actin as distinct entities and the recognition of the fact that early studies on "myosin" had really been carried out on a complex of actin and myosin, *via* actomyosin, studies of the interaction of purified myosin and actin under well controlled conditions have been of paramount interest. Since *in vivo* actin and myosin are located in separate filamentous structures, interactions presumably similar to those obtained on mixing myosin and actin *in vitro* are likely to take place. On mixing myosin and actin at $\mu = 0.5$–0.6 there is a considerable increase in viscosity and birefringence of flow, both being reversed on the addition of ATP. This was interpreted as the dissociation of actomyosin into actin and myosin (*19*) and was confirmed on the basis of light scattering measurements and ultracentrifuge experiments (*176, 177*). It is at lower ionic strength, of the order of 0.05 to 0.15, more closely corresponding to that prevailing in the muscle cell itself, that the most interesting properties of the actin–myosin interaction come to light. Under these conditions myosin ATPase is changed into what one might call the actomyosin enzyme characterized by a high activity even in the presence of Mg^{2+} in contrast to the inhibition of the myosin enzyme (19). Thus actin acts as a powerful modifier of the ATPase activity of myosin immediately suggesting interesting possibilities from the point of view of physiological regulation of energy release.

Actomyosin at low ionic strength forms a fine suspension which on addition of ATP undergoes a characteristic reaction consisting of the aggregation of the particles into a network which eventually shrinks. This process is known as superprecipitation. If an actomyosin solution in 0.6M KCl is extruded in distilled water, a fine thread is formed which contracts when ATP is added. Thus the contractile protein not only splits ATP enzymatically, as first shown by Engelhardt and Ljubimova (14), but itself undergoes a change; Szent-Györgyi (19) considered the above phenomena as models of muscle contraction. We return to this point after considering current molecular theories of contraction.

STRUCTURAL ASPECTS. In addition to the study of actomyosin and reconstitution, studies on systems of somewhat higher complexity have also found wide use as *in vitro* models of muscle contraction. Glycerol-extracted muscle fibers or fiber bundles (211) or myofibrils contain actomyosin in their essentially undisturbed configuration. However, these preparations do not contain the excitatory mechanism of the intact cells nor do they contain the energy supply necessary for contraction. It was found, however, that on the addition of ATP glycerinated fibers contract, or, depending on the experimental conditions, develop tension. Myofibrils exhibit ATPase activity that has the characteristic feature of Mg^{2+} activation observed on reconstituted and natural actomyosin, and this contraction can be observed under the microscope.

H. E. Huxley (212) has added a new dimension to the study of the interaction of actin and myosin by observing that adding fragments of myosin to actin filaments—either polymerized *in vitro* from G-actin or isolated while still attached to the Z disc—produces characteristic structures termed arrowheads. These arrowheads were separated by about 40 nm along an actin filament, and Huxley observed that the arrowheads on any actin filament all point in the same direction. With filaments attached to the Z disc, arrowheads on either side pointed away from the Z disc (or line). This fact indicated a polarity in the actin filaments which, together with the already discussed dipolar structure of the thick myosin filaments, plays an important role in the molecular mechanism of contraction driving the thin filaments toward the center of the thick filaments.

Experiments with HMM S-1 "decorated" actin filaments using 3-D reconstruction of electron micrographs have elucidated the details of the myosin–actin combination resulting in arrowheads (156). The head portions of the myosin molecule attach to the individual actin globules, forming an angle of about 45° with the long axis of the actin filament; in addition there is a "slewing" of the HMM S-1 out of the plane containing the point of attachment and the axis of the filament. The angling of the heads on actin is considered as an expression of the potential driving force in the contraction process (*see* p. 247).

ACTIVATION OF MYOSIN ATPASE BY ACTIN. In the light of the scheme of ATP hydrolysis a drastic reevaluation of views generally held concerning the activation of myosin ATPase by actin has taken place. According to Lymn and Taylor (*70, 213*) the rate of dissociation of actomyosin by ATP is faster than the rate of hydrolysis by actomyosin. Thus activation by actin has to be viewed not as a result of a more rapid hydrolysis of ATP by the actomyosin complex but rather an acceleration of the processes that follow the hydrolysis of ATP on dissociated myosin. This view can be expressed in the following simplified scheme which omits many intermediates in the corresponding scheme for myosin:

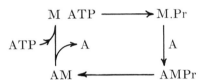

In this view then actin combines with the myosin product complex leading to rapid dissociation of the products, followed by dissociation of the actin–myosin complex by ATP into ATP.myosin and actin. ATP will again be hydrolyzed with the products remaining bound to myosin and the cycle restarted. Detailed measurements on the various rates involved in the cycle of the actomyosin system are not yet available to the same extent that steps of the myosin ATP system have been worked out. Precise information about these rate constants is, of course, very important in interpreting the molecular events that underlie the crossbridge cycle *in situ*.

An interesting new aspect that has emerged from the study of the actin–myosin ATP system and particularly from the actin–HMM–ATP system is the concept of the *refractory state (214)*. This expression has been coined to indicate the existence of the intermediate somewhere in the cycle that is unable to combine with actin unless it first undergoes some further conformational change. The basis of this suggestion is the observation of Eisenberg *et al.* in the actin–HMM–ATP system at $0°C$ that under conditions leading to almost maximal ATP hydrolysis a large portion of HMM is free. Intuitively one would have expected that as the actin concentration increases, more and more of the myosin or HMM will be present, complexed to actin as the ATPase activity approaches V_{max}. In principle, of course, it is possible that in a complex reaction scheme much myosin could remain free at high actin concentrations even without a separate refractory state, simply as a result of the quantitative relation among rate constants. This has been analyzed in some detail by Lymn, but it can be easily shown for a simpler system that if, for instance, the rate constant of the hydrolytic step is of the same order of magnitude as the dissociation step, then the amount of unattached myosin (including the product complex) would be about half of the total present even at extremely high actin concentrations. Eisenberg and his colleagues argue that the experimental values for the relevant rates would exclude such interpretations. Apparently the final verdict is not in yet, and at this point one cannot decide whether additional intermediate states are to be included in the above scheme. More recently Mulhern, Eisenberg, and Kielley (215) have suggested that the reduced activation by actin of the ATPase of SH_1 modified myosin or HMM accompanied by an even greater dissociation of the acto-HMM complex near V_{max} is a strong argument against the view that the persistence of dissociation of acto-HMM at high actin concentrations in the native protein could be explained in simple kinetic terms.

REGULATION OF ACTIN–MYOSIN INTERACTION. This process involves (a) Ca^{2+} and (b) regulatory proteins. The role of Ca^{2+} emerged over a few years from early observations on the inhibition by EDTA of the contraction of glycerinated fibers in the presence of ATP and the inhibition by EDTA of myofibrillar ATPase activity. Various facts such as the release of the inhibitory effect of ATP in excess of Mg^{2+} by minute amounts of Ca^{2+}, the inhibitory effect of an EDTA analog (EGTA) having a much higher affinity for Ca^{2+} than Mg^{2+}, and the demonstration that the ATPase activity and superprecipitation of actomyosin, or syneresis of myofibrils, is a function of the Ca^{2+} concentration, firmly established the Ca^{2+} control of the functionally important actin–myosin interaction [for reviews see Gergely (41), Ebashi, Endo, and Ohtsuki (35), and Weber and Murray (216)]. The most direct demonstration that Ca^{2+} is involved in the excitation–contraction coupling process comes from the work of Ashley and Ridgeway (217). These workers have injected into giant barnacle muscle fiber aequorin, a protein that emits light when it interacts with Ca^{2+}, and have shown that contraction is preceded by a rise in free Ca^{2+} concentration in the cell. The biological control of the Ca^{2+} level in muscle is discussed in Chapter 3 of this volume.

However, the Ca^{2+} requirement for the actin–myosin interaction as discussed above (pp. 233-4) is not present with highly purified proteins, but only if other protein components—*viz.*, tropomyosin and troponin—are present; the latter is itself a complex of three subunits.

Combination of Ca^{2+} with the TnC component of the intact actin–myosin–tropomyosin–troponin system causes a conformational change not only in it (218, 219, 220, 221, 222) but also in the other components, making it possible for myosin and actin to combine. We now turn to some aspects of the mechanism by which this is accomplished. It has been puzzling in view of the localization of tropomyosin and troponin discussed on p. 237 that the combination of Ca^{2+} with a single troponin can affect all actin molecules lying in the range of the tropomyosin molecule with which troponin is associated. An interesting aspect of this long range interaction is the cooperation that exists among myosin molecules attached to actin monomers within the domain of the same tropomyosin molecule. A. Weber and her co-workers have shown in a series of elegant experiments that the affinity of myosin for actin increases once several other myosins are attached within the domain of the same tropomyosin molecule (223). The biological significance of this type of cooperativity is not clear. It may have an important role in the temporal development of tension as successive myosin bridges form linkages with actin or, conversely, on the time course of delay of tension and the cessation of stimulation. On the other hand, according to x-ray evidence (*see below*) only a fraction of all myosin molecules are in contact with actin even

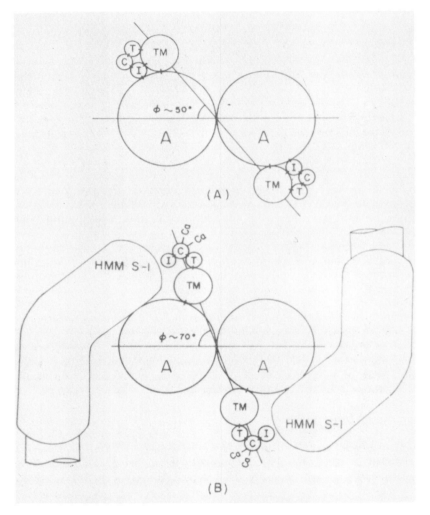

Figure 5. Scheme of regulation of muscle contraction by troponin and Ca^{2+} (227). The relative positions of actin, tropomyosin, and the head of the myosin molecule (HMM S-1) in the model are essentially as proposed by Spudich et al. (224), Haselgrove (225), Huxley (36), and Parry and Squire (226). Key: A, actin; TM, tropomyosin; T, TnC; I, TnI; C, TnC. A, relaxation in the absence of Ca^{2+}; B, activation, $[Ca^{2+}] \sim 1\ \mu M$. Suggested interactions between proteins are indicated by a short connecting line.

under conditions of maximum activation, and thus the probability of many myosins' being attached within the same tropomyosin domain is low.

What is then a possible molecular basis of the regulation of actin and myosin by the calcium troponin interaction? Recent experiments in several laboratories dealing with conformational changes in troponin C and the effect of calcium on the combination between various troponin components, tropomyosin and actin (a detailed description of which would exceed the scope of this chapter), suggest that a possible scheme of regulation can be conceived using some recent information concerning changes in the position of tropomyosin depending on whether muscle is in the active or relaxed state. Under conditions corresponding to activation—*viz.*, in the presence of calcium, tropomyosin molecules move closer to the center of the grooves formed by the long-pitch actin helices of the thin filaments, whereas under conditions corresponding to relaxation tropomyosins are located further away from the grooves. It also appears that the position of tropomyosin in the resting state is such that it would block the attachment of the myosin head to actin (*36, 224, 225, 226*). Thus the activation by calcium, resulting in the movement of tropomyosin, may be attributed to the removal of a steric block at the myosin binding site. The role of the troponin complex in the translocation of tropomyosin associated with actin could be conceived in the following way. The troponin complex is attached to tropomyosin through its TnT component. The linkages between actin and troponin in the resting state occur *via* the TnI component. Combination of calcium with TnC leads to a release of the TnI component from actin (*137, 154, 227, 228*). Thus upon combination tropomyosin would be free to move into the position that would permit the combination of myosin with actin (Figure 5). A somewhat more detailed discussion in terms of changes in the structure (conformation) of the various proteins is given in the next section.

In concluding this section it should be mentioned that in muscles of a number of lower organisms (molluscs), and probably in smooth muscle of higher organisms (*229*), no troponin is found. In these systems Ca^{2+} control is exerted directly through myosin *via* a specialized subunit (*51, 73*).

Molecular Mechanisms and Models of Contraction

Movement of Crossbridges. Several references have been made in this chapter to the crossbridges that form links between the myosin and actin filaments in stimulated muscle. It is generally accepted that the crossbridges are involved in the force generation process which, when the force is unopposed or not sufficiently opposed, leads to shortening. X-ray work on living muscle showing that the well defined pattern of layer lines found in x-ray diagrams of resting muscle disappears on activation suggesting that movements occur in the crossbridges. This is accompanied by what appears to be a transfer of mass from the thick filaments to the thin filaments as deduced from changes in the so-called equatorial reflections (*30, 36*). The latter process would be consistent with links formed by the crossbridges attached to the thin filaments.

Changes in orientation of crossbridges were first observed in insect flight muscle (230). In relaxed muscle of this type both electron microscopic and x-ray analysis showed crossbridges perpendicular to the thick filaments, while in rigor crossbridges formed a roughly 45° angle with the axis of the filaments. This angle roughly corresponds to that deduced from electron micrographs of the combination of actin and subfragment 1 (156). Recent studies on glycerinated fibers utilizing fluorescence depolarization techniques (231, 232, 233) also showed that the contractions induced by ATP involve a change in the attitude of the myosin heads in the direction of that found in rigor. The mechanism of contraction based on crossbridge interactions implies a cyclic process since as the overlap of thick and thin filaments changes over a distance of about 1 μm, crossbridges that have become attached must be detached to allow others to take their place. Each attachment–detachment cycle is thought to involve the hydrolysis of at least one ATP molecule according to the scheme outlined for the hydrolysis of ATP by myosin and actomyosin in solution. The finding of bound ADP, in addition to that bound to actin, in relaxed muscle is consistent with this scheme (234).

Conformational Changes. In considering the molecular driving force causing the movement of crossbridges many investigators have been looking for conformational changes in the protein molecules. Conformational changes in many protein systems are revealed by large changes in α-helix content detectable by measurement of circular dichroism and optical rotation. Such changes have not been found in myosin. However, changes in fluorescence, the mobility of attached spin labels, and optical properties of myosin have been observed on its interaction with ATP (see pp. 240–241). The question, therefore, arises as to whether these changes can be considered as the driving force for crossbridge movement. Unfortunately, no special changes have been observed on interaction with actin so that now it is a matter of speculation what role the conformational changes observed in myosin may play in the process. The movement of the head region of the myosin molecule is predicted in the so-called hinge regions—that is, regions that may show greater flexibility within the myosin structure. The existence of hinge regions has been suggested at the junction of the globular S-1 and the S-2 rod portion as well as at the junction of the S-2 portion with the remainder of the rod. These hinge regions would not only permit flexibility in the crossbridges but would make it possible for the myosin heads to interact with actin at an optimal angle, even as the interfilamentar distance increases with the shortening of the sarcomere, a fact that has been deduced from x-ray diffraction studies. Segmental flexibility of myosin in solution has been deduced from measurements of fluorescence depolarization (235) and more recently from applying a new version of the electron spin resonance technique, saturation transfer spectroscopy (236) and by studies on the thermal stability of various portions of the rod portion of the myosin molecule (232). While some earlier speculation placed the driving force for crossbridge movement in the hinges, H. E. Huxley has introduced (212) the notion that the driving force for the crossbridge motion is developed at the interaction site of actin and myosin.

Conformational changes in some components of regulatory proteins, particularly troponin C, have been referred to above. These conformational changes are induced *in vitro* as troponin C combines with Ca^{2+} and are reflected in changes of various properties including circular dichroism, reactivity of the single sulfhydryl group in TnC, and reactivity in the mobility of attached spin label (*141, 219*). It is puzzling that these conformational changes occur in a Ca^{2+} concentration range that is lower than that required for activation of the actin–myosin interaction and the activation of the actin–myosin ATPase (*238*). Since there are two classes of binding sites in troponin C, it seems that binding of calcium to the higher affinity site produces an easily detectable conformational change. However, Ca^{2+} binding to the remaining sites, which are essential for the activation of the myofibrillar ATPase, has so far failed to produce readily detectable conformational changes. Whether the movement of tropomyosin that is fundamental in allowing the reaction of myosin–actin to proceed is merely the result of the release of a troponin anchor or whether the interaction of calcium with troponin C induces a conformational change in tropomyosin itself and in actin remains to be seen. The occurrence of conformational changes in actin attributable to the changed interaction with tropomyosin is suggested by the fact that once tropomyosin is in the activating position—either because of interaction of Ca^{2+} with TnC or because of the action of crossbridges that induce movement of tropomyosin (*131, 223*)—the activity of actomyosin or acto-HMM can be higher than in the pure, unregulated system.

Studies utilizing quasi-elastic light scattering of laser light have also suggested conformational changes in actin upon combination of Ca^{2+} with Tn bound to TM + actin that may play a role in the molecular mechanism of contraction (*239*).

Relation to Mechanical Properties. In describing the mechanical properties of muscle the presence of elastic elements has always played an important role. These elastic elements have variously been assigned to series and parallel components. Recent work by Huxley and Simmons (*38, 240, 241*) suggests that, at least in single fibers, not only the force generating mechanism but also the elastic elements reside in the crossbridges. The simplest solution would be to place the elastic element in the portion of the myosin molecule connecting the head with the body of the thick filaments—*viz.*, the coiled-coil α-helical S-2 portion of the myosin molecule. It is, however, entirely possible that the elasticity resides in one of the hinges or in S-1. Huxley and Simmons also suggest a novel way of looking at the active elastic component of muscle that has long been known as being involved in tension transients following quick release or quick stretch. These recent experiments utilizing superior recording techniques of extremely high temporal resolution have shown that the immediate tension drop following release (to level T_1) is followed by a rapid return to an intermediate level of tension during a time period of the order of milliseconds, followed by a slow return to essentially the original tension in milliseconds. Huxley and Simmons have suggested that the rapid return to intermediate tension level denoted by T_2 is caused by redistribution of crossbridges among two or more possible angular positions. The crossbridges could thus be in a state of equilibrium among various possible states, and this equilibrium would show a shift on changes in tension. This extension of the original model proposed by A. F. Huxley (2), serving as a basis for the active as well as the passive component, can be shown to account not only for the force velocity properties of muscle (the so-called Hill relation) but also satisfactorily accounts for the transient phenomena observed on rapid changes in length or rapid changes in tension. An alternative view proposed by Podolsky and his colleagues (242, 243) to explain the mechanical transient following quick releases would involve a rapid attachment of new crossbridges. The view proposed by Huxley and Simmons does not envisage a change in the number of crossbridges attached but rather a redistribution among possible positions of crossbridges relative to the thin filaments. Careful measurement of the compliance of muscle following quick stretches or during imposed rapid oscillations may resolve the problem.

In the model outlined by Huxley and Simmons the tendency of the crossbridge upon attachment to move into the new position has been attributed to the existence of discrete modes of attachment characterized by a decreasing set of free energy levels (*see* Figure 6). Alternatively, in terms of a perhaps biochemically more realistic mechanism, one may think of a possibly mutually induced conformational change in actin and myosin leading to the establishment of contact over an increasing area, resulting in the change of angle of the crossbridge and a decrease in free energy. The equilibrium which would be regulated by tension exerted on the head *via* the connecting link would, in

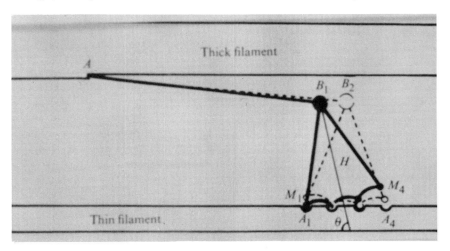

Nature

Figure 6. Diagram showing assumed crossbridge properties (240). The myosin head H is connected to the thick filament by a link AB containing the undamped elasticity which shows up as T_1 in the whole fiber. Full line shows head in position where M_1A_1 and M_2A_2 attachments are made; broken lines show position where M_2A_2 and M_3A_3 attachments are made.

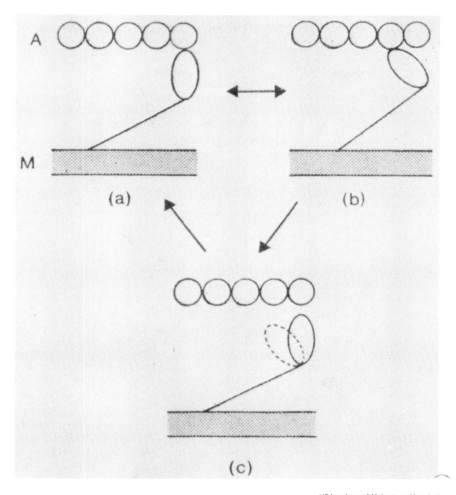

"Disorders of Voluntary Muscle"

Figure 7. Model of interaction of the myosin head with actin based on Huxley (212) and Huxley and Simmons (240). (a): Attachment of myosin to actin; (b) tilting of the myosin head; (c) detachment produced by ATP followed by its hydrolysis. On the basis of in vitro kinetic studies the myosin species present in (a) + (b) carries the ADP.P product complex. Whether the dissociated myosin head can oscillate between the perpendicular or tilted position or whether it is locked in the perpendicular position is not finally settled. Similarly, the precise point at which the product leaves the actin–myosin complex requires clarification. The attached head may oscillate between positions (a) and (b)
(41).

turn, exist among various conformational substrates of the actin–myosin interaction region. A recent proposal, quite independent of the muscle system (244), concerned with the interaction of small molecules with proteins contains suggestions quite similar to this view. The cyclic operation of the crossbridges during contraction is shown in Figure 7.

Energy Transduction

Energy Balance. As discusssed in the first section of this chapter participation of ATP and creatine phosphate is well established, at least for a series of contractions. Although studies on the actomyosin system *in vitro* strongly indicated that the ultimate energy source is ATP, it took more than 10 years after A. V. Hill's challenge to biochemists (*245*) to prove that ATP is in fact broken down in a single contractile event (*246*). There is now also good evidence that there is a chemical counterpart, in terms of the actual breakdown of creatine phosphate and ATP, of the Fenn effect—*viz.*, the increase in the total energy released by muscle when it shortens while doing work. However, Hill's "further challenge to biochemists" (*247*) concerning the precise relation of chemical change to heat and work has still not been fully met.

The original partitioning of the energy released by muscle during contraction in three terms—one corresponding to activation independently of shortening or work done, another proportional to shortening, and a third equivalent to mechanical work—has been questioned by A. V. Hill, father of the idea, himself (*248–251*). The existence of a heat term which depends only on shortening appears doubtful at least over a single full contraction and relaxation cycle (*252, 253*). As pointed out by Wilkie and his colleagues there is no doubt that muscles liberate heat at a higher rate while they are shortening than while they are not. However, this does not mean that the total quantity of heat produced by the shortened muscle will be greater since compensation does set in. A further complication that has recently been uncovered and is as yet not fully resolved, concerns the apparent discrepancy between the total energy liberated in the form of work and heat in the early stage of the contraction when compared with the known chemical reactions and the heat that could result from them on the basis of *in vitro* measurements (*252, 254, 255; see*, however, Ref. *256*). The chemical and energetic changes associated with calcium movement (*see* Chapter 3 of this volume) that control contraction and relaxation may further complicate this picture.

Mechanochemical Energy Transduction. As the search for the mechanism of oxidative phosphorylation has been characterized by the pursuit of the elusive "squiggle," the covalently linked high energy phosphorylated intermediate generated by electron transport, so the connection between ATP hydrolysis and mechanical performance has long been sought in terms of a phosphorylated high energy intermediate (*210*). The earliest attempt seems to be that of Riseman and Kirkwood (*67*) who postulated that covalent attachment of phosphate to myosin or, perhaps, in terms of the then extant knowledge, actomyosin, would keep the muscle relaxed while hydrolysis of the P bond would lead to contraction. About 10 years later H. H. Weber (*257*) attempted to explain the connection between ATP hydrolysis and actin–myosin interaction in terms of stepwise phosphorylation and dephosphorylation of a set of groups located on actin. The current position in the field of oxidative phosphorylation, having led to the recognition that perhaps there exists no phosphorylated intermediate, is somewhat paralleled by that in the field of muscle. While efforts to demonstrate directly a phosphorylated intermediate of myosin have failed (see p. 240), the current thinking discussed in the section on the ATPase mechanism suggests that there are tightly bound intermediates but none, however, covalently bound. Another interesting parallel between oxidative phosphorylation and the hydrolysis of ATP by myosin is that in both systems the negative free energy of ATP hydrolysis in its bound form with the resultant tightly bound ADP and inorganic phosphate, is considerably smaller than in solution. As suggested by Boyer and his colleagues (210, 258) in oxidative phosphorylation it is the removal of the ATP so formed that requires energy derived from electron transport. In the muscle system the tight binding of ATP to myosin may be the key reaction in bringing about the dissociation of the tight actin–myosin complex with the myosin head in the configuration corresponding to high tension in the crossbridge. One can only speculate on the significance for muscle contraction of the low free energy of hydrolysis of ATP in its bound form (see also p. 255). It is, however, a fact that myosin crossbridges in relaxed muscle carry tightly bound ADP and inorganic phosphate (234). In terms of the currently accepted kinetic scheme, myosin in the presence of Mg^{2+} turns over only slowly, the rate-limiting steps being those that follow the rapid hydrolysis of ATP on the surface of the myosin head. Combination with actin leads to rapid liberation of the products, and a new molecule of ATP is required to dissociate the actin–myosin complex. This mechanism avoids waste of ATP in resting muscle in which actin and myosin exist in a dissociated state, and through the accelerated hydrolysis of ATP by actin it allows for a higher rate of energy liberation as the actin–myosin interaction leads to contraction and, if the muscle is loaded, to work. In a continuously working muscle it is the hydrolysis of ATP that provides the energy which does the mechanical work.

Thermodynamic Considerations. Considerable ingenuity and effort have been spent on clarifying the question of energy transfer from ATP to the muscle machine and finally conversion into work (chemomechanical coupling). A great deal of this may appear to be directed at answering pseudo questions, particularly in the light of detailed statistical, mechanical, thermodynamic, and kinetic analysis carried out by Hill and his colleagues (259, 260, 261). While the work obtainable from the hydrolysis of one molecule of ATP is given by the free energy change

$$\Delta\mu = \mu_{ATP} - \mu_{ADP} - \mu_{P_i}$$

at the prevailing concentrations of ATP, ADP, and inorganic phosphate, the actual amount of work obtainable is determined by what one might call the structural details of the muscle machine—or more precisely the crossbridge working part—and cannot be derived directly from kinetic and thermodynamic studies in solution. The simplest free energy diagram involving two states that can illustrate the operation of the crossbridge cycle is shown in Figure 8. This diagram assumes that all detached states of the crossbridge are represented by one state (A_o) and all the attached states are represented by another (A_1). For the latter one has to assume that the free energy is a function of some length parameter, let us say the distance of the head of the crossbridge attached to the actin filament from some conveniently chosen point on the thick filament—e.g., $x = 0$ at the minimum of A_1, viz., at the point where the force is 0. Work in the crossbridge cycle can be obtained only if

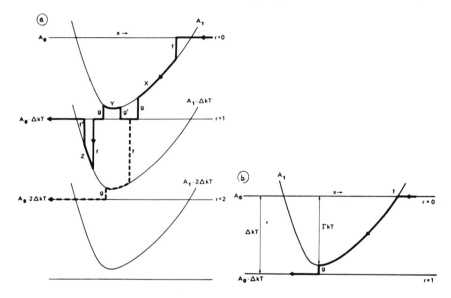

Figure 8. Schematic set of possible transitions in the two-state model of crossbridge actin interaction (261). The parameter r counts the number of ATP molecules hydrolyzed by the crossbridge. A_1 and A_1 are free energy levels of the detached and attached state, respectively. The free energy change of ATP hydrolysis is $-\Delta kT$. Free energy levels are displaced downward by the amount for each molecule of ATP hydrolyzed. Rate constants for crossbridge attachment are f and g', for crossbridge detachment f' and g. A cycle involving an f and g step leads to a change in r—viz., net hydrolysis of ATP. (a): Transitions in a single pass of an actin site by a crossbridge. Heavy path, one net cycle, r = 1. Dashed path, two net cycles, r = 2. (b): Transitions giving optimal efficiency of Γ/Δ.

there is a change in x between the attachment and detachment at the crossbridges. Following Hill, one assumes that the crossbridge carrying the various intermediates formed on hydrolysis of ATP can form an attachment at some value of x. This is a stochastic process determined by the probability parameters—viz., the rate constants—that vary as some function of x. Detachments would also be governed by a probability function (rate of dissociation), again as a function fo x. The amount of work obtainable would be determined by the free energy drop in the attached state between the average point of attachment and detachment. Thus while the maximum free energy would be obtained if attachments always occurred at the point where the line and the curve intersect and detachment always occurred at the minimum of the free energy curve of the attached crossbridge, in fact only smaller amounts can be realized because of the stochastic nature of the process. The free energy level of the detached crossbridge, after hydrolysis of ATP, would be displaced from the previous one by an amount corresponding to the free energy of the hydrolysis of ATP, not standard free energy but actual $\Delta\mu$ determined by the prevailing concentrations. These energy levels are repeated an infinite number of times, each set corresponding to the hydrolysis of one ATP molecule. If one assumes that there are several attached substates among which transitions occur at a finite rate (*see* Huxley and Simmons model), then several curves similar to the one shown must be drawn. In this case the transitions between the states can be represented by a vertical transition from one curve to the other, and the net work will be equal to the net free energy change occurring in the attached states in downward travel along the other curves. In terms of this scheme the significance of the shift to the right of the equilibrium ATP $\rightleftharpoons$ ADP $+$ P_i in the bound state appears to be as follows. If there were a large free energy drop in the hydrolytic reaction on the myosin head, a second horizontal line, correspondingly displaced, would have to be drawn to show that a part of the total free energy available ($\mu_{ATP} - \mu_{ADP} - \mu_{P_i}$) is lost since the drop occurs in an unattached state. If the drop is small or if the two lines are drawn close together, most of the $\Delta\mu$ can, in principle, be utilized in the downward slide to do work.

What would be the energetics in the absence of ATP? It is, of course, thermodynamically possible to get some work—in a single cycle—in the absence of a hydrolyzable substrate; experimental data (*e.g.*, Ref. *262*) furnish evidence for this. What is not possible in the absence of a hydrolyzable substrate is to repeat the crossbridge cycles. Thus, if, in the absence of ATP, the relation of the detached and attached crossbridge free energy levels is the same as with ATP, the work obtainable in one cycle would depend on the free energy difference between a point on the curve at which attachment occurs and the minimum of the attached curve. If, in the absence of ATP, the free energy level of the detached crossbridge does not intersect the free energy curve of the attached crossbridge, then no attachment could occur in the absence of nucleotides. Actually, myosin and its fragments do combine with actin, and tension development can accompany rigor development (*263*); this, in terms of the above simplified scheme, suggests the tendency of the crossbridges to move toward the lowest point on the curve. In fact, the myosin "heads" on actin execute a change in angle in the absence of ATP that seems to be identical with that occurring in active contraction; thus the "energization" of the crossbridge need not be sought in terms of making possible a movement which otherwise would not take place.

Conformational Changes Revisited. A comment may be in order on the relation of conformational changes observed in solution or in structured systems as interactions with nucleotides occur. Conformational changes need not be associated with large changes in free energy although they may be, but they certainly can be associated with changed probabilities of attachment and detachment. Thus the conformational change observed on transition from ATP.myosin in the product–myosin complex may reflect a change in the myosin structure from a form of myosin that cannot attach to actin to one that can. An earlier report by Holmes and his colleagues suggested that hydrolysis of the substrate was necessary to produce the angular rotation of the myosin head necessary for attachment to actin and contraction (264). Recent data indicate that there is no discernible difference in the attitude of detached crossbridges whether or not the bound substrate is hydrolyzed (265). Conformational differences between states that do not greatly differ in free energy may be important in governing free energies of activation of detachment or attachment, resulting in changed rate constants of transition.

Nomenclature

HMM: heavy meromyosin
LMM: light meromyosin
S-1: subfragment 1
S-2: subfragment 2
SDS: sodium dodecyl sulfate
Tn: troponin

Literature Cited

1. Needham, D. M., "Machina Carnis," Cambridge University Press, 1971.
2. Huxley, A. F., *Progr. Biophys.* (1957) **7**, 255.
3. Kalckar, H. M., "Biological Phosphorylation," Prentice Hall, 1969.
4. Kalckar, H. M., *Mol. Cell. Biochem.* (1974) **5**, 55.
5. Cori, F. C., *Mol. Cell. Biochem.* (1974) **5**, 47.
6. Eggleton, P., Eggleton, G. P., *Biochem. J.* (1927) **21**, 190.
7. Eggleton, P., Eggleton, G. P., *J. Physiol. (London)* (1927) **63**, 155.
8. Fiske, C. H., Subbarow, Y., *Science* (1927) **65**, 401.
9. Fiske, C. H., Subbarow, Y., *Science* (1928) **67**, 169.
10. Lundsgaard, E., *Biochem. Z.* (1920) **217**, 162.
11. Lohmann, K., *Biochem. Z.* (1928) **203**, 172.
12. Fiske, C. H., Subbarow, Y., *Science* (1929) **70**, 381.
13. Lohmann, K., *Biochem. Z.* (1934) **271**, 264.
14. Engelhardt, W. A., Ljubimova, M. N., *Nature (London)* (1939) **144**, 668.
15. Needham, J., Kleinzeller, A., Miall, M., Dainty, M., Needham, D. M., Lawrence, A. S. C., *Nature* (1942) **150**, 46.
15a. Kühne, W., Untersuchungen über das Protoplasma und die Contratilität, W. Engelmann, Leipzig, 1864.
16. Huxley, T. H., "The Crayfish," Kegan Paul, Ltd., London, 1880.
17. Muralt, A. von, Edsall, J. T., *J. Biol. Chem.* (1930) **89**, 315.
18. Muralt, A. von, Edsall, J. T., *J. Biol. Chem.* (1930) **89**, 351.
19. Szent-Györgyi, A., *Acta Physiol. Scand.* (1945) **9**, Suppl. XXV.
20. Huxley, H. E., "The Structure and Function of Muscle," Vol. I, p. 302, Geoffrey H. Bourne, Ed., Academic, New York, 1972.
21. Franzini-Armstrong, C., "The Structure and Function of Muscle," Vol. II, Part 2, p. 532, Geoffrey H. Bourne, Ed., Academic, New York, 1973.
22. Adrian, R. H., Peachey, L. D., *J. Physiol.* (1973) **235**, 103.
23. Close, R., *Physiol. Rev.* (1972) **52**, 129.
24. Goldspink, G., "The Structure and Function of Muscle," Vol. I, p. 181, Geoffrey H. Bourne, Ed., Academic, New York, 1972.

25. Kuhn, T., "The Structure of Scientific Revolutions," University of Chicago Press, Chicago, 1962.
26. Huxley, H. E., Hanson, J., *Nature* (1954) **173**, 973.
27. Hanson, J., Huxley, H. E., *Symp. Soc. Exp. Biol.* (1955) **9**, 228.
28. Huxley, A. F., Niedergerke, R., *Nature* (1954) **173**, 971.
29. Hasselback, W., *Z. Naturforsch.* (1953) **8b**, 449.
30. Huxley, H. E., Brown, W., *J. Mol. Biol.* (1967) **30**, 383.
31. Tregear, R. T., Squire, J. M., *J. Mol. Biol.* (1973) **77**, 279.
32. Squire, J. M., *J. Mol. Biol.* (1973) **77**, 291.
33. Morimoto, K., Harrington, W. F., *J. Mol. Biol.* (1974) **83**, 83.
34. Ebashi, S., Endo, M., *Progr. Biophys. Mol. Biol.* (1968) **18**, 123.
35. Ebashi, S., Endo, M., Ohtsuki, I., *Quart Rev. Biophys.* (1969) **2**, 351.
36. Huxley, H. E., *Cold Spring Harbor Symp. Quant. Biol.* (1972) **37**, 361.
37. Gordon, A. M., Huxley, A. F., Julian, F. J., *J. Physiol.* (1966) **184**, 170.
38. Huxley, A. F., *J. Physiol.* (1974) **243**, 1.
39. Lowey, S., Slayter, H. S., Weeds, A. G., Baker, H., *J. Mol. Biol.* (1969) **42**, 1.
40. Lowey, S., "Subunits in Biological Systems," Part A, p. 201, S. N. Timasheff and G. D. Fasman, Eds., Marcell Dekker, New York, 1971.
41. Gergely, J., "Disorders of Voluntary Muscle," 3rd ed., p. 102, J. N. Walton, Ed., Churchill Livingstone Ltd., Edinburgh, 1974.
42. Young, D. M., *Ann. Rev. Biochem.* (1969) **38**, 913.
43. Gergely, J., "The Physiology and Biochemistry of Huscle as a Food," p. 349, E. J. Briskey, R. G. Cassens, B. B. Marsh, Eds., University of Wisconsin Press, Madison, 1970.
44. Stone, D., Perry, S. V., *Biochem. J.* (1972) **128**, 106.
45. Balint, M., Sreter, F. A., Gergely, J., *Arch. Biochem. Biophys.* (1975) **168**, 557.
46. Balint, M., Sreter, F. A., Wolf, I., Nagy, B., Gergely, J., *J. Biol. Chem.* (1975) **250**, 6168.
46a. Gazith, J. S., Himmelfarb, S., Harrington, W. F., *J. Biol. Chem.* (1970) **245**, 15.
47. Sarker, S., Sreter, F. A., Gergely, J., *Proc. Natl. Acad. Sci. USA* (1971) **68**, 946.
48. Lowey, S., Risby, D., *Nature* (1971) **234**, 81.
49. Weeds, A. G., Pope, B., *Nature* (1971) **234**, 85.
50. Frank, G., Weeds, A. G., *Eur. J. Biochem.* (1974) **44**, 317.
51. Kendrick-Jones, J., *Nature* (1974) **249**, 631.
52. Weeds, A. G., Taylor, R. S., *Nature* (1975) **257**, 54.
53. Sreter, F. A., Gergely, J., Luff, A. L., *Biochem. Biophys. Res. Commun.* (1974) **56**, 84.
54. Sreter, F. A., Romanul, F. C. A., Salmons, S., Gergely, J., "Exploratory Concepts in Muscular Dystrophy," Vol. II, A. Milhorat, Ed., *Excerpta Med. Intern. Congr. Ser. No.* **333**, 338, 1975.
55. Collins, J. H., *Fed. Proc.* (1975) **34**, 1804.
56. Starr, R., Offer, G., *J. Mol. Biol.* (1974) **81**, 17.
57. Trayer, I. P., Harris, C. I., Perry, S. V., *Nature* (1968) **217**, 452.
58. Kuehl, W. M., Adelstein, R. S., *Biochem. Biophys. Res. Commun.* (1970) **39**, 956.
59. Barany, M., *J. Gen. Physiol.* (1967) **50**, 197.
60. Sreter, F. A., Seidel, J. C., Gergely, J., *J. Biol. Chem.* (1966) **241**, 5772.
61. Huszar, G., Elzinga, M., *Nature* (1969) **223**, 834.
62. Kuehl, W. M., Adelstein, R. S., *Biochem. Biophys. Res. Commun.* (1969) **37**, 59.
63. Yamaguchi, M., Sekine, T., *Biochem. J.* (1966) **39**, 24.
64. Kimura, M., Kielley, W. W., *Biochem. Z.* (1966) **345**, 188.
65. Kielley, W. W., Bradley, L. B., *J. Biol. Chem.* (1956) **218**, 653.
66. Tonomura, Y., "Muscle Proteins, Contraction and Cation Transport," University Park Press, Baltimore, 1973.

67. Riseman, J., Kirkwood, J., *J. Amer. Chem. Soc.* (1948) **70,** 2820.
68. Sartorelli, L., Fromm, H. J., Benson, R. W., Boyer, P. D., *Biochemistry* (1966) **5,** 2877.
69. Taylor, E. W., *Ann. Rev. Biochem.* (1972) **41,** 577.
70. Taylor, E. W., *Curr. Top. Bioenerg.* (1973) **5,** 201.
71. Perrie, W. T., Smillie, L. B., Perry, S. V., *Biochem. J.* (1973) **135,** 151.
72. Pires, E., Perry, S. V., Thomas, M. A. W., *FEBS Lett.* (1974) **41,** 292.
73. Lehman, W., Kendrick-Jones, J., Szent-Györgyi, A. G., *Cold Spring Harbor Symp. Quant. Biol.* (1972) **37,** 319.
74. Huxley, H. E., *J. Mol. Biol.* (1963) **7,** 281.
75. Szent-Györgyi, A. G., Cohen, C., Philpott, D. E., *J. Mol. Biol.* (1960) **2,** 133.
76. Nakamura, A., Sreter, F. A., Gergely, J., *J. Cell. Biol.* (1971) **49,** 883.
77. Cohen, C., Lowey, S., Harrison, K. G., Kendrick-Jones, J., Szent-Györgyi, A. G., *J. Mol. Biol.* (1970) **47,** 605.
78. Harrington, W. F., Burke, M., *Biochemistry* (1972) **11,** 1448.
79. Pepe, F. A., *Progr. Biophys. Mol. Biol.* (1971) **22,** 17.
80. Pepe, F. A., Drucker, B., *J. Cell. Biol.* (1972) **52,** 255.
81. Straub, F. B., "Studies Inst. Med. Chemistry," Vol. II, p. 3, University of Szeged, 1942.
82. Elzinga, M., Collins, J. H., Kuehl, W. M., Adelstein, R. S., *Proc. Natl. Acad. Sci. USA* (1973) **70,** 2687.
83. Hama, H., Maruyama, K., Noda, H., *Biochim. Biophys. Acta* (1965) **102,** 149.
84. Martonosi, A., Gouvea, M., Gergely, J., *J. Biol. Chem.* (1960) **235,** 1707.
85. Straub, F. B., Feuer, G., *Biochim. Biophys. Acta* (1950) **4,** 455.
86. Laki, K., Bowen, W. J., Clark, A., *J. Gen. Physiol.* (1950) **33,** 437.
87. Maruyama, K., Gergely, J., *Biochem. Biophys. Res. Commun.* (1961) **6,** 245.
88. Kasai, M., *Biochim. Biophys. Acta* (1969) **172,** 171.
89. Barany, M., Finkelman, F., Therattil-Anthony, T., *Arch. Biochem. Biophys.* (1962) **98,** 28.
90. Tonomura, Y., Oosawa, F., *Ann. Rev. Biophys. Bioeng.* (1972) **1,** 159.
91. Mommaerts, W. F. H. M., *J. Biol. Chem.* (1952) **198,** 469.
92. Asakura, S., *Biochim. Biophys. Acta* (1961) **52,** 65.
93. Asakura, S., Taniguchi, M., Oosawa, F., *Biochim. Biophys. Acta* (1963) **74,** 140.
94. Kakol, I., Weber, H. H., *Z. Naturforsch.* (1965) **20,** 977.
95. Moos, C., Estes, J. E., Eisenberg, G., *Biochem. Biophys. Res. Commun.* (1966) **23,** 347.
96. Kitagawa, S., Drabikowski, W., Gergely, J., *J. Gen. Physiol.* (1967) **50,** 1083.
97. Kitagawa, S., Drabikowski, W., Gergely, J., *Arch. Biochem. Biophys.* (1968) **125,** 706.
98. Pragay, D., *Naturwissenschaften* (1957) **44,** 397.
99. Grubhofer, N., Weber, H. H., *Z Naturforsch.* (1961) **16,** 435.
100. Hayashi, T., Rosenbluth, R., *Biochem. Biophys. Res. Commun.* (1962) **8,** 20.
101. Nagy, B., Jencks, W. P., *Biochemistry* (1962) **1,** 987.
102. Lehrer, S. S., Kerwar, G., *Biochemistry* (1972) **11,** 1211.
103. Cooke, R., *J. Supramol. Structure* (1975) **3,** 146.
104. Kuschinsky, G., Turba, F., *Biochim. Biophys. Acta* (1951) **6,** 426.
105. Drabikowski, W., Gergely, J., *J. Biol. Chem.* (1963) **238,** 640.
106. Bailin, G., Barany, M., *Biochim. Biophys. Acta* (1967) **140,** 208.
107. Elzinga, M., Collins, J. H., *Cold Spring Harbor Symp. Quant. Biol.* (1972) **37,** 1.
108. Lusty, C. J., Fasold, H., *Biochemistry* (1969) **8,** 2933.
109. Lehrer, S. S., Nagy, B., Gergely, J., *Arch. Biochem. Biophys.* (1972) **150,** 164.
110. Stone, D. B., Prevost, S. C., Botts, J., *Biochemistry* (1970) **9,** 3937.
111. Burley, R., Seidel, J., Gergely, J., *Arch. Biochem. Biophys.* (1971) **146,** 597.
112. Burley, R., Seidel, J., Gergely, J., *Arch. Biochem. Biophys.* (1972) **150,** 792.
113. Pollard, T. D., Weihing, R. R., *Critical Rev. Biochem.* (1974) **2,** 1.

114. Huszar, G., Elzinga, M., *Biochemistry* (1971) **10,** 229.
115. Hanson, J., Lowy, J., *J. Molec. Biol.* (1963) **6,** 46.
116. Drabikowski, W., Gergely, J., *J. Biol. Chem.* (1962) **237,** 3412.
117. Martonosi, A., *J. Biol. Chem.* (1962) **237,** 2795.
118. Bailey, K., *Biochem. J.* (1948) **43,** 271.
119. Woods, F., *J. Biol. Chem.* (1967) **242,** 2859.
120. Cohen, C., Szent-Györgyi, A. G., *J. Amer. Chem. Soc.* (1957) **79,** 248.
121. Cohen, C., Holmes, K. C., *J. Mol. Biol.* (1963) **6,** 423.
122. Hodges, R. S., Sodak, J., Smillie, L. B., Jurasek, L., *Cold Spring Harbor Symp. Quant. Biol.* (1972) **37,** 299.
123. Yamaguchi, M., Greaser, M. L., Cassens, R. G. *Ultrastruct. Res.* (1974) **33,** 48.
124. Cummins, P., Perry, S. V., *Biochem. J.* (1973) **133,** 765.
125. Caspar, D. L. D., Cohen, C., Longley, W., *J. Mol. Biol.* (1971) **41,** 87.
126. Cohen, C., Caspar, D. L. D., Parry, D. A. D., Lucas, R. M., *Cold Springs Harbor Symp. Quant. Biol.* (1971) **36,** 205.
127. Ebashi, S., Ebashi, F., *J. Biochem. (Tokyo)* (1964) **55,** 604.
128. Ebashi, S., Kodama, A., *J. Biochem (Tokyo)* (1965) **58,** 107.
129. Ebashi, S., Ohtsuki, I., Mihashi, K., *Cold Spring Harbor Symp. Quant. Biol.* (1972) **37,** 215.
130. Katz, A. M., *J. Biol. Chem.* (1964) **239,** 3304.
131. Eaton, B. L., Kominz, D. R., Eisenberg, E., *Biochemistry* (1975) **14,** 2718.
132. Hartshorne, D. J., Dreizen, P., *Cold Spring Harbor Symp. Quant. Biol.* (1972) **37,** 225.
133. Hartshorne, D. J., Mueller, H., *Biochem. Biophys. Res. Commun.* (1968) **31,** 647.
134. Hartshorne, D. J., Theiner, M., Mueller, H., *Biochem. Biophys. Acta* (1969) **175,** 301.
135. Schaub, M. C., Perry, S. V., *Biochem. J.* (1969) **115,** 993.
136. Greaser, M., Gergely, J., *J. Biol. Chem.* (1971) **246,** 4226.
137. Hitchcock, S. E., Huxley, H. E., Szent-Györgyi, A. G., *J. Mol. Biol.* (1973) **80,** 825.
138. Greaser, M. L., Yamaguchi, M., Brekke, C., Potter, J., Gergely, J. *Cold Spring Harbor Symp. Quant. Biol.* (1972) **37,** 235.
139. Greaser, M., Gergely, J., *J. Biol. Chem.* (1973) **248,** 2125.
140. Wilkinson, J. M., Grand, R. J. A., *Proteins of Contractile Systems, Fed. Eur. Biochem. Soc., Meet., 9th, 1974,* **31,** 137.
141. Potter, J. D., *Arch. Biochem. Biophys.* (1974) **162,** 436.
142. Perry, S. V., Cole, H. A., Head, J. F., Wilson, F. J., *Cold Spring Harbor Symp. Quant. Biol.* (1972) **37,** 251.
143. van Eerd, J. P., Kawasaki, Y., "Calcium Binding Proteins," p. 153, W. Drabikowski, H. Strzelecka-Golaszewska, E. Carafoli, Eds., Elsevier, Amsterdam, 1974.
144. Winter, M. R. C., Head, J. F., Perry, S. V., "Calcium Binding Proteins," p. 109, W. Drabikowski, H. Strzelecka-Golaszewska, E. Carafoli, Eds., Elsevier, Amsterdam, 1974.
145. Drabikowski, W., Barylko, B., Dabrowska, R., Nowak, E., Szpacenko, A., "Calcium Binding Proteins," p. 69, W. Drabikowski, H. Strzelecka,Golaszewska, E. Carafoli, Eds., Elsevier, Amsterdam, 1974.
146. Hamoir, G., *Biochem. Soc. Symp.* (1968) **6,** 8.
147. Pechere, J. F., *C. R. Acad. Sci.* (1974) **278,** 2577.
148. Lehky, Pl, Blum, H. E., Stein, E. A., Fischer, E. H., *J. Biol. Chem.* (1974) **249,** 4332.
149. Kretsinger, R. H., *Nature (New Biol.)* (1972) **85,** 240.
150. Kretsinger, R. H., Nockolds, C. E., *J. Biol. Chem.* (1973) **248,** 3313.
151. Collins, J. H., Potter, J. D., Horn, M. J., Wilshire, G., Jackman, N., *FEBS Lett.* (1973) **36,** 268.
152. Collins, J. H., *Biochem. Biophys. Res. Commun* (1974) **58,** 301.

153. Weeds, A. G., McLachlan, A. D., *Nature* (1974) **252,** 646.
154. Margossian, S. S., Cohen, C., *J. Mol. Biol.* (1973) **81,** 409.
155. Wilkinson, J. M., Perry, S. V., Cole, H. A., Trayer, I. P., *Biochem. J.* (1972) **127,** 215.
156. Moore, P. B., Huxley, H. E., DeRosier, D. J., *J. Mol. Biol.* (1970) **50,** 279.
157. Hanson, J., *Quart. Rev. Biophys.* (1968) **1,** 177.
158. Cohen, C., Caspar, D. L. D., Johnson, J. P., Nauss, K., Margossian, S. S., Parry, D. A. D., *Cold Spring Harbor Symp. Quant. Biol.* (1972) **37,** 287.
159. Ebashi, S., Ebashi, F., *J. Biochem. (Tokyo)* (1965) **7,** 58.
160. Maruyama, K., *Biochim. Biophys. Acta* (1965) **102,** 542.
161. Seraydarian, K., Briskey, E. J., Mommaerts, W. F. H. M., *Biochim. Biophys. Acta* (1967) **133,** 399.
162. Briskey, E. J., Seraydarian, K., Mommaerts, W. F. H. M., *Biochim. Biophys. Acta* (1967) **133,** 424.
163. Briskey, E. J., Seraydarian, K., Mommaerts, W. F. H. M., *Biochim. Biophys. Acta* (1967) **133,** 412.
164. Goll, D., Mommaerts, W. F. H. M., Reedy, M. K., Seraydarian, K., *Biochim. Biophys. Acta* (1969) **175,** 174.
165. Maruyama, K., Ebashi, S., *J. Biochem. (Tokyo)* (1965) **13,** 58.
166. Maruyama, K., *J. Biochem.* (1971) **69,** 369.
167. Kawamura, M., Maruyama, K., *Biochim. Biophys. Acta* (1972) **267,** 422.
168. Maruyama, K., Abe, S., Ishii, T., *J. Biochem.* (1975) **77,** 131.
169. Pepe, F. A., *J. Histochem. Cytochem.* (1975) **23,** 543.
170. Etlinger, J. D., Zak, R., Fischman, D. A., *J. Cell. Biol.* (1976) **68,** 123.
171. Morimoto, K., Harrington, W. F., *J. Biol. Chem.* (1972) **247,** 3052.
172. Turner, D. C., Wallimann, T., Eppenberger, H. M., *Proc. Natl. Acad. Sci. USA* (1973) **70,** 702.
173. Masaki, T., Takaiti, O., *Biochem. J. (Tokyo)* (1972) **71,** 355.
174. Trinick, J., Lowey, S., *Biophys. J.* (1976) **16,** 199a.
175. Offer, G., Moos, C., Starr, R., *J. Mol. Biol.* (1973) **74,** 653.
176. Weber, A., *Biochim. Biophys. Acta* (1956) **19,** 345.
177. Gergely, J., *J. Biol. Chem.* (1956) **220,** 917.
178. Nihei, T., Tonomura, Y., *J. Biochem.* (1959) **46,** 305.
179. Warren, J. C., Stowring, L., Morales, M. F., *J. Biol. Chem.* (1966) **241,** 309.
180. Muhlrad, A., Fabian, F., Biro, N. A., *Biochim. Biophys. Acta* (1964) **89,** 186.
181. Offer, G. W., *Biochim. Biophys. Acta* (1964) **89,** 566.
182. Seidel, J. C., *Biochim. Biophys. Acta* (1969) **180,** 216.
183. Sekine, T., Kielley, W. W., *Biochim. Biophys. Acta* (1964) **81,** 336.
184. Reisler, E., Burke, M., Himmelfarb, S., Harrington, W. F., *Biochemistry* (1974) **13,** 3837.
185. Seidel, J. C., Gergely, J., *Cold Spring Harbor Symp. Quant. Biol.* (1972) **37,** 187.
186. Wagner, P. D., Yount, R. G., *Biochemistry* (1975) **14,** 5156.
187. Ohnishi, H., Ohtsuka, E., Ikehara, M., Tonomura, Y., *J. Biochem.* (1973) **74,** 435.
188. Haugland, R. P., *J. Supramol. Struct.* (1975) **4,** 338.
189. Muhlrad, A., Lamed, R., Oplatka, A., *J. Biol. Chem.* (1975) **175,** 250.
190. Blum, J. J., *Arch. Biochem. Biophys.* (1955) **55,** 486.
191. Tonomura, Y., Inoue, A., *J. Mol. Cell. Biochem.* (1974) **5,** 127.
192. Boyer, P. D., *Curr. Top. Bioenerg.* (1967) **2,** 49.
193. Weber, A., Hasselbach, W., *Biochim. Biophys. Acta* (1954) **15,** 237.
194. Tokiwa, T., Tonomura, Y., *J. Biochem.* (1965) **57,** 616.
195. Lymn, R. W., Taylor, E. W., *Biochemistry* (1970) **9,** 2975.
196. Seidel, J. C., Gergely, J., *Biochem. Biophys. Res. Commun.* (1971) **44,** 826.
197. Cheung, H. C., *Biochim. Biophys. Acta* (1969) **194,** 478.
198. Morita, F., *J. Biol. Chem.* (1967) **242,** 4501.
199. Werber, M. J. W., Szent-Györgyi, A. G., Fasman, G. D., *Biochemistry* (1972) **11,** 2872.

200. Trentham, D. R., Bardsley, R. G., Eccleston, J. F., Weeds, A. G., *Biochem. J.* (1972) **126**, 635.
201. Koretz, J. F., Hunt, T., Taylor, E. W., *Cold Spring Harbor Symp. Quant. Biol.* (1972) **37**, 179.
202. Bagshaw, C. R., Trentham, D. R., *Biochem. J.* (1974) **141**, 331.
203. Bagshaw, C. R., Eccleston, J. F., Eckstein, F., Goody, R. S., Gutfreund, H., Trentham, D. R., *Biochem. J.* (1974) **141**, 351.
204. Malik, M. N., Martonosi, A., *Arch. Biochem. Biophys.* (1972) **152**, 243.
205. Arata, T., Inoue, A., Tonomura, Y., *Biochem. J.* (1974) **133**, 323.
206. Bagshaw, C. R., Trentham, D. R., *Biochem. J.* (1973) **133**, 323.
207. Morales, M. F., Botts, J. (1956), "Currents in Biochemical Research," D. E. Green, Ed., p. 609, Interscience, New York.
208. Mannherz, H. G., Schenck, H., Goody, R. S., *Eur. J. Biochem.* (1974) **48**, 287.
209. Wolcott, R. G., Boyer, P. D., *Biochem. Biophys. Res. Commun.* (1974) **57**, 709.
210. Boyer, P. D., *FEBS Lett.* (1975) **50**, 91.
211. Szent-Györgyi, A., *Biol. Bull.* (1949) **96**, 140.
212. Huxley, H. E., *Science* (1969) **164**, 1356.
213. Lymn, R. W., Taylor, E. W., *Biochemistry* (1971) **10**, 4617.
214. Eisenberg, E., Dobkin, L., Kielley, W. W., *Proc. Natl. Acad. Sci. USA* (1972) **69**, 667.
215. Mulhern, S., Eisenberg, E., Kielley, W. W., *Biochemistry* (1975) **14**, 3863.
216. Weber, A., Murray, J. M., *Physiol. Rev.* (1973) **53**, 612.
217. Ashley, C. C., Ridgeway, E. B., *J. Physiol.* (1970) **105**, 209.
218. Tonomura, Y., Watanabe, S., Morales, M., *Biochemistry* (1969) **8**, 2171.
219. Ebashi, S., Ohnishi, S., Abe, S., Maruyama, K., *J. Biochem. (Tokyo)* (1974) **75**, 211.
220. Potter, J. D., Seidel, J. C., Leavis, P., Lehrer, S. S., Gergely, J., "Calcium Binding Proteins," W. Drabikowski, H. Strzelecka-Golaszewska, E. Carafoli, Eds., p. 129, Elsevier, New York, 1974.
221. van Eerd, F. P., Kawasaki, Y., *Biochemistry* (1973) **12**, 4972.
222. Mani, R. S., McCubbin, W. D., Kay, C. M., *FEBS Lett.* (1975) **52**, 127.
223. Bremel, R. D., Murray, J. M., Weber, A., *Cold Spring Harbor Symp. Quant. Biol.* (1972) **37**, 267.
224. Spudich, J. A., Huxley, H. E., Finch, J. T., *J. Mol. Biol.* (1972) **72**, 619.
225. Haselgrove, J. C., *Cold Spring Harbor Symp. Quant. Biol.* (1972) **37**, 341.
226. Parry, D. A. D., Squire, M. M., *J. Mol. Biol.* (1973) **81**, 409.
227. Potter, J. D., Gergely, J., *Biochemistry* (1974) **13**, 2697.
228. Hitchcock, S. E., *Eur. J. Biochem.* (1975) **52**. 255.
229. Bremel, R. D., *Nature* (1974) **252**, 405.
230. Reedy, M. K., Holmes, K. C., Tregear, R. T., *Nature* (1965) **207**, 1276.
231. Steiger, G. J., Ruegg, J. C., Boldy, K. M., Lubbers, D. W., Breull, W., *Cold Spring Harbor Symp. Quant. Biol.* (1972) **37**, 377.
232. Nihei, T., Mendelson, R., Botts, J., *Proc. Natl. Acad. Sci. USA* (1974) **71**, 274.
233. Remedios, D. C. dos, Millikan, R. G. C., Morales, M. F., *J. Gen. Physiol.* (1972) **59**, 103.
234. Marston, J. B., Tregear, R. T., *Nature (New Biol.)* (1972) **23**, 235.
235. Mendelson, R. A., Morales, M. F., Botts, J., *Biochemistry* (1973) **12**, 2250.
236. Thomas, D. D., Seidel, J. C., Hyde, J. S., Gergely, J., *Proc. Natl. Acad. Sci. USA* (1975) **72**, 1729.
237. Burke, M., Himmelfarb, S., Harrington, W. F., *Biochemistry* (1973) **12**, 701.
238. Potter, J. D., Gergely, J., *J. Biol. Chem.* (1975).
239. Ishiwata, S., Fujime, S., *J. Mol. Biol.* (1972) **68**, 511.
240. Huxley, A. F., Simmons, R. M., *Nature* (1971) **233**, 533.
241. Huxley, A. F., Simmons, R. M., *Cold Spring Harbor Symp. Quant. Biol.* (1972) **37**, 669.

242. Podolsky, F. J., Nolan, C., *Cold Spring Harbor Symp. Quant. Biol.* (1972) **37**, 661.
243. Julian, F. J., Sollins, K. R., Sollins, M. R., *Cold Spring Harbor Symp. Quant. Biol.* (1972) **37**, 685.
244. Burgen, A. S. V., Roberts, G. C. K., Feeney, J., *Nature* (1975) **253**, 753.
245. Hill, A. V., *Biochim. Biophys. Acta* (1950) **4**, 4.
246. Cain, D. F., Davies, D. R., *Biochem. Biophys. Res. Commun.* (1962) **8**, 361.
247. Hill, A. V., *Biochem. Z.* (1966) **1**, 345.
248. Hill, A. V., *Proc. Roy. Soc. Ser. B.* (1964) **159**, 297.
249. Hill, A. V., *Proc. Roy. Soc. Ser. B.* (1964) **159**, 319.
250. Hill, A. V., *Proc. Roy. Soc. Ser. B.* (1964) **159**, 589.
251. Hill, A. V., *Proc. Roy. Soc. Ser. B.* (1964) **159**, 596.
252. Gilbert, C., Kretzschmar, K. M., Wilkie, D. R., *Cold Spring Harbor Symp. Quant. Biol.* (1972) **37**, 613.
253. Woledge, R. C., *Progr. Biophys. Molec. Biol.* (1971) **22**, 37.
254. Woledge, R. C., *Cold Spring Harbor Symp. Quant. Biol.* (1972) **37**, 629.
255. Curtin, N. A., Gilbert, C., Kretzschmar, K. M., Wilkie, D. R., *J. Physiol.* (1974) **238**, 455.
256. Homsher, E., Rall, J. A., Wallner, A., Ricchiuti, N. V., *J. Gen. Physiol.* (1975) **1**, 65.
257. Weber, H. H., "The Motility of Muscle and Cells," Harvard University Press, Cambridge, Mass., 1958.
258. Cross, R. L., Boyer, P. D., *Biochemistry* (1965) **14**, 392.
259. Hill, T. L., *Prog. Biophys. Mol. Biol.* (1974) **28**, 267.
260. Hill, T. L., *Prog. Biophys. Mol. Biol.* (1975) **29**, 105.
261. Hill, T. L., Eisenberg, E., Chen, Y. D., Podolsky, R. J., *Biophys. J.* (1975) **15**, 335.
262. Kuhn, H. J., *Experientia* (1973) **29**, 1086.
263. White, D. S. C., *J. Physiol.* (1970) **208**, 583.
264. Mannherz, H. G., Barrington-Leigh, J., Holmes, K. C., Rosenbaum, G., *Nature (New Biol.)* (1973 **241**, 226.
265. Goody, R. S., Holmes, K. Cl, Mannherz, H. G., Barrington-Leigh, J., Rosenbaum, G., *Biophys. J.* (1975) **15**, 687.

Index

A

A band	224
Absorption and emission of light in photosynthesis	181
Absorption spectrum of chlorophyll a	181
Acetaldehyde by NAD^+, oxidation of	10
Acid, lactic	222
Actin	232
F-	232
G-	232
activation of myosin ATPase by	244
interaction of myosin with	242
—myosin ATPase	249
ATPase interactions with	237
interaction, Ca^{2+}-dependent regulation of	234
interaction, regulation of	245
—tropomyosin—troponin system, intact	245
α-Actinin	237
β-Actinin	237
Action spectra	194
Activation of myosin ATPase by actin	244
Active transport	23, 55
and corresponding osmotic flow	89
systems, studies on	93
Activity, high ATPase	163
Actomyosin	223
Adenine nucleotide transport	160
Adenylate cyclase system	57
ADP/ATP exchange	100
Aerobic photosynthesis	173
Affinity	7, 26
Aldosterone	28
Amino acid composition and sequence of membranes	45
Ammonia	205
Analyses, method for rapid kinetic	159
1-Anilino-8-naphthalene sulfonate (ANS)	68
Animal and microbial membranes, lipid composition of some	44
Anion	
pump model	157
transport	157
transporters, mitochondrial	160
(ANS), 1-anilino-8-naphthalene sulfonate	68
Apolar membrane—protein associations	49
Apparatus, light saturation of the photosynthetic	214
Aqueous intra- and extracellular compartments	50
Arsenate	150
(*Athiorhodaceae*), purple non-sulfur bacteria	196
ATP	202
/2e ratio	203
enzyme-bound	242
exchange, ADP/	100
hydrolysis	10
intermediate steps in the hydrolysis of	240
and muscle contraction	222
synthetic reaction mediated by	9
ATPase	10
by actin, activation of myosin	244
actin—myosin	249
activity, high	163
inhibition of K^+-	239
interactions with actin, myosin	237
isolation of the Na/K—	97
Na^+–K^+ sensitive	64
proton-translocating	145
purification of Ca—	110
reaction sequence for Ca—	114
of the sarcoplasmic reticulum, Ca—	108
sites, ouabain-sensitive	45
Aurovertin	142

B

Bacteria	
(*Athiorhodaceae*), purple non-sulfur	196
(*Chlorobacteriaceae*), green sulfur	196
(*Thiorhodaceae*), purple sulfur	196
Bacterial photosystem	196
Bacteriochlorophyll a	196
Balance, energy	252
Band	
A	224
I	224
M	224

Behavior of muscles, mechanical	32
Benson cycle, Calvin–	209
Benzoquinone	187
Bilayer, phosphatide	47
Bile pigments	180
Binding	
calcium	235
to the enzyme, ion	113
site of myosin, metal–	239
substrate	99
Biochemistry of transport	90
Biomembranes	38
protein, phospholipid and cholesterol content of some	44
Biothermodynamics	1
Bonding requirements	98
Bound ATP, enzyme–	242
Boyer's conformational coupling hypothesis	148
Bulk flow	15

C

Calcium	
–ATPase, purification of	110
–ATPase, reaction sequence for	114
–ATPase of the sarcoplasmic reticulum	108
binding	235
-dependent regulation of actin–myosin interaction	234
pump	108
reconstitution of the	112
transport	154
mechanism of	115
and muscle relaxation	108
Cardioactive steroids and the Na/K pump, interaction between	97
Calvin–Benson cycle	209
Capture, early events of energy	179
Carbohydrates in membranes	51
Carbon compounds, production of reduced	209
Cardiac muscle	229
Carotenoids	180
Carrier catalyzed transport, criteria for	158
Cation	
interactions, monovalent	103
transport	
monovalent	155
multivalent	153
system	56
Cells, motile	58
Cellular membrane systems	89
Cellular recognition	57
Centers	
identification of membrane associated redox	125
iron–sulfur	129
mitochondrial redox	131
photochemical conversion	174
sequence of redox	129
Chain, organization of the respiratory	134
Chain, respiratory	125

Changes in membrane protein conformation, functionally related	63
Changes in membrane protein molecules, conformational	248
Chemical	
hypothesis of electron transfer	203
reactions	7
in oxidative phosphorylation	123
transport phenomena and the influence of	11
Chemiosmotic hypothesis of electron transfer	203
Chemomechanical coupling	253
Chlorophylls	179
Chlorophyll a, absorption spectrum of	181
Chloroplast	177
Cholesterol content of some biomembranes, protein, phospholipid and	44
Chromatic rate transients	193
Citrate, translocation of	53
Classification of membrane proteins	46
Compartmentalization and organization of functional macromolecules	53
Compartments, aqueous intra- and extracellular	50
Competition between NADH oxidation and succinate oxidation	138
Complete coupling	24
Complex, enzyme–substrate	83
Complex, Golgi	38
Composition, general features of membrane	40
Compounds, production of reduced carbon	209
Conductance, tissue	26
Conferring protein (OSCP), oligomycin sensitivity	164
Conformation, functionally related changes in membrane protein	63
Conformational	
changes in the protein molecules	248
coupling hypothesis, Boyer's	148
hypothesis of electron transfer	203
Contraction	
ATP and muscle	222
mechanism of muscle	221
molecular mechanisms and models of	247
Conversion centers, photochemical	174
Conversion efficiency, energy	212
Cooperative lattice model	73
Counter transport	84
Coupling	
chemomechanical	253
complete	24
factors, properties of specific	162
hypothesis, Boyer's conformational	148
between reactions and flows	12
stationary state	13
and stoichiometry	9
C_3 plants	210
C_4 plants	210
C protein	237

Criteria for carrier-catalyzed transport	158
Crossbridges	225
movement of	247
Curie–Prigogine principle	13
Cyclase system, adenylate	57
Cycle, Calvin–Benson	209
Cycle, pentose	209
Cyclic phosphorylation	202
Cytochrome	
b_{559}	189
b_{563}	185
c oxidase	133
f	185
oxidase, a-hemes of	134
Cytochromes, hemes of	129

D

DAD	188
DCMU	190
DCPIP	187
Dehydrogenase, NADH–	134
2-Deoxy-D-glucose	28
Dephosphorylation, enzyme	100
Dephosphorylation, K-dependent	101
Diffraction studies, x-ray	59
Diffusion	
facilitated	83
Fick's law of	79
models for facilitated	84
simple	79
single-file	15
Discs, Z	224
Dissipation function	11
Dissipation by irreversible processes	2
Distance, interfilamental	248
Distribution of light energy	192
Distribution, quantum	194
Donnan equilibrium, Gibbs–	81
Double beam spectrophotometer	125
Double hits	201
Drag, solvent	80

E

Early events of energy capture	179
EDTA	238
Effectiveness of energy utilization	22
Efficiency, energy conversion	212
Electrical properties of membranes, permeability and	54
Electrochemical gradient	85
Electrogenic pumps	86
Electrokinetics	12
Electron	
flow, in-membrane-	135
flow, reversed	150
paramagnetic resonances to study mitochondrial redox centers	127
spin resonances (ESR)	67, 93
transfer	
chemical hypothesis of	203
chemiosmotic hypothesis of	203
conformational hypothesis	203

Electron (*Continued*)	
transport	
inhibitors of	132
light-driven	176
and oxidative phosphorylation, scheme for	126
Electrons, transfer of	192
Electro-osmotic work	24
Electrophoresis, gel	98
Electrophoresis, SDS-polyacrylamide gel	230
Emission of light in photosynthesis, absorption and	181
Endoplasmic reticulum	38
Energy	
balance	252
capture, early events of	179
conversion	19
efficiency	212
distribution of light	192
expenditure without performance of work	21
linked reactions	149
transduction	252
inhibitors of	141
mechanochemical	252
sites	137
transfer	185
inhibitors	142
utilization, effectiveness of	22
Enhancement	193
Entropy	
Enzyme	
-bound ATP	242
dephosphorylation	100
ion binding to the	113
phosphorylation	100
–substrate complex	83
Epithelia, sodium transport in	27
Epithelial membranes	26
Equation	
Gibbs	3, 4
Hill	32
Michaelis–Menten	83
Equations, phenomenological	6, 10
Equilibrium	8
Gibbs–Donnan	81
(ESR), electron spin resonance	93
N-Ethylmaleimide (NEM)	91
Evolution in photosynthesis, oxygen	199
Evolution, reaction steps involved in O_2	201
Exchange, ADP/ATP	100
Exchange reactions	149
Extracellular compartments, aqueous intra- and	50
Extrusion, rate of Na	94

F

Facilitated diffusion	83
models for	84
F-actin	232

Factors, properties of specific coupling	162
Ferredoxin	185
Ferricyanide	187
Fick's law of diffusion	79
Filament formation, mechanism of	231
Filaments	
relation of molecular structure of myosin to thick	231
sliding	226
thick	224
Flexibility of myosin, segmental	248
Flow	7
active transport and corresponding osmotic	89
in-membrane-electron	135
level	22
reversed electron	150
water	80
Flows, coupling between reactions and	12
Flows and forces, relationships between	6
Fluorescence	181
Flux ratio and isotope interaction	16
Fluxes, unidirectional	88
FMN	187
Forces, relationships between flows and	6
Formation, mechanisms of filament	231
Free energy, Gibbs	8
Freeze-fracturing of membranes	40
Fructose	9
Functionally related changes in membrane protein conformation	63
Functional macromolecules, compartmentalization and organization of	53
Functions, some major membrane	53

G

G-actin	232
Gel electrophoresis	98
SDS-polyacrylamide	230
General aspects of membrane morphology	39
Ghosts	88
Gibbs	
–Donnan equilibrium	81
equation	3, 4
free energy	8
Globular protein mosaic models (LGPM), lipid–	71
Globular proteins, high-resolution x-ray analyses of	39
Glucose	9
Glycerophosphatides	41
Glycoproteins, major membrane	52
Glycoproteins from mitochondria	154
Golgi complex	38
Gradient, electrochemical	85
Gradient, pH	206
Gramicidin	205
Green sulfur bacteria (*Chlorobacteriaceae*)	196

H

Hatch–Slack–Kortschak pathway	210
Heavy meromyosin	228
H_2 photoevolution	210
Hemes of cytochromes	129
a-Hemes of cytochrome oxidase	134
High ATPase activity	163
High-resolution x-ray analyses of globular proteins	39
Hill equation	32
Homeostasis	123
Hydrolysis of ATP, intermediate steps in the	240
Hydrophobic membrane interior	116
Hypothesis	
Boyer's conformational coupling	148
of electron transfer	
chemical	203
chemiosmotic	203
conformational	203
H zone	224

I

I band	224
Identification of membrane associated redox centers	125
Immune recognition	57
Information translation	57
Ingen–Housz, Jan	173
Inhibition of K^+–ATPase	239
Inhibition, ouabain	102
Inhibitor, phosphorylation	205
Inhibitors	
of electron transport	132
of energy transduction	141
energy transfer	142
metabolite transport	142
In-membrane-electron flow	135
Inner mitochondrial membrane in oxidative phosphorylation, role of the	147
Inogen	222
Intact actin–myosin–tropomyosin–troponin system	245
Interaction	
with actin, myosin ATPase and	237
Ca^{2+}-dependent regulation of actin–myosin	234
monovalent cation	103
of myosin with actin	242
between the photosystems	192
regulation of actin–myosin	245
Interfilamental distance	248
Interior, hydrophobic membrane	116
Intermediary steps in oxidative phosphorylation	143
Intermediate steps in the hydrolysis of ATP	240
Intra- and extracellular compartments, aqueous	50
Ion binding to the enzyme	113
Ionophore	205
-bearing lattices	56
versatile	56

INDEX 267

Ion transport	153
IR spectroscopy	62
Iron–sulfur centers	129
Isotope interaction	17, 19
flux ratio and	16
Isotope kinetics	15

K

K^+	
—ATPase, inhibition of	239
—ATPase, isolation of the Na/	97
—dependent dephosphorylation	101
—phosphatase activity	101
pump, interaction between cardioactive steroids and the Na/	97
sensitive ATPase, Na^+–	64
Kinetic(s)	
analyses, method for rapid	159
isotope	15
Michaelis–Menton	8
Kortschak pathway, Hatch–Slack–	210

L

Labels, spin	67
Labilization-stabilization NMR	66
Lactic acid	222
Lattice model, cooperative	73
Lattices, ionophore-bearing	56
Law of diffusion, Fick's	79
Level of flow	22
(LGPM), lipid–globular protein mosaic models	71
Light	
-driven electron transport	176
energy, distribution of	192
meromyosin	228
in photosynthesis, absorption and emission of	181
saturation of the photosynthetic apparatus	214
Linearity	7
Linked reactions, energy	149
Lipid	
composition of some animal and microbial membranes	44
—globular protein mosaic models (LGPM)	71
matrix, influence of the	98
Lipoprotein, N-terminal portions of murein	50
Low-affinity substrate sites	108
Luminescence	181

M

M band	224
M protein	237
Macromolecules, compartmentalization and organization of functional	53
Major membrane glycoproteins	52
Manganese	199
Matrix, influence of the lipid	98
Matrix space	124
Maximum turnover rate	214
Measurements, redox–potential	130
Mechanical behavior of muscles	32
Mechanism	
of Ca, transport	115
of filament formation	231
of muscle contraction	221
sliding filament	224
Mechanisms	
and models of contraction, molecular	247
oscillating-pore	86
in photosynthesis	172
Mechanochemical energy transduction	252
Mehler reaction	201
Membrane	
-associated redox centers, identification of	125
characteristics	11
composition, general features of	40
-electron flow, in-	135
functions, some major	53
glycoproteins, major	52
interior, hydrophobic	116
morphology, general aspects of	39
in oxidative phosphorylation, role of the inner mitochondrial	147
potentials	82, 207
protein	
associations, apolar	49
conformation, functionally related changes in	63
classification of	46
structure, some physical approaches to	59
structure, spectroscopic techniques to elucidate	60
systems, cellular	89
transport	78
unit	39
Membranes	
amino acid composition and sequence of	45
carbohydrates in	51
epithelial	26
freeze-fracturing of	40
lipid composition of some animal and microbial	44
permeability and electrical properties of	54
semipermeable	80
Menten equation, Michaelis–	83
Meromyosin, heavy	228
Meromyosin, light	228
Metabolic poisons	87
Metabolite transport inhibitors	142
Method for rapid kinetic analyses	159
Methods for identifying transporter systems	158
Metal-binding site of myosin	239
Methylamine	204

Michaelis–Menten equation	83
Michaelis–Menten kinetics	8
Microbial membranes, lipid composition of some animal and	44
Midpoint redox potentials	138
Mitochondria	124
gylcoproteins from	154
Mitochondrial	
anion transporters	160
membrane in oxidative phosphorylation, role of the inner	147
redox centers	131
electron paramagnetic resonances to study	127
Model	
anion pump	157
cooperative lattice	73
proton pump	157
Shaw	96
transport	106
Models	
of contraction, molecular mechanisms and	247
for facilitated diffusion	84
(LGPM), lipid–globular protein mosaic	71
Moieties, sugar	52
Motile cells	58
Molecular mechanisms and models of contraction	247
Molecular structure of myosin to thick filaments, relation of	231
Molecules, conformational changes in protein	248
Monovalent cation interactions	103
Monovalent cation transport	155
Morphology, general aspects of membrane	39
Mosaic models (LGPM), lipid–globular protein	71
Movement of crossbridges	247
Multivalent cation transport	153
Murein lipoprotein, N-terminal portions of	50
Muscular contraction	32
Muscle	
cardiac	229
contraction, ATP and	222
contraction, mechanism of	221
relaxation, calcium transport and	108
structure of striated	227
ultrastructure of	224
Muscles, mechanical behavior of	32
Myelin	38
Myofibril-bound TnC	235
Myofibrillar proteins	228
Myosin	223, 228
with actin, interaction of	242
ATPase	
actin–	249
by actin, activation of	244
and interactions with actin	237
B	223

Myosin (Continued)	
interaction, Ca^{2+}-dependent regulation of actin–	234
interaction, regulation of actin–	245
metal-binding site of	239
segmental flexibility of	248
to thick filaments, relation of molecular structure of	231
–tropomyosin–troponin system, intact actin–	245

N

Na^+	
extrusion, rate of	94
/K-ATPase, isolation of the	97
/K pump, interaction between cardioactive steroids and the	97
/K^+ sensitive ATPase	64
net outward transport of	94
transport in epithelia	27
NAD^+, oxidation of acetaldehyde by	10
NADH-dehydrogenase	134
NADH oxidation and succinate oxidation, competition between	138
NADP	185
NET analysis	24
NET (nonequilibrium thermodynamics)	1
Nicotinamide nucleotide transhydrogenation	152
NMR	152
labilization–stabilization	66
Noncyclic photophosphorylation	202
Nonequilibrium thermodynamics (NET)	1
Nuclear pores	53
Nucleotide transhydrogenation, nicotinamide	152
Nucleotide transport, adenine	160

O

Oligomycin	103, 142
sensitivity conferring protein (OSCP)	164
Oligosaccharide structure	51
Onsager relations	11
Onsager symmetry	31
Optimal quantum yield	212
Organization of functional macromolecules, compartmentalization and	53
Organization of the respiratory chain	134
Oscillating-pore mechanisms	86
(OSCP), oligomycin sensitivity conferring protein	164
Osmotic flow, active transport and corresponding	89
Ouabain inhibition	102
Ouabain-sensitive ATPase sites	45
Oxalacetate	133
Oxidase, a-hemes of cytochrome	134
Oxidase, cytochrome c	133

Oxidation
  of acetaldehyde by NAD^+ 10
  competition between NADH
    oxidation and succinate 138
  substrate 30
Oxidative phosphorylation 10, 28, 137
  chemical reactions in 123
  intermediary steps in 143
  role of the inner mitochondrial
    membrane in 147
  scheme for electron transport and .. 126
Oxygen
  evolution in photosynthesis 199
  evolution, reaction steps involved in 201
  titrations 129

P

P_{700} 185
P_{890} 198
Packages, separate 194
Paramagnetic resonances to study
  mitochondrial redox centers,
  electron 127
Particles, submitochondrial 141
Pathway, Hatch–Slack–Kortschak 210
Pentose cycle 209
Permeability coefficients 15
Permeability and electrical properties
  of membranes 54
pH gradient 206
Phagocytosis 53
Phenomenological equations 6, 10
Phosphatase activity, K– 101
Phosphate potentials 138
Phosphatide bilayer 47
Phospholipid and cholesterol content
  of some biomembranes, protein, 44
Phosphorescence 181
Phosphorylation 64
  chemical reactions in oxidative 123
  cyclic 202
  enzyme 100
  inhibitor 205
  intermediary steps in oxidative 143
  oxidative 10, 28, 137
  role of the inner mitochondrial
    membrane in oxidative 147
  scheme for electron transport and
    oxidative 126
  substrate level 10
Photochemical conversion centers 174
Photoevolution, H_2 210
Photophosphorylation, noncyclic 202
Photorespiration 210
Photosystem 185
  bacterial 196
Photosystems, interactions between
  the 192
Photosynthesis
  absorption and emission of light in 181
  aerobic 173
  mechanisms in 172
Photosynthesis (*Continued*)
  oxygen evolution in 199
  rate-limiting step of 215
  Z scheme of 174
Photosynthetic
  apparatus, light saturation of the .. 214
  pigments, structure of the major 179
  structures 177
  unit 184
Physical approaches to membrane
  structure, some 59
Pigments, structure of the major
  photosynthetic 179
Pinocytosis 53
Plants, C_3 210
Plants, C_4 210
Plasma membrane permeability to
  protons, K^+, and Na^+ 54
Plastocyanin 185
Plastoquinone 189
Plots, substrate–velocity 116
PMS 187, 188
Poisons, metabolic 87
Polyacrylamide gel electrophoresis,
  SDS– 230
Polyvinylmethylpyridinium bromide .. 19
Pore mechanisms, oscillating- 86
Pores, nuclear 53
Potassium (*see* K^+)
Potential measurements, redox– 130
Potential, membrane 207
Potentials
  membrane 82
  midpoint redox 138
  phosphate 138
Prigogine principle, Curie– 13
Probes, spectroscopic 67
Process, sulfate transport 90
Production of reduced carbon
  compounds 209
Properties of specific coupling factors 162
Protein
  associations, apolar membrane– 49
  C 237
  conformation, functionally related
    changes in membrane 63
  M 237
  molecules, conformational changes
    in the 248
  mosaic models (LGPM), lipid–
    globular 71
  (OSCP), oligomycin sensitivity
    conferring 164
  phospholipid and cholesterol
    content of some biomembranes 44
Proteins
  classification of membranes 46
  high-resolution x-ray analyses of
    globular 39
  myofibrillar 228
  regulatory 233
Proton pump model 157
Proton-translocating ATPase 145
Pull phenomena, push and 193

Pump
  Ca 108
  interaction between cardioactive
    steroids and the Na/K 97
  model, anion 157
  model, proton 157
  reconstitution of the Ca 112
Pumps, electrogenic 86
Purification of Ca–ATPase 110
Purple non-sulfur bacteria
  (*Athiorhodaceae*) 196
Purple sulfur bacteria
  (*Thiorhodaceae*) 196
Push and pull phenomena 193

Q

Quantum
  distribution 194
  yield 192
    optimal 212
(Q) ubiquinone 133

R

Rapid kinetic analyses, method for .. 159
Rate
  -limiting step of photosynthesis 215
  maximum turnover 214
  transients, chromatic 193
Ratio, ATP/2e 203
Reaction
  Mehler 201
  sequence for Ca–ATPase 114
  steps involved in O_2 evolution 201
  velocity 7
Reactions
  and flows, coupling between 12
  chemical 7
  energy linked 149
  exchange 149
  mediated by ATP, synthetic 9
  in oxidative phosphorylation,
    chemical 123
  transport phenomena and the
    influence of chemical 11
Receptors, surface 38
Recognition, cellular 57
Recognition, immune 57
Reconstitution of the Ca pump 112
Redox
  centers
    electron paramagnetic resonances
      to study mitochondrial 127
    identification of membrane
      associated 125
    mitochondrial 131
    sequence of 129
  -potential measurements 130
  potentials, midpoint 138
Reduced carbon compounds,
  production of 209
Refractory state 244

Regulation of actin–myosin
  interaction 245
  Ca^{2+}-dependent 234
Relation of molecular structure of
  myosin to thick filaments 231
Relationships between flows and
  forces 6
Relaxation, calcium transport and
  muscle 108
Regulatory proteins 233
Resonance (ESR), electron spin 67, 93
Resonances to study mitochondrial
  redox centers, electron
    paramagnetic 127
Respiratory chain 125
  organization of the 134
Reticulum, Ca–ATPase of the
  sarcoplasmic 108
Reticulum, endoplasmic 38
Retinal rod outer segments 59
Reversed electron flow 150
Rod outer segments, retinal 59

S

Sarcomeres 227
Sarcoplasmic reticulum, Ca–ATPase
  of the 108
Saturation of the photosynthetic
  apparatus, light 214
Scheme for electron transport and
  oxidative phosphorylation 126
SDS-polyacrylamide gel
  electrophoresis 230
Segmental flexibility of myosin 248
Semipermeable membranes 80
Sensitivity conferring protein
  (OSCP), oligomycin 164
Sequence of redox centers 129
Shaw transport model 96, 106
Simple diffusion 79
Single-file diffusion 15
Sites, energy transduction 137
Sites, low-affinity substrate 108
Slack–Kortschak pathway, Hatch– 210
Sliding filament mechanism 224, 226
Sodium (*see* Na)
Solvent drag 80
Space, matrix 124
Spectra, action 194
Spectrophotometer, double beam 125
Spectroscopic probes 67
Spectroscopic techniques to
  elucidate membrane structure 60
Spectroscopy, IR 62
Spectrum of chlorophyll *a*,
  absorption 181
Sphingolipids 42
Spill-over 195
Spin labels 67
Spin resonance (ESR), electron 67, 93
Stabilization NMR, labilization- 66
State, refractory 244
Static head 21

INDEX
271

Stationary state coupling 13
Steroids and the Na/K pump, interaction between cardioactive 97
Stoichiometric ratio 26
Stoichiometry, coupling and 9
Striated muscle, structure of 227
Structure
  of the major photosynthetic
    pigments 179
  oligosaccharide 51
  some physical approaches to
    membrane 59
  spectroscopic techniques to
    elucidate membrane 60
  of striated muscle 227
Structures, photosynthetic 177
Studies, x-ray diffraction 59
Submitochondrial particles 141
Substances that influence transport 27
Substrate
  binding 99
  complex, enzyme– 83
  level phosphorylation 10
  oxidation 30
  sites, low-affinity 108
  –velocity plots 116
Succinate oxidation, competition
  between NADH oxidation and 138
Sucrose, synthesis of 9
Sugar moieties 52
Sulfate transport process 90
Sulfur
  bacteria
    (Athiorhodaceae), purple non- .. 196
    (Chlorobacteriaceae), green 196
    (Thiorhodaceae), purple 196
  centers, iron- 129
Surface receptors 38
Synthesis of sucrose 9
Synthetic reactions mediated by ATP 9
System
  adenylate cyclase 57
  cation transport 56
  intact actin–myosin–tropomyosin–
    troponin 245
  troponin–tropomyosin 237
Systems
  cellular membrane 89
  methods for identifying transporter 158
  studies on active transport 93

T

T tubules 109
Techniques to elucidate membrane
  structure, spectroscopic 60
Terminal portions of murein
  lipoprotein, N- 50
Thermodynamics (NET),
  nonequilibrium 1
Thick filaments 224
  relation of molecular structure of
    myosin to 231

(Thiorhodaceae), purple sulfur
  bacteria 196
Tissue conductance 26
Titrations, oxygen 129
TnC, myofibril-bound 235
Transduction
  energy 252
  inhibitors of energy 141
  mechanochemical energy 252
  sites, energy 137
Transfer
  chemical hypothesis of electron 203
  chemiosmotic hypothesis of electron 203
  conformational hypothesis of
    electron 203
  of electrons 192
  of energy 185
  inhibitors, energy 142
Transhydrogenation, nicotinamide
  nucleotide 152
Transients, chromatic rate 193
Translation, information 57
Translocating ATPase, proton– 145
Translocation of citrate 53
Transport
  active 23, 55
  adenine nucleotide 160
  anion 157
  biochemistry of 90
  Ca^{2+} 154
  and corresponding osmotic flow,
    active 89
  counter 84
  criteria for carrier catalyzed 158
  inhibitors of electron 132
  inhibitors, metabolite 142
  ion 153
  light-driven electron 176
  mechanism of Ca 115
  membrane 78
  model, Shaw 106
  monovalent cation 155
  multivalent cation 153
  and muscle relaxation, calcium 108
  of Na, net outward 94
  and oxidative phosphorylation,
    scheme for electron 126
  phenomena and the influence of
    chemical reactions 11
  process, sulfate 90
  substances that influence 27
  system, cation 56
  systems, studies on active 93
Transporter systems, methods for
  identifying 158
Transporters, mitochondrial anion 160
Tropomyosin 233
  system, troponin– 237
  troponin system, intact actin–
    myosin– 245
Troponin 234
  C 249
  system, intact actin–myosin–
    tropo–myosin 245

Troponin (*Continued*)
—tropomyosin system 237
Turnover rate, maximum 214

U

Ubiquinone (Q) 133
Ultrastructure of muscle 224
Uncouplers 204
Unidirectional fluxes 88
Unit membrane 39
Unit, photosynthetic 184

V

VanNiel equation 173
Velocity plots, substrate— 116
Velocity, reaction 7
Versatile ionophores 56
Viologen 187

W

Water flow 80

X

X-ray analyses of globular proteins,
  high-resolution 39
X-ray diffraction studies 59

Y

Yield, quantum 192, 212

Z

Z discs 224
Z scheme of photosynthesis 174
Zone, H 224

Date Due

Due	Returned	Due	Returned
DEC 17 1986	DEC 22 1986		
APR 28 1988	APR 13 1988		
FEB 25 1992	FEB 04 1992		
NOV 18 1992			
DEC 06 1993	DEC 08 1993		
APR 20 1997	APR 18 1997		